ELECTRON MICROSCOPY IN FORENSIC, OCCUPATIONAL, AND ENVIRONMENTAL HEALTH SCIENCES

ELECTRON MICROSCOPY IN FORENSIC, OCCUPATIONAL, AND ENVIRONMENTAL HEALTH SCIENCES

Edited by

Samarendra Basu
New York State Police Crime Laboratory
Albany, New York

and

James R. Millette
McCrone Environmental Services, Inc.
Norcross, Georgia

PLENUM PRESS • NEW YORK AND LONDON

Library of Congress Cataloging in Publication Data

Electron microscopy in forensic, occupational, and environmental health sciences.

"Proceedings of a symposium...part of the Electron Microscopy Society of America/Microbeam Analysis Society joint annual meeting, held August 5-9, 1985, in Louisville, Kentucky" — T.p. verso.
   Bibliography: p.
   Includes index.
   1. Forensic pathology — Technique — Congresses. 2. Industrial hygiene — Technique — Congresses. 3. Environmental health — Technique — Congresses. 4. Electron microscopy — Congresses. 5. Particles — Identification — Congresses. I. Basu, Samarendra. II. Millette, James R. III. Electron Microscopy Society of America. IV. Microbeam Analysis Society.
RA1063.4.E43   1986   614'.1   86-25290
ISBN-13: 978-1-4684-5247-1   e-ISBN-13: 978-1-4684-5245-7
DOI: 10.1007/978-1-4684-5245-7

Proceedings of a symposium on Forensic, Occupational and Environmental
Health Sciences, part of the Electron Microscopy Society of America/Microbeam
Analysis Society Joint annual meeting, held August 5-9, 1985, in Louisville, Kentucky

© 1986 Plenum Press, New York
Softcover reprint of the hardcover 1st edition 1986
A Division of Plenum Publishing Corporation
233 Spring Street, New York, N.Y. 10013

DEDICATION

    This volume is dedicated to all the desperate children who need our active support. They are hungry, homeless, malnourished, crippled, and disease-ravaged, and they have no hope. The volume is also dedicated to all the children who are forgotten and deserted, abused, retarded, handicapped, or blind. Between the time this volume was conceived and its publication, thousands of children will have died, having lost the battle against starvation and disease. Hundreds of children will commit crimes against themselves before they have the chance to learn that we care for them - that they are very special to someone. We want all children to know this about scientists - that while we strive to solve the puzzles of science, it is the children who are foremost to our hearts. The purpose of forensic and health sciences is one, to create a better world for everyone to live in. Our children can depend on us for that.

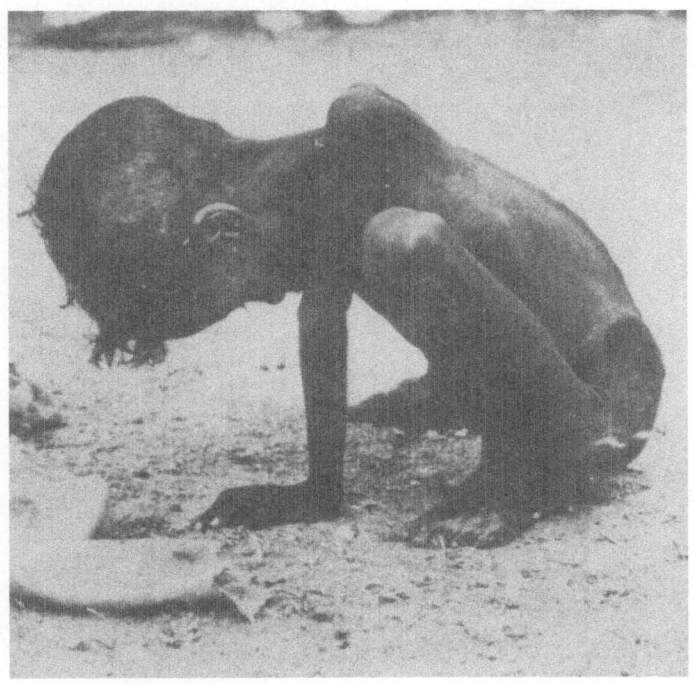

They have no food, no water, and no hope - victims of
natural environmental disaster (Reprinted with permission
from David Livingstone Missionary Foundation Letter
from Africa).

The editors have decided to donate their royalties to children's charities in the United States, including the African Famine Relief Organization. The charities selected are the following:

David Livingstone Missionary Foundation (Africa)
The Pearl S. Buck Foundation, Inc.
Meals for Millions/Freedom from Hunger Foundation
Christian Appalachian Project
New York Easter Seal Society
Siloam International
Epilepsy Foundation of America
National Federation of the Blind
Christian Children's Fund
Save the Children Fund

In due time the list may be augmented to include other charities in the United States and abroad, especially in Africa and Asia, which provide benevolent services to orphans and abused children. The editors are grateful to be able to make this small contribution and hope that others will join in the effort.

Samarendra Basu, Ph.D.

James R. Millette, Ph.D.

PREFACE

    Forensic, Occupational and Environmental Health Sciences are identification sciences dealing with criminal and delicate societal problems for which scanning electron microscopy (SEM) with energy dispersive x-rays (EDAX) and analytical transmission electron microscopy (TEM) are providing increasingly definitive solutions. This is particularly true in the area of particulates. However, electron microscopists working independently in these three related fields are often called upon to identify particulates which have been well characterized by microscopists in one of the other fields of study. Exchange of information on particulate identification and techniques for analyzing new unusual samples has been difficult between the three disciplines. For example, automated search and identification of particulates has become a standard procedure in environmental and occupational health. The procedure, however, has yet to find its way into forensic laboratories for analysis of gunshot residue particles. This volume represents a beginning for collaboration and exchange of ideas between such areas of study with diverse interests but similar analytical problems. It is virtually certain that this effort will also interest other electron microscopists in some novel scientific problems with criminal and societal relevance.

    The volume presents full-length articles of several invited speakers and participants at the "Forensic, Occupational and Environmental Health Sciences" Symposium held as a part of the EMSA-MAS Joint Annual Meeting, 5-9 August 1985, Louisville, Kentucky. Extended 2-4 page abstracts of these presentations have been published in the EMSA Proceedings (1985). These abstracts, however, are not long enough to permit detailed descriptions of methodologies and substantial results that are necessary for evaluation of each work by the scientific community. This volume is essentially a small expanded segment from one session of the 1985 - EMSA/MAS Joint Meeting. A few articles, written by researchers of international standing who were on the editors' (Symposium Organizer and Co-Chairman) original list, are also included.

    The present volume consists of three sections dealing with outstanding and novel applications of electron microscopy in forensic science, environmental health and occupational health. The included topics of forensic interest are the mechanism of gunshot residue deposition and its probing characteristics to reconstruct shootings, bulb filament and globe failure in car and plane accidents, hair comparisons by analysis of biological elements and by x-ray line scans for spatial variations of sulfur and absorbed copper ions, forensic uses of deflection modulation and x-ray dot maps, evaluation of staple impressions, impressions of cutting tools and bite mark analysis.

    The topics of occupational health are particulate contents of human lung in black lung diseases, alpha quartz in respirable coal mine dust, asbestos measurements in work places and ambient atmospheres and the effect of nickel on ultrastructure of the heart muscle.

The topics of environmental health are respirable aerosols of sulfate (acid rain precursors) reaching Whiteface Mountain, New York, airborne lead particulates in an urban industrial environment, evidence of hepatotoxicity by microanalysis of subcellular ions, and liver tissue preparation by a modified cryoultramicrotomy kit.

Although in terms of techniques and their applications, this volume has set forth definitive standards in areas in which significant advancements have been made that are contributing to Criminal Justice system and to health organization systems, the opinions expressed in individual articles should be considered to be those of the authors. Their opinions do not necessarily reflect the views of the sponsoring organizations.

SPECIAL ACKNOWLEDGMENTS

        Several chapters in this book are largely an elaboration of invited
and contributed papers presented by the contributors at the Forensic,
Occupational and Environmental Health Sciences Symposium, sponsored by the
Electron Microscopy Society of America at the 1985 joint national meetings
of that Society and the Microbeam Analysis Society in Louisville,
Kentucky, August 5-9. Extended abstracts of those presentations have been
published by San Francisco Press, Inc. We should like to express our
thanks to San Francisco Press, Inc., for releasing their copyrights to
those abstracts. We also express our gratitude to all contributors of
this book.
        Because the progress in our fields has to move on hand in hand with
public interest, scientists and publishers of science must come forward
and work as a team. We, therefore, commend Plenum Publishing Corporation
for their efforts to act as the publisher of this volume.
        We would like to specially acknowledge the financial support the
Symposium (1985) received from the Electron Microscopy Society of America.
Donations from AMRay Inc., Cambridge Instruments, Inc., and Ernest F.
Fullam, Inc. toward the Symposium (1985) are also acknowledged.
        Thanks are also due to our individual employers, New York State
Police (SB) and McCrone Environmental Services, Inc. (JRM), for support
during the preparation of this volume. Special thanks from the principal
editor (SB) are due to the 1985 - EMSA Program Committee headed by
Professor George D. Pappas, to the 1985 - EMSA Local Arrangements
Committee headed by Dr. George H. Herbener and Dr. Alan G. Farman, to Dr.
Charles E. Lyman (EMSA), to Donald O. Chesworth, the Superintendent of New
York State Police, to Colonel C. R. Baker, the Deputy Superintendent of
Administration, New York State Police, and to Lt. W. M. Dale, the
Assistant Director, New York State Police Crime Laboratory, for their
active interest in the Symposium (1985).
        The principal editor (SB) wishes to express his deep appreciation to
Mary Beth Larmour for her professional, secretarial assistance in
organizing the Symposium (1985). Grateful acknowledgment from this
editor (SB) is rendered to Robert Miazga for photography and to Sgt. A.
Dooney for his assistance in preparing posters at the Symposium (1985).
        Finally, we thank Ellen Raynor-Enco, the editor of Plenum Publishing
Corporation, for her assistance in seeing this book to completion.

                                        Samarendra Basu, Ph.D
                                        James R. Millete, Ph.D

CONTENTS

OCCUPATIONAL HEALTH

THE MECHANISM OF GUNSHOT RESIDUE DEPOSITION AND ITS PROBING CHARACTERISTICS TO RECONSTRUCT SHOOTINGS*

S. Basu

New York State Police Crime Laboratory
Albany, New York 12226

ABSTRACT

The significance of the predominant distribution of gunshot residue (GSR) spheroids, mainly on the back of the hand, has been explained by fundamental studies of the mechanism of GSR deposition and its probing characteristics to reconstruct shootings. These studies were based upon Glue-lift collection of GSR for examination in the scanning electron microscope with energy dispersive X-ray analysis. By performing "muzzle blast block" and "trigger-blast block" experiments with selected shotguns, it has been shown that in routine test firings of these weapons the breech GSR contribute the most to the GSR deposits on the shooter's hands. Muzzle-blast GSR seldom settle from air, or by a blow-back, onto the shooter's hands. The breech GSR are predominantly smooth, regular spheroids, whereas muzzle-blast GSR are mostly irregular and accompany bullet residues and bullet fragments. Since collections were made within a few seconds after each firing, the previous observation suggests that breech GSR are blown instantaneously onto the immediate surfaces of the shooting weapon, and on any hand in contact with, or in close proximity to the weapon. This forced deposition is an advantageous as well as cautioning phenomenon. If the gun is precleaned and the ammunition and the hand grasps remain unchanged, reproducible GSR counts can be obtained from single firings, providing that all deposited particles containing one, two or all three characteristic elements (Pb, Sb, Ba) of GSR are taken into account. This density distribution is a function of the weapon and the ammunition, and is therefore the basic criterion necessary for reconstruction of a shooting crime. Within each such distribution, the elements, sizes, and the morphologies of residue particles may vary significantly from shot to shot. Residue particles may also be transferred by contact from a freshly fired gun to a non-firing hand. In fact, more GSR may result in this way on the palm of a hand than on the shooter's hand if specific areas of a weapon (e.g. cylinder and barrel of a revolver) are handled. If a shooting is attended or encountered at a very close distance (inches), similar GSR distribution may result on the shooter's and non-shooter's hands. However, if collections are made at the crime scene from undisturbed hands and the shooting incident is duplicated by test firing the suspect weapon with the ammunition in evidence, significant interpretation of the GSR collections may come out even in difficult cases. In such cases, the density distribution and the GSR characteristics are to be given

---

*Symposium invited paper presented at the 43rd Annual Meeting of the Electron Microscopy Society of America held jointly with the Microbeam Analysis Society, Louisville, Kentucky, 5-9 Aug. 1985.

strong consideration. The most important interpretation of all is that the positive indication of GSR, mainly on the back of the hand, implies a shooting hand only if it is a proven case of an unattended shooting.

INTRODUCTION

Because of its strong evidential value in the investigation of a shooting crime, gunshot residue (GSR) detection on the hands of suspects or victims has long been of major interest to many forensic laboratories. Of all techniques currently in use, scanning electron microscopy with energy dispersive X-rays (SEM-EDX) is the most definitive in testing for GSR.[1-4] The Glue-Lift collection of GSR and the author's testimony based upon GSR analysis in the SEM-EDX have been accepted well in the Court of Law (NY).

Following its development in 1979, the Glue-Lift[1] collection technique has been a research tool aimed at improving the credibility of GSR examination in the SEM-EDX. The technique has aided the understanding that the rigid conditions involved in GSR formation, such as steep rises in temperature and pressure and their equilibria, make GSR particles unique products of primer explosion.[2] The elemental distribution and structural features of GSR, both in its interior and exterior, are characteristic of condensates.[1,2] The Glue-Lift collection was essential to the reconstruction of shootings in self-inflicted deaths.[3] This indicated for the first time the potential value of this technique. In fact, since its introduction to casework, this collection technique has provided useful interpretation of the GSR distributions on a suspect's or victim's hands.

The interpretation of a GSR distribution requires a clear understanding of the mechanism of GSR deposition and its probing characteristics. Knowing these characteristics was crucial to reconstruction of shootings.[3] The experiments developed in this report are intended to give a foundation to a simple concept, or the postulate, that when a firearm is discharged, GSR are blown instantaneously onto the immediate surfaces of the shooting weapon, and on any hand in contact with, or in close proximity to, the weapon. This forced deposition is essentially a blasting process:[5] it is unlike the settling of lead aerosols (diameters $\leq$ .42 μm) from the air containing gunsmoke.[6]

If GSR deposition is time-dependent, the evidential value of GSR will surely be questioned in the court of law. The problem has been attacked here from a different angle, that is, the GSR on the shooting hand can be shown to have originated from the nearest GSR emanating port, i.e. the breech(s) of trigger housing. Therefore, shooting experiments were undertaken by blocking out "muzzle-blast GSR" and then "trigger-blast GSR" (breech residues) in alternate series of test firings with the same weapon(s) and ammunition(s), and then determining whether the breech or the muzzle contributed the most to the GSR deposits on the shooter's hand(s). The mechanics by which GSR emanate from "close-breech" weapons and deposit on both hands of the shooter remains unexplored (See Discussion with reviewer, in ref. 1). Therefore, shotguns and rifles were preferred for the block tests.

The block tests revealed that a systematic deposition of GSR occurs with only pre-cleaned weapons. This developed into a criterion called "density distribution".[3] Since this criterion is the major consideration to the reconstruction of shootings,[3] it has been analyzed in this report for its consistency and pitfalls as well.

The block tests have been instrumental to determine unambiguously the distinctive features of breech GSR and muzzle-blast GSR. Previously this evidence assisted in distinguishing the hand-on-muzzle from the hand-on-trigger in self-inflicted shooting deaths.[3] Therefore, examples of interpretation, based upon block tests, density distribution and GSR characteristics, will be presented.

This report has appeared as an extended abstract in the proceedings of EMSA/MAS (1985).[7]

MATERIALS AND METHODS

GSR Collection

The GSR were collected from four specific areas of hands by Glue-Lift.[1,2] (See Fig. 2, in ref. 3 for areas of the back of hands and the palms). A field-ready GSR-Kit is shown in Fig. 1. The collectors were four half-an inch (13 mm) diameter polished carbon planchets (width 1/8 in. or 3 mm) each containing a spread, thin layer (thickness 2 μm) of Carter®'s rubber cement used as diluted (4 times) with toluene. With a finger inserted into a thimble base, the attached planchet was gently touched five times (without retouching the same spot) to a specific area: either the back of a hand or the palm. The advantages of GSR collection by Glue-Lift[1] are several. Briefly, (1) a thin glue layer rejects skin cells, debris and hairs, etc., as it collects GSR (Size 0.3-55 μm) and small-sized particulates from the surfaces of these materials; (2) the collected particles hardly imbed into the glue since the glue layer is thin; (3) the substrate (graphite) is polished, conductive and is recommended for energy dispersive x-ray analysis; and (4) the GSR collections (disks) do not require carbon coating.

GSR Search and Classification

With the forefinger inserted into a base thimble, four circles (diameter 1.5

FIG. 1.   New York State Police GSR Kit. Four Glue-Lift[1,2] disks (polished carbon, 13 mm diameter) are mounted with adhesive tapes on 13 mm sanded top of "hat-like" red plastic thimbles (Caplug®). Each thimble (#1) is marked on side for the collection area of hand and is covered from top with an unsanded thimble (#2). Each set is allowed to stand inside a cylindrical box (3), and is held in place by the tight fitting cap of the box (ht. 2.54 cm). The four boxes (#3) are enclosed in a square-like plastic box (ht. 3.2 cm) (#4). This box (#4), the collection procedure (#5) and a pair of polythene gloves (#6) are packaged into a rectangular paper box (#7, Kit).

mm) were marked on the attached collector disk, with the sharpened rim of a retracted ballpoint pen. The planchet was peeled off the base thimble with a surgical blade (scalpel) and then mounted on a 13-mm diameter aluminum pin mount with carbon cement, or with a small cut piece (3 mm X 3 mm) of Scotch® transfer tape. After each examination at 20 kV in the SEM-EDX (AMR®1000, EDAX®707A), the carbon planchet was returned to its place in the kit.

The marked circles were first recorded for full-screen views of the circle boundaries with a reverse-biased (-100 V) Everhart-Thornley detector.[8] This back scattered electron mode provides uniform contrast of details at a low magnification (e.g. 60X) for the circles. (See also GSR Distribution.) The same detector was used with the usual positive bias (+300 V) to the grid for identification of topography and shape of GSR by secondary electrons.[1] When the density of GSR was too high (>100 per circle), several sectors of each circle were photographed at a high magnification (500 X to 200 X). An annular solid-state backscatter detector ("Quad"),[9] mounted around the incident electron beam on the final pole piece, was used to search for GSR within the circles and their sectors. Since backscattered electrons with higher energy (>10keV) are collected from a large solid angle, these images are responsive to atomic contrast, or material contrast.[10] In fact, due to heavy metallic composition (Pb, Sb, Ba) of GSR, these particles appear as bright dots. Each such particle was first analyzed for its elements with the energy dispersive x-ray analyzer (EDAX®707A), using a concentrated beam (partial field) of incident electrons. When the GSR elements (Pb, Sb, Ba) were detected, the particle was verified for its condensate and/or molten spheroidal shape.[1,2] The confirmed particle was marked on the circle, or on its sector, by a number. A tabulation was made of all detected particles according to their elements, sizes and morphologies. All spheroids containing either three elements (Pb, Sb, Ba), two elements (PbSb/SbBa/BaPb), or only one element (Pb/Sb/Ba) were counted. The classification of GSR into "full-GSR" (three element-containing), "binaries" (two element-containing) and "monomers" (one element-containing) has been explained elsewhere.[3] The counts, due to all condensate spheroids in four circles, were added to obtain the density (ρ) of GSR distribution, or the number of GSR per 7.1 mm^2 of a collection disk.[3] This distribution did not take into account bullet fragments, irregular and crystalline environmental particles, which often contain one (e.g. Ba, Pb) or two elements of GSR.

Block Experiments With a Shotgun or a Rifle

The blockades were heavy-duty plastic sheets (thickness 0.15 mm) used as screens in the "muzzle-blast block" (Fig. 2) and similar quality plastic bags to enclose the weapon in the "trigger-blast block" (Fig. 3). Each test firing was preceded by control collections from pre-cleaned hands (the back of hands and palms) of the test shooter.[1,3] Also, the plastics were examined as cut pieces (1.8m by 2.7m) and were installed vertically inside a clean firing range, and the screen was taped to the side walls, to the ceiling and to the floor. The muzzle of a pre-cleaned shotgun or rifle was inserted about 5 cms through a pre-cut hole on the screen, which (hole) was then sealed to the muzzle, and the weapon was fired once, keeping normal shooting positions of the shooter's hands (pre-cleaned) on the weapon (Fig. 2). GSR particles were collected in a few seconds after the firing by Glue-Lift.[1] Each test firing was preceded by control collections from the precleaned hands of the shooter. The unclean weapon was used for the next firing, or the weapon was cleaned inside and on its outside,[3] depending upon the shown sequence of six single firings in Fig. 2. The same sequence of firings was followed in the "trigger-blast block" (Fig. 3). In this test the plastic enclosure was heat-sealed at the end that contained the butt of the weapon, and at the other end the bag was sealed to the surface of the muzzle within about 5 cms from the exit (Fig. 3). After a firing, pieces of the plastic were cut from regions which were on top of the breeches and these were examined in the SEM-EDX. These studies added evidence to the fact that the breech-GSR were in fact barricaded by the plastic bag, and the plastic was impervious to emanating GSR.

FIG.(s) 2 and 3. Tests for GSR deposition in "muzzle-blast block" (2) and "trigger-blast block" (3) shootings with a shotgun (Reprinted with permission from San Francisco Press, Inc., ref. 7).

GSR Capture

The results of the block tests, concerning the spread of GSR and their (GSR) elemental and morphologic characteristics, needed additional support by a means that would not involve collection from the shooter's hands; rather, GSR would be captured by suitable targets and then analyzed. This test was incorporated into the "muzzle-blast" block by affixing Glue-Lift[1,2] disks juxtaposed on either side of the screen (not shown in Fig.(s) 2 and 3). The disks were at a vertical distance (radius) of about 1/2 ft. (15.2 cm) from the axis of the muzzle (or the bullet exit). Since in normal shootings of selected shotguns and rifles the supporting hand on the barrel indicates residue particles,[1] the same as the shooting hand (hand on the trigger), one would like to know if breech residues would be ejected toward the screen (distance 0.9 m from the trigger guard) and would possibly deposit on the Glue-Lift disks on the screen. The ejection chamber, which appears between the two hands of the shooter in Fig. 2 (top), and a hammer-gap (trigger-breech), were the major breeches of the shotgun (pump action) in these various tests. The Glue-Lift disks on the muzzle-side of the screen were necessary for capturing muzzle-blast residues which would strike the glue layer at a glancing angle.

GSR Emanation Test by Residue Color

A confirmation of the GSR emanation from breeches of shotguns and rifles was achieved by using lint-free white cloths to enclose the weapon in "trigger-blast" shootings, and developing residue color due to lead (Pb) on the test cloths with sodium rhodizonate.[11,12]

Consistency of ρ

The consistency of ρ, i.e., the reproducibility of GSR counts per shot, was examined by test-firing a pre-cleaned weapon, with the same ammunition in each firing, while maintaining the same hand positions on the weapon. Each test firing was preceded by control collections from the pre-cleaned hands of the test-shooter. The test was undertaken through several firings (2 shots to 6 shots) with a given firearm (See ref. 3 for list of weapons and ammunition).

Forced Deposition of Breech GSR

Since GSR would be collected immediately after a test firing, a consistency in density (ρ) distribution would imply an instantaneous, forced deposition process. The surface structures of the shooting hand would be immaterial to this process. This examination has been made by comparing GSR counts on the bare hand versus gloved-in hand of the test shooter. Clean polyethylene gloves were used in each test firing. Also, the same collection spots (5 touches) were chosen on both the bare hand and the gloved-in hand. Each test firing was preceded by control collections from the pre-cleaned gloved hand. The gun in evidence was also pre-cleaned prior to each test firing.

GSR Transfer by Contact

The gun, the test shooter's two hands and the non-shooter's hands were all pre-cleaned prior to each test firing in this experiment. After a shooting, the test-shooter placed the fired gun on a clean table and the non-shooter was asked to pick up the weapon. The two were not allowed any activity that might transfer GSR from the palm of hand to the back of hand and vice versa. The non-shooter held the gun in his palm(s) for 15 seconds to one minute under various conditions, as described in Results.

Reconstruction of Shooting

The technique for reconstruction of suicide shootings uses a target to simulate the human body and has been described in a previous report.[3]

RESULTS

"Muzzle-Blast Block" and "Trigger-Blast Block" with a Shotgun

The two "block" experiments were performed by 18-to-20 firings with a pump action shotgun and a pump action rifle. The main results were confirmed by additional firings with a single shot double barrelled, hinge frame shotgun and a 30-30 caliber lever action rifle. Typical results, as obtained with a pump action shotgun (Winchester®Model 1200, 12 gauge; ammunition Remington®Express, 12 gauge, 2 3/4 inch, or 7 cm, 0-0 buckshot) are shown in Fig.(s) 2 and 3.

The main observations were the following: (1) Breech GSR invariably deposited on the shooter's hands in the "muzzle-blast block" shootings. This deposition was consistent with pre-cleaned weapons (1st and 6th firings, Fig. 2). (2) In the corresponding firings (1st and 6th) in the "trigger-blast block" there was no GSR deposition on the shooter's hands (Fig. 3). (3) In those firings (2nd-5th) in the "muzzle-blast block" in which the weapon was unclean, the counts due to "full-GSR" and "binaries" did not increase at all. Yet the deposition was sporadic because the net counts increased as much as five to ten times, due primarily to settling of contaminant Pb particles (diameter 0.1-0.5 µm) of the weapon. Cleaning the test weapon was therefore essential in the reconstruction of suicides.[3] In fact, when pre-cleaned shotguns or rifles were test fired (no "block"), the GSR counts from the back of the shooter's hands agreed with the GSR counts obtained in the 1st and 6th firings of the "muzzle-blast block" (Fig. 2). Since GSR, each containing Pb, Sb and Ba, did not settle from the air or by a blow-back onto the shooting hand (Fig. 3), the main conclusion from the block tests is that the GSR deposits on the shooter's hands were mainly the breech deposits.

"Muzzle-Blast Block" and "Trigger-Blast Block" with a Rifle

The trend in results of "muzzle-blast block" with pre-cleaned and unclean rifles was similar to the previous results obtained with shotguns (Fig. 2). A slight difference was observed only in the "trigger-blast block", in which, whether the rifles were pre-cleaned or unclean, a few lead (Pb) spheroids (diameter 0.1-0.5 µm) were consistently found on the support-hand (back of hand) of the test shooter on the muzzle. These were presumably bullet particles. The trigger hand of the shooter did not exhibit these (Pb) particles.

GSR Capture

The main results obtained with GSR targets (Glue-Lift disks attached to screen) in the "muzzle-blast block" shootings were the following (Table 1 and 2). (1) The net capture of residue particles on either kinds of target increased several times in the 3rd, 4th and 5th firings, in which the weapon was not cleaned (cf. 1st firing with the pre-cleaned weapon). The increments of residues in Table 1 were due mainly to "full-GSR" (cf. net counts 46, 42 full-GSR inclusive, Table 1). The increments of residues in Table 2 were due primarily to lead (Pb) spheroids (cf. net counts 74, 66 Pb-monomer inclusive, Table 2). (2) The captured breech residue particles (Table 2) were predominantly regular spheroids with rather smooth surface topography (64 out of 74 GSR), whereas, the captured muzzle-blast residues (Table 1) were predominantly irregular spheroids (42 out of 46 GSR). The observation in (2) has been considered in another section See GSR Characteristics).

The observation in (1) was consistent with the observed deposition of lead (Pb) particles on both the shooter's hands (cf. Fig. 2 vs Table 2); that is, the capture of these particles (Pb) was also sporadic. Since the attached Glue-Lift disk on the "muzzle-blast block" screen was only 1/2 ft (15.2 cm) away from the axis of the muzzle, the captured particles were obviously ejected at sharp angles, grazing the barrel of the shotgun (See the illustration on top of Fig. 2). This seems to suggest that these high velocity lead (Pb) particles were most likely bullet-derived. Note, that lead (Pb) particles also contribute to the net deposit on the support-hand of the test shooter on the barrel (Fig. 2).

TABLE 1. Deposition of Muzzle-Residues On a Glue-Lift Disk on the Muzzle Side of the Screen in "Muzzle-Blast Block"

Experiment[a,b]	# of GSR particles per 4 circles (7.1 mm^2) having different elemental compositions, ρ				
	Full GSR's (PbSbBa)	Binaries (PbSb/ PbBa/BaSb)	Monomers (Pb/Sb/Ba)	Range of Diameters (μm)	Total GSR in 4 Circles
1st shot with a pre-cleaned shotgun	2	1	1	0.5-6	4
2nd shot with the unclean shotgun	2	2	1	1 - 5	5
3rd, 4th and 5th shots, that is, total 3 shots with the un-clean shotgun	42	0	4	0.5-6	46 4 regular spheroids + 42 irregular spheroids)

[a]Distance (vertical) of disk from muzzle = 15 cm.
[b]Shotgun and ammunition same as in Fig.(s) 2 and 3.

TABLE 2. Deposition of Trigger-Blast Residues On a Glue-Lift Disk on the Shooter's Side of the Screen in "Muzzle-Blast Block"

Experiment[a,b]	# of GSR particles per 4 circles (7.1 mm^2) having different elemental composition, ρ				
	Full GSR's (PbSbBa)	Binaries (PbSb/ PbBa/BaSb)	Monomers (Pb/Sb/Ba)	Range of Diameters (µm)	Total GSR in 4 Circles
1st shot with a pre-cleaned shotgun	1	0	4 (Pb mainly)	0.5-1.0	6
2nd shot with the unclean shotgun	1	0	5 (Pb mainly)	0.2-7	6
3rd, 4th and 5th shots, i.e. total 3 shots with the un-clean shotgun	8	0	66 (Pb mainly)	0.2-10	74 (64 regular spheroids + 10 irregular spheroids)

[a] Distance (vertical) of disk from muzzle = 15 cm. (opposite to the disk in Table 1).
[b] Shotgun and ammunition same as in Fig.(s) 2 and 3.

GSR Emanation from Shotguns and Rifles

Four weapons (12 and 20 gauge single-shot double barrelled hinge frame shotguns, 12 gauge pump action shotgun, and a 30.06 Springfield caliber pump action rifle) were used to confirm the pattern of GSR emanation from the breeches of these weapons. The residue color was the strongest at the site of cloths encompassing the ejection chamber (pump action shotgun, Fig.(2) 2 and 3), hinge and hammer-gap (single or double barrelled hinge frame shotguns). Superimposed on these colors, there was a faint persistent color representing arcs of residues, which unravelled the length (top) of a shotgun. This confirmed that breech residues are ejected, both forward and backward, at sharp angles grazing the surfaces of these weapons and depositing by forced deposition onto both of the shooter's hands on a shotgun (or a rifle).

The plastic bags used in the "trigger-blast block" experiments indicated that the GSR deposits on plastics (5 cm. by 5 cm) in the areas of ejection chamber and hammer-gap (ea. approx. 180 GSR/4 circles) were about 3 times more than the deposits in the area of trigger-pull, i.e., under the gun (60 GSR/4 circles).

Consistency of ρ

The forced deposition of GSR and its characteristics prevail also in the shooting of open-breech weapons (handguns). If a revolver (.38 caliber Smith & Wesson®) is pre-cleaned and the ammunition and grasp of firing hand on the weapon remain unchanged, the same amount of residue is deposited per firing on the back of the firing hand (Fig. 4, line on top). This density distribution

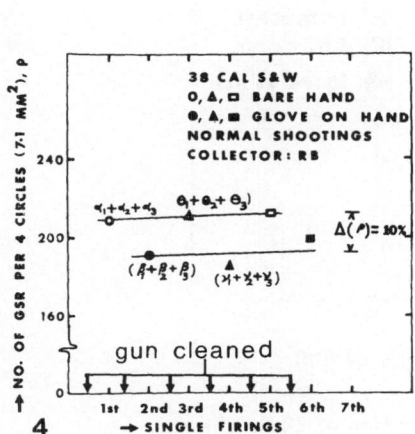

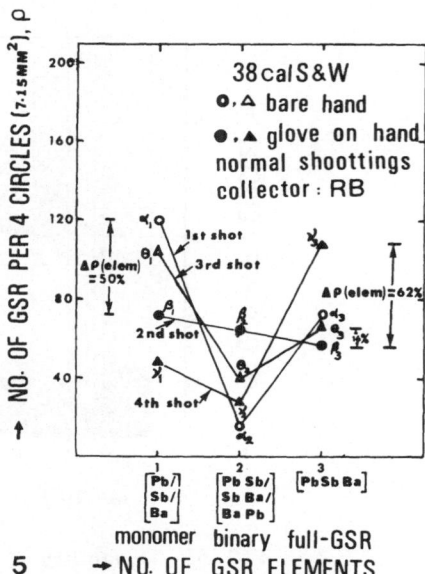

FIG.(s) 4 and 5. Consistency of density (ρ) distribution (4) <u>versus</u> variation of elemental distribution (5) of GSR in single firings with a pre-cleaned
revolver (ammunition .38 spl, S & W® jacketted bullets). The elemental data in Fig. 5 are based upon first 4 shots in Fig. 4. (Reprinted with permission from San Francisco Press, Inc., ref. 7).

(ρ), which takes into account all deposited particles containing one, two and
three characteristic elements of GSR (Pb, Sb, Ba) changes only a little (10%
less) when a smooth plastic glove is on the shooting hand (Fig. 4, lower plot).
Since surface textures of hands have little influence on density distribution,
the forced deposition of GSR must be like a blasting process. The consistency
of density distribution (ρ) has been verified by test firing 12 revolvers (0.22-0.45
calibers), 5 pistols, and 7 shotguns (12-20 gauge) (<u>See</u> ref. 3 for a partial list
of these weapons). Within each such distribution, the elements (Fig. 5), the sizes
(Fig. 6) and the morphologies (Fig. 7) of residue particles may vary significantly
from one shot to another. The variation in observed morphologies (viz. 15% to
20%) was much less compared to the variations in elements (viz. 50% to 62%)
and in sizes. Among the observed morphologies with the revolver (.38 cal. <u>S &
W®</u>) used in Fig.(s) 4-7, regular spheroids ("S") and nodular spheroids ("NS")
accounted for 60% to 70% of total GSR counts per shot. The remaining 30%
to 40% of GSR counts was due to irregular spheroids ("IS") and peeled-orange
spheroids ("POS"). The hollow spheroids ("HS") and hollow peeled-orange spheroids
("HPO") were not detected on the shooting hand with this revolver (.38 cal.) (For
these various GSR morphologies, <u>See</u> Basu and Ferriss[1]).

The regular spheroids ("S", "NS" and "POS") from the shooting hand (hand
on trigger) possessed remarkably smooth surface topography, which made them
distinct from the irregular spheroids ("IS").

The size distribution studies (Fig. 6) indicate that density distribution criterion should not fail unless very large GSR (diameters 10-30 μm, 30-100 μm,
etc.) were formed. The maximum probability of this failure is 16%, or one out
of six or seven firings, and this drawback has been overcome by additional test
shots.[3]

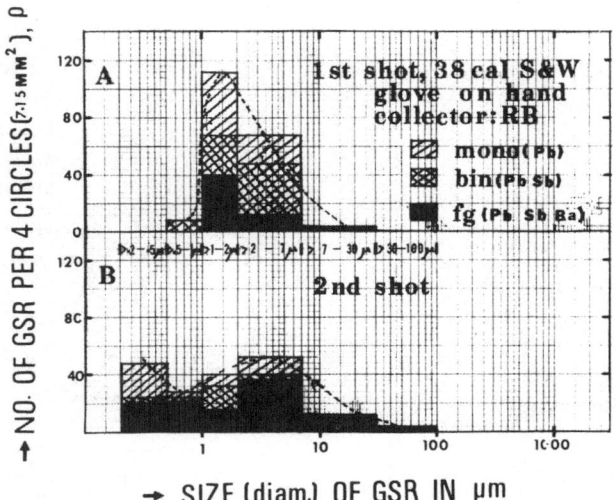

FIG. 6.   Variability of size distribution of GSR. The six
size groups were - > .2 to .5 μm, > .5 to 1 μm, > 1
to 2 μm, > 2 to 7 μm, > 7 to 30 μm and > 30 to
100 μm. The peak diameters were 1.6 μm in the
first shot (A) and 3.7 μm in the second shot (B).
The gun and ammunition were the same as in
Fig. 4. (Reprinted with permission from San
Francisco Press, Inc., ref. 7).

GSR Transfer

The disadvantage of forced deposition is that residues are blown in onto
the escape ports and onto the immediate surfaces of the weapon, and on any
hand in contact with the weapon, or in close proximity of the breeches. If a non-
shooter picks up a fired gun with one hand, residues are readily transferred to
the palm of this hand. Two revolvers (ea. .38 caliber), two semi-automatic pistols
(.22 and .25 caliber) and two shotguns (12 gauge and 20 gauge) were used to obtain
estimates of transferred GSR to the palm of two non-shooter's hands (one per
person) under controlled conditions. These conditions were then varied. The variables were the number of firings, the specific area of a gun that has been touched
and the number of touches or rubbing activities to wipe off residues from the
surfaces of the gun. Table 3 contains representative results obtained with two
.38 caliber revolvers. Because no activity of the shooter and of the non-shooter
was allowed that might cause transfer of GSR from the back of the hand to the
palm, and vice versa, the level of GSR density (ρ) on the back vs palm was opposite
between the shooter and the non-shooter (compare: Table 3, row 1a and 1b, etc.).
The degree of GSR transfer was only slight (12% - 29%) when the non-shooter
picked up the fired gun holding the trigger in the same manner as the shooter
normally does to fire a revolver (See Table 3, rows 1b, 2b and 3b). When the barrel
and the cylinder were touched, the transferred residues exceeded the hand deposit
of the shooter. When those areas (barrel and cylinder) were handled, or touched
many times, the cumulative yield of GSR on the palm increased considerably
(compare: Table 3, row 2c and 3c). The GSR deposits on the back of the hand
of the shooter did not increase in proportion with the number of firings, because
a portion of the hand deposits (> 28%) was always blown off by the residue blast
in the successive firing. Krishnan et al.,[13] Guin,[14] and Goleb and Midkiff[15] also
observed this effect using the techniques for bulk analysis, such as, atomic absorption and neutron activation analysis. This effect is also true for the deposits
on the surface of a gun, but residues certainly build up on the escape ports (e.g.

10

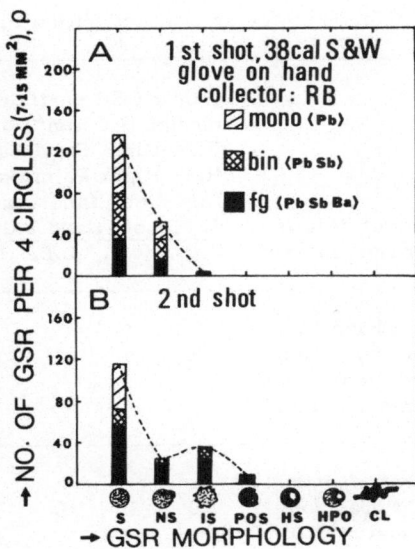

FIG. 7. Variation of GSR morphology from shot to shot. Notations: S: Spheroid, NS: Nodular spheroid, IS: Irregular spheroid, POS: Peeled orange spheroid, HS: Hollow spheroid, HPO: Hollow peeled orange, and CL–cluster.[1] Variation ($\Delta \rho$) in each morphologic category ranged from 15% to 20%. All these spheroids except IS presented smoother topography.

cylinder gaps) unless the weapon is cleaned. Remote parts of the gun (e.f. butt) received very few particles from the trigger blast, because the latter was partially blocked by the shooter's grasp of the weapon. Touching this area did not increase the yield of transferred GSR, even when the weapon was fired several times (compare: Table 3, rows 1b and 2b).

GSR Characteristics

The results of two "block" tests have found meaningful applications in casework when GSR were collected at the crime scene. The muzzle-blast residues are predominantly irregular spheroids (about 90%), whereas the trigger-blast residues are more regular spheroids (about 80%) (Table 1 vs Table 2) (compare: Wolten et al.[5]). This contrast is more marked than the shot-to-shot variation in morphology of breech residues (Fig. 7). The "muzzle-blast GSR" also accompany a large distribution of bullet-derived GSR, bullet particles (Pb, PbCu, etc.) and bullet fragments. These characteristics have been used (1) to distinguish the

11

TABLE 3. GSR Transfer by Contact With a Fired Weapon

Experiment[a]		Experiment details (shooter/non-shooter	ρ or # GSR particles/4 circles (7.1 mm^2) of R.B.-Right Hand Back; R.P.-Right Palm Back; L.B.-Left Hand Back, & L.P.-Left Hand Palm				Transfer of GSR: Non-Shooter's Hand ×100% Shooter's Hand
			R.B.	R.P.	L.B.	L.P.	
1	a.	Shooter's hand (R), tight grasp, 1 firing with pre-cleaned gun (no barrier at muzzle)	156	0	0	0	
	b.	Non-shooter's hand (R), tight grasp on trigger and rear of gun for 15 seconds.	0	36	0	0	23%
2	a.	Same as 1a except – 3 succesive firings	339	0	0	0	
	b.	Same as 1b	0	42	0	0	12%
	c.	Non-shooter's hand (R), tight grasp on cylinder and barrel for 15 seconds	0	504	0	0	150%
3	a.	Shooter's hands (R&L), one simulative suicide firing, that is, firing against a target at muzzle; tight grasp of R-hand on trigger and rear of gun with L-hand support under R-hand	88	0	49	0	
	b.	Same as 1b and 2b	0	40	0	0	29%
	c.	Same as 2c plus rubbing or several touches on cylinder and barrel with palm	0	268	0	0	195%

[a]Gun and Ammunition: For expt.(s) in (1) and (2): .38 caliber Smith and Wesson® revolver; Remington®158 gr round nose.
    For expt.(s) in (3) .38 caliber Smith and Wesson ® revolver; Winchester® Western 145 Lubaloy.

hand of suicide victims on the trigger from their hand on the muzzle (Fig.(s) 8-12); (2) to identify the bullet entrance (wound) (Fig. 13), and (3) to distinguish the hands of the homicide victims from the shooter's hands in combined homicide-suicide cases.

Usually in suicide shootings with a shotgun, if the supporting hand on the muzzle is within the "cut-off" distance of rebound GSR (ca. contact shootings), the hand deposit appears to be enriched in lead monomers and irregular GSR particles. By contrast, the hand on the trigger represents mainly regular, smooth GSR of trigger-blast (breech residues) (cf. Fig.(s) 8 and 10 vs Fig. 9; Fig. 11 vs Fig. 12).

The secondary electron mode is essential for determination of the shape and topography of GSR, but this imaging mode fails in the analysis of GSR in scrapings from the area of the bullet wound. Because the GSR particles imbed into tissues and are obstructed from view by blood, the imaging due to material contrast by backscattered electrons is more useful (Compare: Fig. 13A vs 13B).[10] The entrance wound was in close proximity (inches) of a discharged firearm (muzzle), and so a dense distribution of irregular, regular and hollow GSR spheroids of muzzle-blast was found. These GSR appear as bright particles in the backscattered electron image (Fig. 13B). The approximate diameter of the GSR in

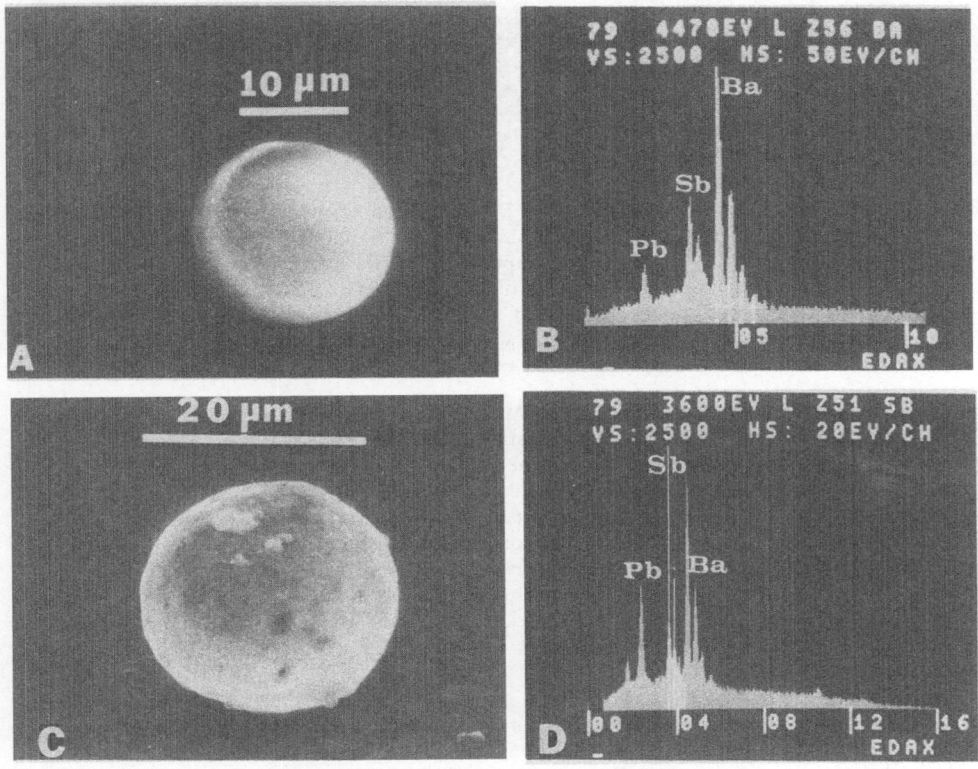

FIG. 8. Secondary electron images and EDX spectra of two large "full-GSR" (Pb SbBa) which represent regular, smooth GSR from the back of firing hands of two suicide victims. A & B: A "Peeled orange" GSR, diameter 18.5 μm, original magnification 2,100X, handgun (.38 Spl. S & W®) suicide; C & D: A GSR with holes (bottom), diam. 23 μm, original magnification 2,100X, shotgun (12 ga. Remington® pump action) suicide. Specimen not carbon coated.

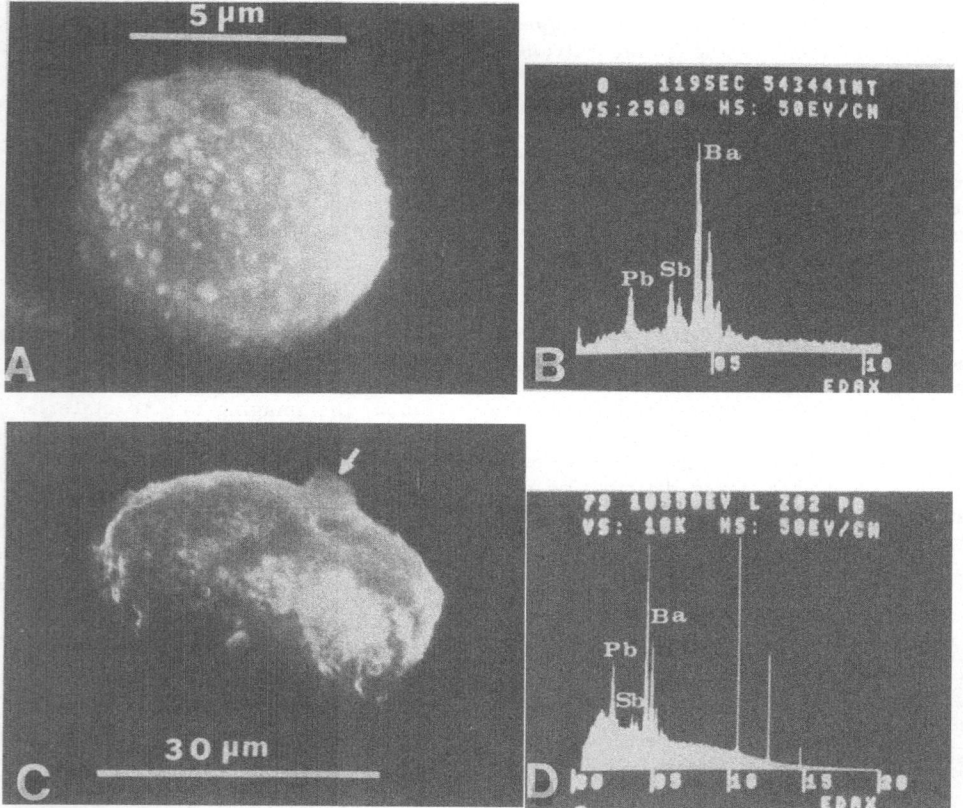

FIG. 9.   Secondary electron images and EDX spectra of two large "full-GSR" which
demonstrate irregular topography of muzzle-blast GSR. These GSR rebound from the wound(s) and deposited on the muzzle hand(s) of victims.
The weapons were the same as in Fig. 8. A & B: GSR diameter 7.1 μm,
original magnification 9,500X, C & D: GSR diameter 32.5 μm, original
magnification 2,400X. The arrow shows a "stalk" of the GSR droplet
(frozen). Specimens not carbon coated.

Fig. 13B ranged from 0.5 to 18 μm. Many of these GSR were associated with
the elements of bullet coating and casing (e.g. tin, copper, etc.), although their
major elements were lead (Pb), antimony (Sb) and barium (Ba). Ueyama et al[16]
first made a systematic analysis of GSR in muzzle deposits at different muzzle-
to-target (solid) distances, using the SEM-EDX. This subject is of course not
new and there are many publications on the determination of muzzle-to-target
distances (See Ueyama et al for references prior to 1980). The result in Fig.
13B is, however, new as it indicates that backscattered electron mode, coupled
with EDX, is able to make a fast discrimination between the bullet entrance
wound and the bullet exit wound. If it was an exit wound only, bullet residues
would have been found.

In combined homicide-suicide cases (attended shootings), the hands of the
victim represented irregular, bullet-derived GSR of muzzle-blast in association
with bullet fragments and bullet residues (Pb). In contrast, at least one hand
(usually right hand) of the murderer contained mainly regular (R) and nodular
(NS) GSR spheroids. These were far from being irregular.

GSR Distribution

The interpretation of a shooting crime requires a strong scrutiny of the GSR

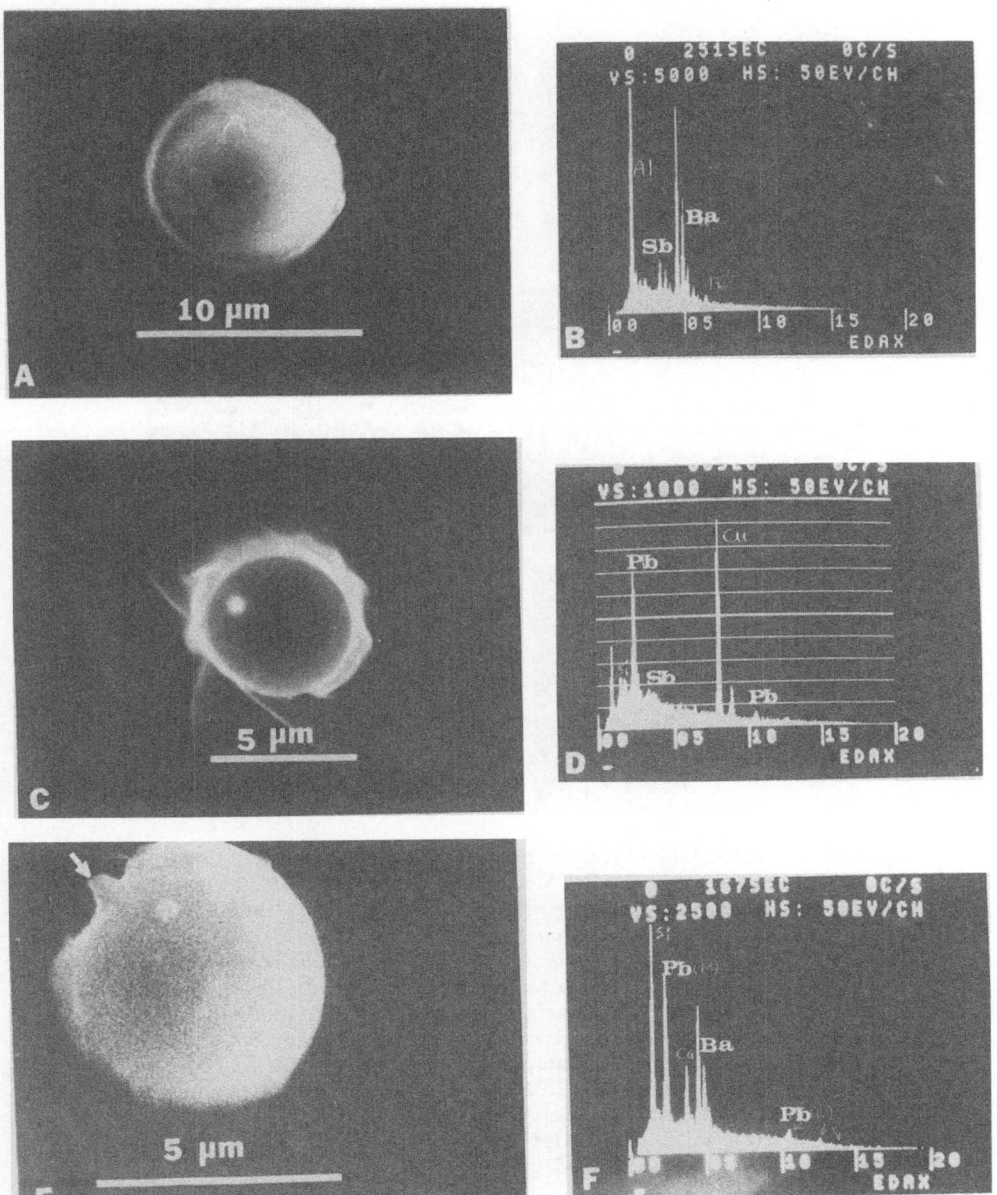

FIG. 10. Secondary electron images and EDX spectra of regular, smooth binary GSR. All spheroids are from the firing hands (hands on trigger). The spheroids in A, B and E, F are smooth, primer-GSR. The spheroid in C, D is bullet-derived since it contains the element copper (Cu) of bullet coating. A, B: SbBa diameter 9 μm, original magnification 5,000X, a shotgun (16 ga. Ithaca®) suicide; C & D: PbSbCu, diameter 5 μm, original magnification 6.500X, a handgun suicide (.22 caliber H & R® revolver, 22 long-rim fire, 150 grain cartridge); E, F: PbBa, diameter 5.5 μm, original magnification 11,000X, a test firing with a 20 ga. Savage/ Stevens® hinge frame shotgun with 2 3/4 in. #4 shell. Arrow shows a "stalk" of GSR spheroids. Specimens not carbon coated.

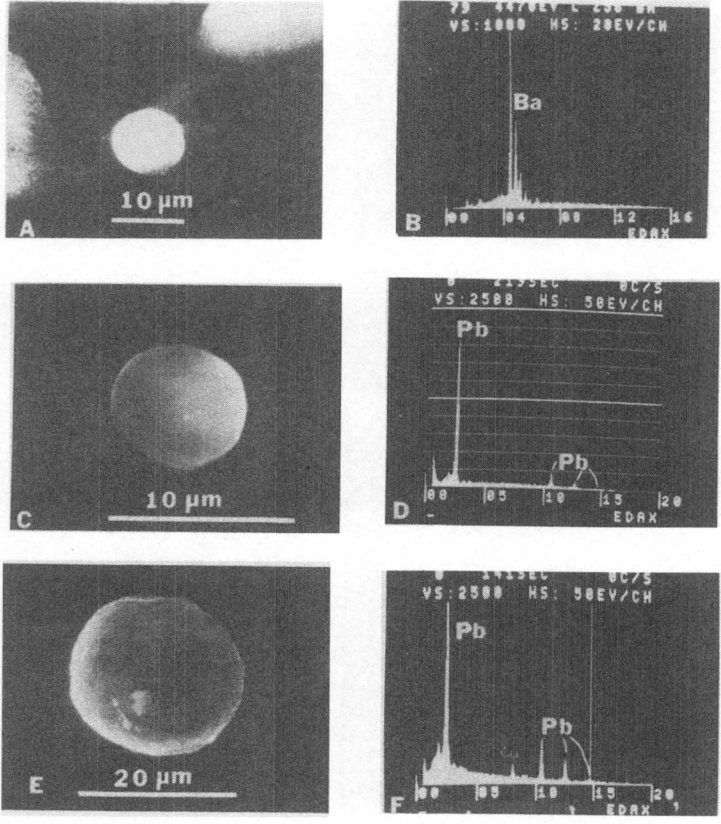

FIG. 11. Secondary electron images and EDX spectra of three
large mono-GSR ("Monomers") which represent smooth,
regular topography of "trigger–blast" GSR (breech residues) from the firing hands. A & B: Ba, GSR diameter
10 μm, original magnification 2,000X, a shotgun (12 ga.
Remington®, hinge frame) suicide; C & D: PbCu, GSR
diameter 7.3 μm, original magnification 5,200X, a
handgun (.22 cal. H & R® revolver) suicide; E & F: Pb,
GSR diameter 22 μm, original magnification 2,000X,
a handgun suicide (.22 cal. H & R® model 929 revolver,
.22 long rifle caliber CCI® stinger). Note that the particle in C, D is bullet–derived since it contains Cu.
Specimen uncoated.

distribution for the four collections from the back of the hands and the palms.
The potential variables that operate on density distribution at the instant of
the shooting are the "specifics of hand positions" on the gun. The relative density of GSR deposits on the back vs palm of a hand can be taken as a shadow
graph of the grasp of the hand (cf. Table 3).

Fig. 14 is an illustration of this subject. The case refers to a self–inflicted
death with a .38 caliber revolver (S&W®), which was first examined by a firearms
examiner and then used for the test firings in Fig.(s) 4–7. The observed GSR
distribution (#GSR per 4 circles per collector) on the submitted collectors (RB
= 47, RP = 3; LB = 2104 and LP = 3) was consistent with the residue distribution
normally found in handgun suicides with the right hand on the trigger (Fig. 14
A and B) and left hand on the cylinder, or around the muzzle (Fig. 14 C and D).
The GSR characteristics were supportive to this end; namely, a mixture of regular

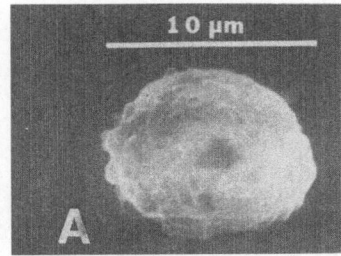 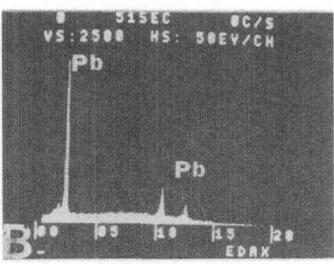

FIG. 12. Secondary electron image and EDX spectrum of a
large mono-GSR (Pb) from the muzzle-hand of a
suicide victim with a 20 ga. Savage/Stevens® single-shot hinge frame shotgun. GSR diameter 10 μm,
original magnification 5,700X, uncoated specimen.

and irregular GSR and an enormous amount of lead (Pb) spheroids were on the
left hand back, whereas the deposit on the right hand back was enriched in regular
GSR spheroids. The heavy GSR deposit on the left hand back (Fig. 14C) indicated
that this area of the hand was an obstruction to the residue blast. Reconstructive
firing(s) against a simulated target confirmed that the left hand of the victim
was around the cylinder of the revolver. (See also Fig. 10 in Basu et al[3] for other
potential hand positions and the results of these firings.) The GSR distribution
(#GSR per 4 circles per collector) obtained on reconstructive firing was as follows:
RB = 47, RP = 0, LB = 2184 and LP = 80 (Fig. 14 E and F). The left hand in these
shootings was a gloved, plastic hand. The back of this hand obstructed the blasted
residues through the flash gap between the cylinder and barrel. Consequently,
this area of the hand had a heavy deposit (Fig. 14 F). An examination of the plastic
(left) hand and the submitted color photograph of the victim's left hand revealed
that both hands had black patches of residues in an identical area (back). This
evidence of cylinder blast added confidence in reporting of the most probable
hand positions of the victim on the submitted weapon. The circles (one per collector) in Fig. 14 A–F indicate the quality of average Glue-Lift collections made
by trained officers. The few dirt particles (arrows) shown for collections from
the victim's hands could have been avoided if "Glue-Lift" touches were slightly
more gentle (cf. Fig. 14 E and F). The GSR particles have been shown by arrow
heads, except in panels C and F (Fig. 14). These fields were overcrowded with
small-sized GSR and lead spheroids of diameters below 1.5 μm. Higher magnifications of the GSR distribution in such circles have been shown elsewhere (See
Fig. 11D in Basu et al[3]; Fig.(s) 2 C and E and Fig. 6 in Basu and Ferriss [1]).

The block tests can also be accessory to interpretation of unusual GSR distribution. For example, if no GSR is detected on the back of the hand(s), despite
collections having been made at the crime scene, then possible blockades of
"trigger-blast GSR", such as towels, cloths and hankerchiefs, etc., should be
found at the crime scene, along with the weapon, and these should be tested
for residue color.

When the GSR distribution on the back of the hands vs the palms are very
similar with the victim and with the suspect, the case refers to a close encounter
(Fig. 15). The density (ρ) distribution are very much alike on the shooter's and
the non-shooter's (by-stander) hands because of the close proximity of these
hands to the weapon. However, the contrast in GSR morphology distinguishes
the shooting hand from the non-shooting hand. Therefore, the positive indication
of GSR, mainly on the back of the hand (Fig. 15, shaded area) implies a shooting
hand, only if it is a proven case of unattended shooting. Infact, in homicide cases,
in which the victims were reportedly at a distance from the shooting weapons,
significant amounts of regular GSR spheroids were obtained from the back of
the hand(s) of the suspects. The maximum delay in sampling in such cases had
been approximately 2 hours.

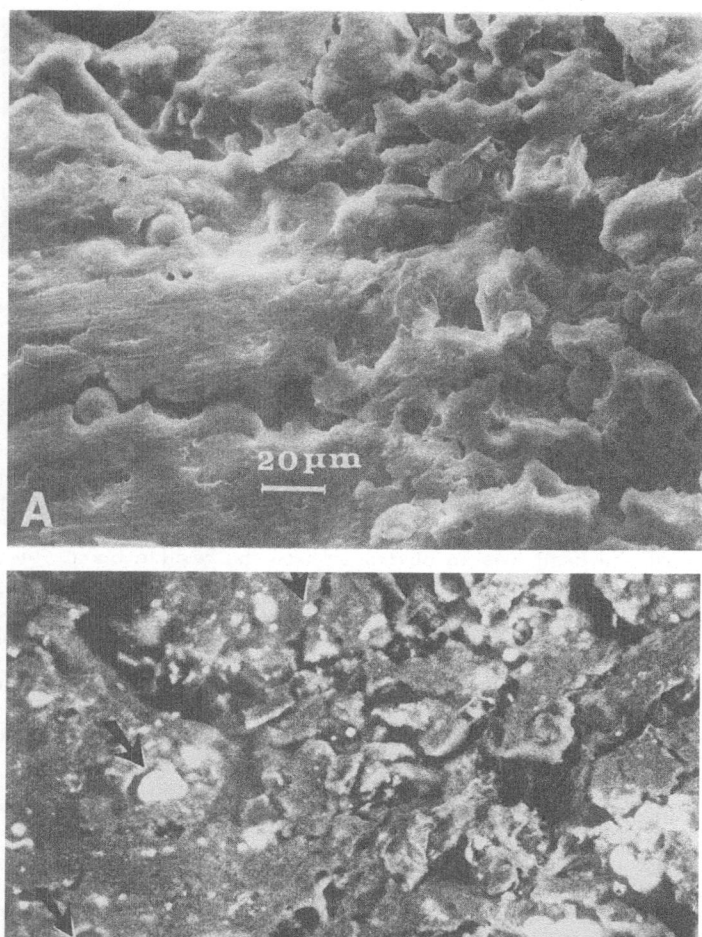

FIG. 13. GSR in tissue scrapings from bullet entrance wound.
A: secondary electron image and B: backscattered
electron image of the same field of view with an annular solid state detector.[9] Uncoated specimen mounted on a Glue-Lift disk. Original magnification: 500X.
Electron potential: 20 kV. Arrows in 'B' show GSR.

DISCUSSION

The discharge of GSR upon firing a weapon is due to detonation of primer,
followed by the controlled burning of gunpowder.[1] The GSR are formed prior
to explosion of gunpowder. Because the discharge takes place under high pressure
of burned powder, a forced deposition (GSR) is the result. Wolten et al[5] first
speculated that GSR may deposit onto the shooter's hands by a blasting process.

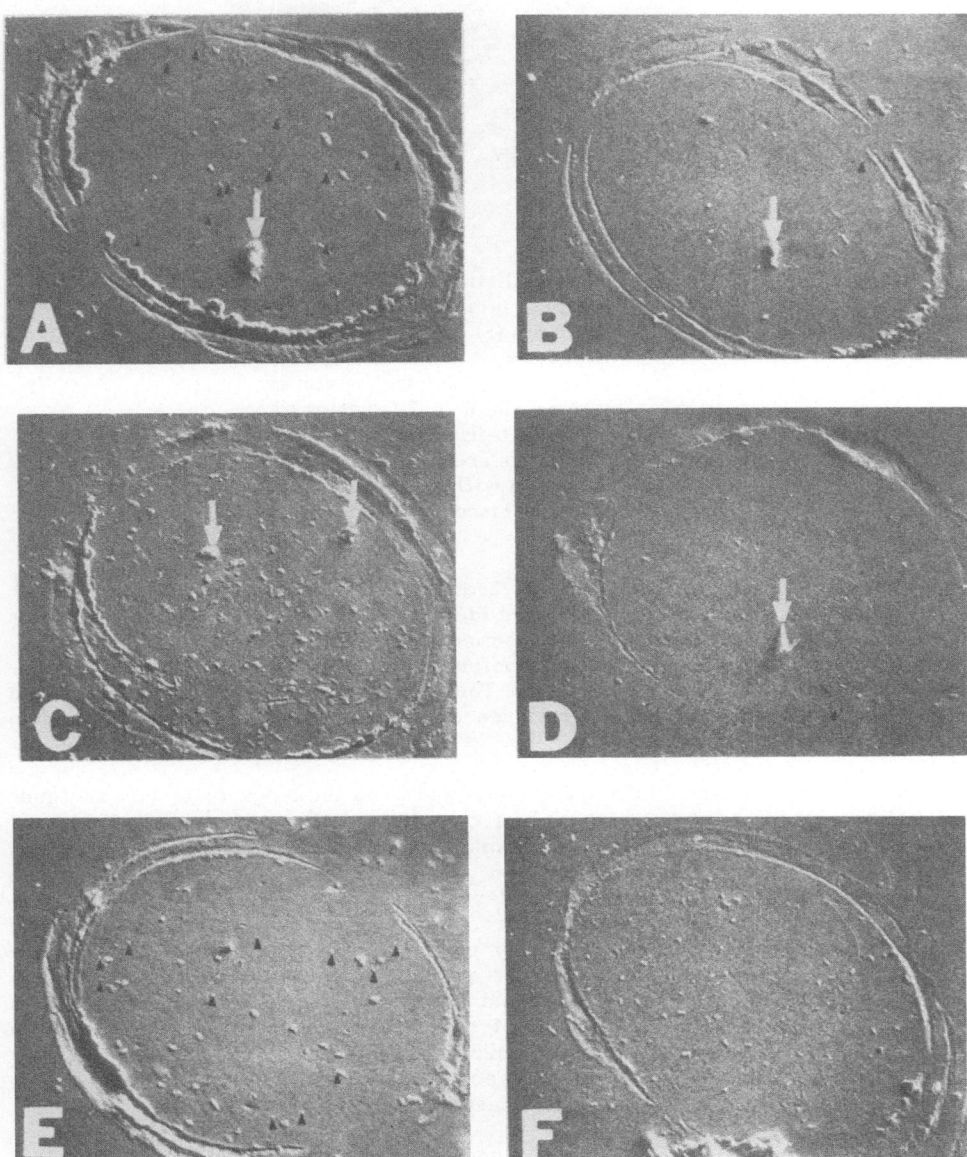

FIG. 14. Backscattered electron images of GSR in circles. A,B: from a suicide
victim's right hand back (A) and palm (B); C,D: from the same victim's
left hand back (C) and palm (D); E,F: from a test-shooter's right hand
back (E) and from the back of a plastic left hand, around cylinder (revolver) in a reconstructive firing (F). Weapon and ammunition same
as Fig. 4.

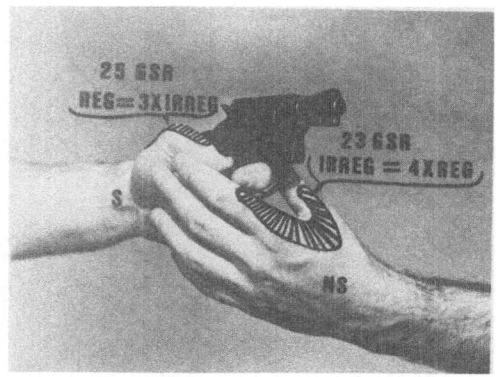

FIG. 15. Density distribution (#GSR/4 circles, 7.1 mm^2) of GSR on the shooter's (S) and non-shooter's (NS) back of hands in close encounter of a shooting revolver (.38 cal. S&W®). The ratios of regular-to-irregular GSR's on the two hands are contrasting. (Reproduced with permission from San Francisco Press, Inc., ref. 7).

The results of the block-tests (Fig.(s) 2 and 3) and the density distribution (ρ) test (Fig.(s) 4-7) are consistent with the idea. The breeches of shotguns and rifles are not necessarily leak-proof to the micron-size GSR. Knowing precisely how the breech residues deposit onto the shooter's hands, especially with these "closed-breech" weapons, had been a mystery. This is the first time it has been shown with these weapons that breech residues contribute the most to the deposition on both hands of the shooter.

The more fundamental information obtained from the block-tests with shotguns and rifles is that the primer gas-mixture containing GSR is subject to a backward thrust ("trigger-blast"), due to the exploding nitrocellulose (gunpowder). This enforces this gas-mixture to be ejected via breeches. The ejected gas mixture would spread backward and sideways, depending upon the size, location and the structure of breech. The detonation front, comprising the primer-gas mixture that strikes the gunpowder, is obviously pumped forward into the barrel, where it mixes with the burned and unburned powder, and the vapors of the etched bullet. However, due to rifling and spinning, the bullet does not emerge at once. With a shotgun, the size and amount of pellets are also a hindrance to residue flow. This forces a portion of the residue gas to escape tangentially forward through the breeches alongside the weapon. The residual gas mixture leaves the muzzle as the bullet emerges ("muzzle-blast").

The residues of two blasts should not superimpose on the shooting hand(s), except in self-inflicted, contact shootings, in which the muzzle-blast residues may rebound from the area of the wound.[3] However, the hand has to be within the "cut-off" distance of rebound GSR. If the GSR on the shooting hand contains bullet-derived elements (e.f. Cu, Sn, etc.) it does not mean that they originated from the muzzle. Such GSR are found with rimfire and rifle caliber (.22 cal.) cartridges (cf. Fig(s) 10 and 11, C and D). These are usually regular, smooth spheroids if they originate from breeches. The muzzle-blast GSR are generally irregular (cf. Fig.(s) 9 and 12). Because these GSR spend more time in the barrel in the presence of the hot gas-mixture of burned and unburned powder, they pass through various metastable states that are due to boiling, fragmenting,

20

or etching, until they escape from the muzzle and freeze in their existing forms.[2]

The forced deposition of GSR with a pre-cleaned weapon is usually systematic. The rate of capture of GSR on the shooter's hands is not influenced by the elements in them, but by the pressure of the incoming particles and by the "through-put" of residue gas through the leaks of the gun. This makes the ammunition (primer content) and the gun-structure more important than the substrate structure (skin or plastic, etc.). The substrate surface properties, such as roughness, surface tension, attachment coefficient and temperature, etc. are less important in forced deposition. Because the leaks are accessible to GSR with a cleaned gun, a consistent deposition is the result of each firing. Whether a GSR would be a mono-GSR (Pb, Sb or Ba), binary-GSR (PbSb; SbBa; BaPb) or a full-GSR (PbSbBa), depends on the probability of mixing of the element gases (Pb, Sb, Ba). The density distribution ( ) of GSR is less variable than other distributions based upon GSR elements, sizes and morphologies (Fig.(s) 4-7). At present, this density distribution of GSR and the spatial distribution of GSR on the firing hand[17] are the most fundamental criteria required for reconstruction of a shooting crime.[3] The size distribution studies (Fig. 6) indicate that density distribution criterion should not fail unless very large GSR are found. This drawback can be overcome by additional test shots.

The forced deposition of GSR gives rise to the residue (density) contrast on the back of the hand vs the palm and vice versa. The disadvantage is that a heavy amount of GSR could be transferred to a non-shooter's hand, due to physical contact with the most contaminated areas of the fired weapon. Our previous studies of reconstruction of suicides[3] were not affected because the observed amounts on the palms of the victim's hands were about 6% in excess of the expected amounts. However, the interpretation of a crime could be difficult if GSR are found only on the palms of the hands, and no GSR are found on the back of the hands. The possible blockade(s) of "trigger-blast" residues have been searched at the crime scene of such cases.

The experiments in this report had to be limited to only single firing tests, so that the deposition mechanism of GSR can be revealed. Whether any difference in conclusion should be found in multiple firing situations, would have to be determined. The residues on multiple firings do not necessarily add up or increase with the number of firings. (Table 3, row 2) (cf. Krishnan et al,[13] Guinn,[14] and Goleb and Midkiff.[15])

ACKNOWLEDGMENTS

Appreciation is expressed to the New York State Police Crime Laboratory for sponsoring GSR research. Thanks are due to Robert Miazga for his assistance in photography. This work would not have been possible without the assistance of the following firearms examiners: Sgt. Charles E. Boone, Jr., Sgt. A. Spencer, Jr. and Sgt. D. J. Denio, Jr., and the cooperation of many trained officers of the New York State Police and various police departments in Upstate New York. Thanks are also due to Mary Larmour for her editing assistance and preparing the typescript, and to Lt. Peter F. Kelly of the New York State Police Crime Laboratory and Dr. J. R. Millette for reviewing the manuscript.

REFERENCES

1. S. Basu and S. Ferriss, A refined collection technique for rapid search of gunshot residue particles in the SEM, Scanning Electron Microsc. I:375 (1980).
2. S. Basu, Formation of gunshot residues, J. Forensic Sci., 27:72 (1982).
3. S. Basu, S. Ferriss and R. Horn, Suicide reconstruction by glue-lift of gunshot residue, J. Forensic Sci., 29:843 (1984).
4. A. Zeichner, H. A. Foner and M. Dvorachek, Evaluation of concentration techniques for small populations of gunshot residue (GSR) particles using calibrated GSR suspensions, in: Proc. 43rd Annual Meeting of the Electron

Microscopy Society of America, G. W. Bailey, ed., San Francisco Press, Inc., San Francisco, pp. 130,131 (1985).

5. G. M. Wolten, R. S. Nesbitt, A. R. Calloway, G. L. Loper and P. F. Jones, Final report on particle analysis for gunshot residue detection, Report ATR-77 (7915)-3, The Aerospace Corporation, El Segundo, Calif., pp. 46-47, 49-53, 93-105. Sept. 1977.

6. V. R. Matricardi and J. W. Kilty, Detection of gunshot residue particles from the hands of a shooter, J. Forensic Sci., 22:725 (1977).

7. S. Basu, The mechanism of gunshot residue deposition and its probing characteristics to reconstruct shootings, in: Proc. 43rd Annual Meeting of of the Electron Microscopy Society of America, G. W. Bailey, ed., San Francisco Press, Inc., San Francisco, pp. 104-107 (1985).

8. T. E. Everhart and R. F. M. Thornley, Wide-band detector for micro-microampere low-energy electron currents, J. Sci. Instr. 37:246 (1960).

9. E. D. Wolf and T. E. Everhart, Annular diode detector for high angular resolution pseudo-kikuchi patterns, Scanning Electron Micros. I:41 (1969).

10. H. Niedrig, Physical background of electron backscattering, Scanning, 1:17 (1978).

11. J. S. Bashinski, J. E. Davis and C. Young, Detection of lead in gunshot residues on targets using the sodium rhodizonate test, Assoc. Firearm and Tool Mark Examiners J. 6 (No. 4): 5 (1974).

12. Gunshot residues and shot pattern tests, published by the Federal Bureau of Investigation, U.S. Department of Justice. Reprints from the FBI Law Enforcement Bulletin, Sept. 1970, pp. 1-7.

13. S. S. Krishnan, K. A. Gillespie and E. J. Anderson, Rapid detection of firearm discharge residues by atomic absorption and neutron activation analysis, J. Forensic Sci., 16:144 (1971).

14. V. P. Guinn, Applications of nuclear science in crime investigation. Ann.Rev. Nucl. Sci., 24:361 (1974).

15. J. A. Goleb and C. R. Midkiff, Jr., The determination of barium and antimony in gunshot residue by flameless atomic absorption spectroscopy using a tantalum strip atomizer, Appl. Spectrosc., 29:44 (1975).

16. M. Ueyama, R. L. Taylor and T. T. Noguchi, SEM/EDS analysis of muzzle deposits at different target distances, Scanning Electron Microsc. I:367 (1980).

17. H. Gansau and U. Becker, Semi-automatic detection of gunshot residue by SEM-EDX, Scanning Electron Microsc., I:107 (1982).

SPATIAL VARIATIONS OF SULFUR AND ABSORBED COPPER IONS IN HUMAN HAIR

BY X-RAY LINE SCANNING*

S. Basu

New York State Police Crime Laboratory
Albany, New York   12226

ABSTRACT

The x-ray line scanning technique in the scanning electron microscope
has the inherent ability to produce useful parameters of comparative hair
identification. Since sulfur is the major endogeneous, inorganic element
of hair keratin and this element is a stable (i.e. unwashable) constituent
of hair cuticles, the line scanning distribution of sulfur in segments of
hair has the potential to serve as a reference for comparative assessment
of the modes of distribution of other minor elements of hair. A secondary
electron image of the sample area is superimposed over the elemental line
scans. The method has been tested with the use of hairs containing absorbed cupric ions from a soaking solution. A seemingly important parameter of hair identification is offered by a sudden, characteristic rise in
the uptake of sulfur within endodermal portion of the root of hair, which
is followed by a gradual increase in the intensity of sulfur along the
length of hair shaft. The rate of incorporation of sulfur measured at the
root of hair ("sulfur index") varies remarkably from one individual to
another. Conventional microscopy test for hair comparison can be significantly aided with the uses of these elemental parameters examined in the
SEM-EDX.

INTRODUCTION

Forensic hair comparison requires parameters of obvious interpersonal
specificity. Such parameters should not be confined to hair anomalies
associated with certain diseases[1] and cosmetic treatments.[2] The hairs
from different parts of an individual's head (scalp) are rarely equivalent
in terms of macroscopic and microscopic characteristics,[3,4] and concentration distribution of many endogeneous, inorganic elements of hair except
sulfur.[5,6] Seta[7] has recently suggested that the concentration distribution of these elements (e.g. potassium, calcium, chlorine, zinc and phosphorus, etc.) in scalp hairs is dependent on physiological status of body.
Interestingly, these inorganic elements of hair are leached out of
the hair structure during washing in distilled water and in standard hair
washing solvents (ethanol and diethyl ether, etc.).[6] Sulfur (S), the major inorganic element (5% w/w) of hair,[6-9] remains unaffected under the

---

*Presented at the Joint Annual Meeting of EMSA/MAS, Louisville,
Kentucky, 5-9 August, 1985.

same condition.[6] Sulfur is known to occur as -S-S- cross-linkages between cystine residues of adjacent peptide chains of eukeratin (α- Keratin).[8] Human hair is composed almost entirely of this keratin (eukeratin).[9] Sulfur is also more abundant in cuticles and cortex than in medulla.[6] This at once suggests that if we are able to obtain superimposed x-ray line scans due to S and to a minor element of hair across the width of hair, or along its length, a comparison of the shapes of these elemental line scans will suggest whether that minor element is distributed uniformly in hair like sulfur, or if that element was applied externally to hair.

Such parameters of comparative hair identification are not only important in forensic investigation but are also needed for diagnostic purposes. The project is intended for x-ray excited fluorescence spectroscopy with a scanning electron microscope (SEM) which will include, in due course, analysis of trace elements. The present studies were conducted by a careful application of the analog,[10] or ratemeter approach, to energy dispersive x rays of sulfur and a minor element (copper) of hair, with a view to determine the merit of the line scanning approach. There is seemingly a skepticism about the origin of metallic elements, such as, copper, zinc, calcium, sodium and potassium, etc., which are occasionally found in elevated amounts in scalp hairs of some individuals. This problem has been avoided here by incorporating copper ions from a solution of Cu $(NO_3)_2$ (pH 4.5) into scalp hairs which have had no prior indication of copper, and maintaining the Cu-uptake as measured after washing and drying the hairs to about 0.6% which is encountered in some individual's hair.

The present report also contains studies of spatial variation of sulfur starting from the proximal end of hair up to a distance of about 1.5cm on the hair shaft. To the knowledge of this author there has not been any such study. These studies have indicated a new parameter of hair identification.

This report is largely an elaboration of a paper which was presented at the 1985 Joint Annual Meetings of the Electron Microscopy Society of America and the Microbeam Analysis Society in Louisville, Kentucky.[11]

MATERIALS AND METHODS

Collection

Sample hairs were plucked from the heads of several individuals who seldom applied any hair formula or cosmetics to their hair. The plucking of hair was accomplished with the use of 3C diamond honed forceps washed in diethyl ether (anhydrous), prior to use. Only hairs in the anagen phase of the hair growth cycle were selected for incorporation of copper ions in view of the potential interest of the correlation in time, of features observed in various single hairs. The plucked hairs with the root attached were examined at 70X magnification in a stereo-microscope to verify the growth cycle phase. Hairs in the anagen phase were taken to be those with a plump and pigmented root, with the white root sheath attached and a curling up of the shaft at the proximal end.[12] These hairs were examined in the SEM-EDX at different orientations. The selected hairs were those which indicated no x-ray intensity due to Cu (K_α) at 8040 ev (energy).

Both plucked and combed hairs were employed for studies of spatial variation of sulfur. The scalps of several donors were combed with a bristled hair brush. The selected hairs (6 to 9 per head), each characterized by a transparent, knobby root at the proximal end of hair and no attached sheath cells were collected in a glassine envelope per scalp. These were samples of hairs in the telogen (i.e. dead or resting) phase in the growth cycle of the hair.12

24

Washing

Successive washings of hairs were effected in diethyl ether, acetone, ether, acetone, ether, acetone and finally ether, for periods of five minutes per step with slight agitation.[13] Hairs were then air dried at room temperature. It was necessary to encapsulate each hair in a twenty-lambda (20λ) micro-pipette. The storage of hair without contaminating the hair shaft and the subsequent soaking experiment in tracer solution were thus facilitated.

Inactive Tracer

A fully saturated solution with a maximum cupric ion concentration in the pH range of 4.5 - 5.0 was prepared by dissolving 61gr Cu $(NO_3)_2$ $3H_2O$ in 100 ml deionized distilled water at $40^{\circ}C$. This solution was diluted to a strength of about 9.7% w/v of Cu $(NO_3)_2$ and its pH was adjusted to about 4.5 with 1 N Na OH. This was the final solution for inactive tracer (Cu-ions).

Soaking in Tracer

About 100 ml of the final tracer (Cu-ions) solution was taken in a 250 ml beaker. The hairs were adjusted inside the micro-pipettes (20λ) so that the tip of hair root extended a few millimeters below each micro-pipette tube, and these ends of the micro-pipettes were inserted vertically into the tracer solution. The capillary rise of the bulk liquid into each micro-pipette (20λ) was to a distance of 2.6cm. The total soaking distance on individual hairs varied from 2.9cm to 3.4cm. This capillary method of soaking gave an access to having a control over the soaking distance on individual hair.

The hairs were incubated in the tracer solution for a period of 48 hours at room temperature ($25^{\circ}C$). After this incubation, the hairs were washed and air dried using the previously described procedure. The hairs were stored for a 24 hr period prior to mounting and analysis in the SEM-EDX.

Mounting

Sample hairs were cut to 1-2mm segments from the root up to a distance of 1.5cm. These were attached in known arrays on diluted rubber cement layers on a polished carbon planchet (Fig. 1). The hairs in evidence (matching test) were cut to about 1.5cm segments which included the root of each hair. All specimens were coated with evaporated carbon.

SEM-EDX

The specimens were examined at 20 kV electron potential using a grounded specimen stage in an SEM (AMRay 1000). Elemental studies were performed with an energy dispersive x-ray spectrometer in conjunction with an analyzer (EDAX-707A) and a computer (EPIC module-609) that identified the element peaks. The x-ray line scanning studies were conducted with a R-C ratemeter (EDAX-352).

These various studies of hairs were preceeded by a thorough search for the sources of spurious metallic emissions in the SEM (AMRay 1000) chamber. The brass surfaces of a "quad" backscatter detector which was installed around the aperture of the final lens were the main sources of x rays due to copper (Cu) and zinc (Zn). The major source of x rays due to aluminum (Al) was the collimator window of the x-ray detector. Since these spurious emissions were caused by high energy (> 10 keV) backscattered electrons reaching those surfaces from the specimen mount, those surfaces were heavily coated with evaporated carbon (thickness $\geq$

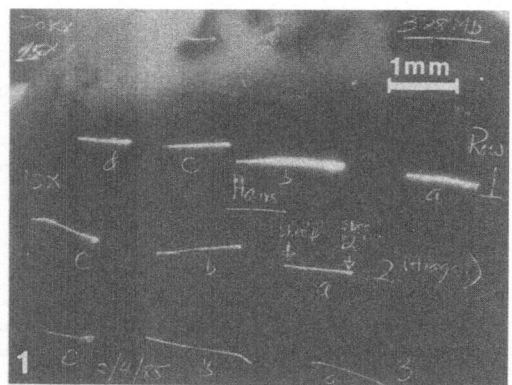

Fig. 1 Cut segments of a hair from the root end up to a length of 1.5cm. Sample mount is a polished carbon planchet (diameter 13mm) with a spread layer of diluted rubber cement. Scale bar: 1mm.

10 µm). The solar windows (four) of the backscatter detector and their leads to a chain of amplifiers were carefully covered with small pieces (circular) of masking tapes. These tapes were removed after the required cycles (several) of carbon evaporation. Commercial grade carbon paints and inks of many marking pens contain ferromagnetic elements (iron and cobalt) and silicon and, therefore, these materials were not used for coating.

The x-ray studies were performed keeping the specimen (tilt: 45°) containing the point of x-ray emission in-line with the axis of the x-ray detector (working distance 12mm in AMR 1000). Also, the specimens were examined in close proximity (3cm) of the collimator window of the x-ray detector. This increased the x-ray take off angle and collection (cf. Fig. 16 in Ref. 14).

Copper Uptake

Each hair segment was irradiated with 20 kV electrons at an emission current of 75-90 µa to produce an integral x-ray count rate of 1200-1500 cts/sec. Either partial field or larger magnifications up to 1300X were used in the scanning mode to obtain a full view of the width of hair on a display screen, while giving the above count rate of total x rays. Several segments of each hair were examined in this manner and the representative x-ray emission spectra due to control (untreated) hair and hair containing Cu-ions (washed and dried) were superimposed (Fig. 2). An estimate of the uptake of copper was obtained by a comparison of the background free x-ray counts due to S and Cu at their K lines and by taking into consideration that S concentration is about five percent (w/w) of scalp hairs.[6,8,9] The brass surfaces of the "quad" (backscatter) detector were not entirely coated with carbon, particularly in the areas under the masking tapes. Therefore the Cu-uptake was further checked first by covering the "quad" detectors with an annular gold (coated) foil and then with an annular glassine paper (uncoated) in two successive experiments. The foil and the paper were cut to the same size and were held onto the "quad" detector with pieces of transfer tape such that the foil, or the paper, would not be in the way of the scanning electron beam. Characteristic x-ray spectra were recorded with and without the above foil

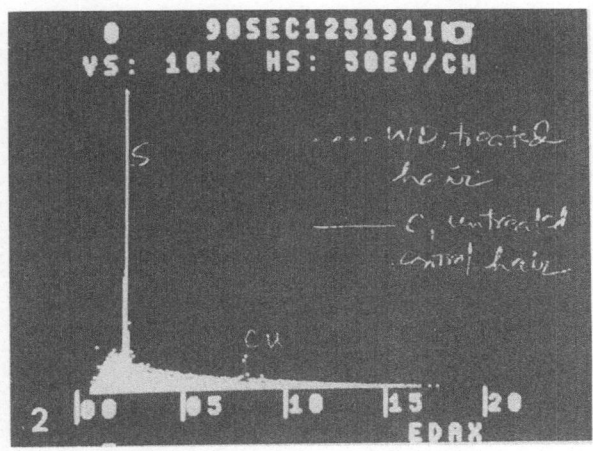

Fig. 2 Characteristic (EDS) x-ray spectra.
Dotted line (WD): Anagen (scalp) hair
soaked in tracer (cupric ions) solution
and then washed and dried. Solid line
(C): anagen hair, untreated control,
from the same individual's head.

(or the paper) and these were superimposed. Despite the appearances of
small gold peaks with the gold foil-in, the Cu (K_α) peaks were identical
in the two cases.

Line Scanning

For line scanning of hair with Cu (K_α) x rays, the analyzer (EDAX
707A) window was set at 8040 ev which corresponded to K line of copper.
The energy window was adjusted to two-thirds of half-band width of the Cu
(K_α) peak so that an acceptable count rate (25-100 cps) was obtained. The
count rate with S (K_α) x rays at 2310 ev was highly variable depending
upon the area of hair under observation (i.e. the root of hair or away
from the root).

The precautions taken in the line scanning studies were the same as
in the x-ray mapping of elements.[15] The SEM x-ray line scans using an
analog, or a R-C ratemeter, has both advantages and disadvantages.[10] The
main advantage is that line scans due to multiple elements and to
secondary electrons can be superimposed on the normal SEM image.[10] This
avoids the required first-order correction for geometry effects.[10] The
main disadvantage is that several problems might arise as a result of
frequent lack of correspondence between time constant of the R-C circuit,
count rate and scan speed. The major problem is "tailing" or masking of
small changes when the time constant for the R-C circuit is too high. The
other problem is due to high count rate (the dead time effect) when the
signal reaches an immediate saturation instead of rising (or decreasing)
in proportion with the count rate (i.e. intensity). These and other
problems of conventional (R-C) analog line scan were minimized by proper
combinations of count rate, slower scan speed and ratemeter settings
(scale factors for integration).

The R-C ratemeter (EDAX-352) allowed line scanning in two different modes called 'set-up' and 'operate' modes. In the set-up mode the signal contained noise due to fluctuation in background counts. The 'noise' was suppressed in the 'operate' mode. The line scans in this report were obtained in the 'operate' mode.

Depending upon how the local axis of hair is inclined to the axis of the x-ray detector, the x-ray line scan may be shifted relative to the position of hair in the secondary electron image. The shift may occur in both orthogonal directions (x and y). This effect was avoided by a slight rotation of the sample disk and an adjustment of the working distance (i.e. height).

RESULTS AND DISCUSSION

Cu-uptake

Fig. 2 shows representative x-ray emission spectra of the Cu-treated hair and a control, untreated hair. These spectra are shown in superimposition. The hair containing absorbed Cu-ions exhibited K (K_α, K_β) and L peaks of copper (Cu) and K (K_α, K_β) peaks of sulfur (S). The control hair showed mainly the K (K_α, K_β) peaks of sulfur (S). The S (K_α) peak was more intense than the Cu (K_α) peak because S is the major inorganic element of hair keratin. The Cu-uptake, measured relative to known sulfur content (5% w/w) of scalp hair,[6,8,9] was in the range of 0.3% to 1% (w/w) (approx.). This spectral analysis has proved that Cu-ions were incorporated in detectable amounts into sample hairs.

The Cu (K_α) peak at 8040 ev was about 3 to 4 times the background radiation at this energy. The atomic number of Cu (Z=29) is much higher than that of S (Z=16) and other organic elements (C, O, H, etc.) of hair. Therefore, the Cu (K_α) intensity was free from any contribution of x-ray (Cu) fluorescence.[16]

Characteristics of Absorbed Copper (ions)

The line scans of Cu (K_α) and S (K_α) x rays have been studied at different areas of the same anagen hair containing incorporated cupric ions. The scans were obtained across the width of hair at the root and at various distances away from the root. These are shown sequentially in Fig.(s) 3-5.

The shapes of these perpendicular line scans were influenced by density effect (i.e. changes in Cu-concentration) and irregular hair topography near the root (Fig. 3). Slightly above the root the surfaces of hair started to be smoother and the density effect (changes in Cu-concentration) continued to prevail (Fig.{s} 4,5). The line scans due to secondary electrons assisted in this evaluation (Fig.{s} 4B and 5A). When the absorbed Cu-ions were present uniformly across the width of hair, the line scans due to S (K_α) and Cu (K_α) x rays were similar (Fig. 4A and B). These were round or dome-shaped (hemispherical) with a maximum intensity corresponding to the middle of the width of hair. Examination of control hairs (no Cu^{2+}) showed that line scans of S (K_α) across hair were also round or dome shaped.

In other segments of the same hair the shapes of line scans due to Cu-ions and S were interchangeable depending upon which edge (right or left) of hair had relatively more copper (Fig. 5A vs Fig. 5B). Thus sulfur distribution was relatively unvarying across hair segments while copper (ion) distribution was varying except in some places where Cu-ions were uniformly distributed (Fig. 4B). These hairs were throughly washed prior to these examinations. Therefore the shown line scan in Fig. 4B was indicative of Cu-ions having been strongly absorbed into hair cuticles.

The intensity of S (K_α) in each sample area was several folds more than that of Cu (K_α) (See the scale factors for Cu and S in captions to Fig.(s) 3 to 5). The peak intensity of Cu (K_α) was about 50 - 60 cts/sec

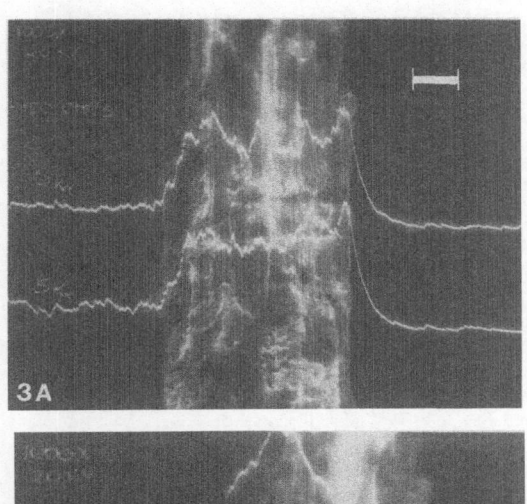

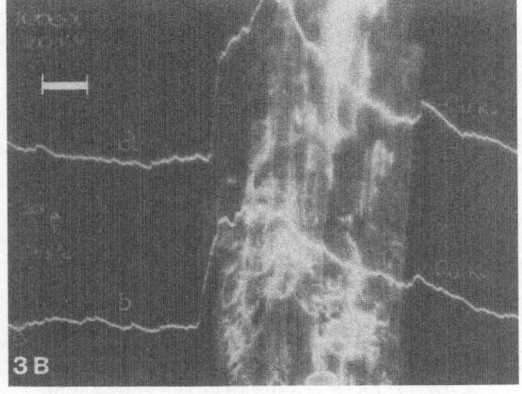

Fig. 3 X-ray line scans across the width
       of a hair at the root. Anagen hair
       soaked in tracer (cupric ions) solu-
       tion, and then washed and dried. A:
       S (Kα) intensity; B: Cu (Kα) inten-
       sity. Scale Factors: 0.5 x 10^3 cps
       (full scale) for S; 10^2 (full scale)
       for Cu. Bar 10 μm, 20 kV. (Re-
       produced with permission from San
       Francisco Press, Inc., Ref. 11)

(Fig.(s) 3-5) whereas the peak intensity of S (Kα) increased from 150
cts/sec (Fig. 3A) to about 300 cts/sec (Fig. 5B). In otherwords the intensity of S (Kα) increased approximately two-folds within a distance of 2
to 3mm from the root of hair. In fact these intensity values first gave
the hint that perhaps the S (Kα) x rays have an increased trend along the
length of hair away from the root of hair.

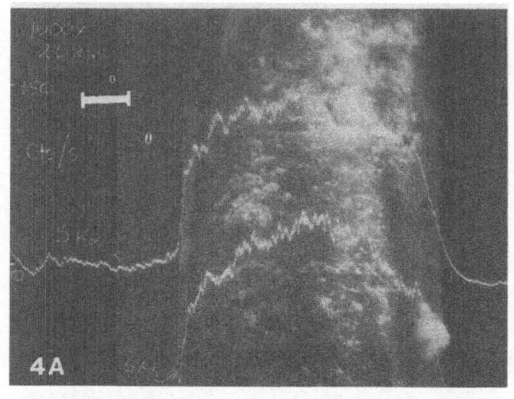

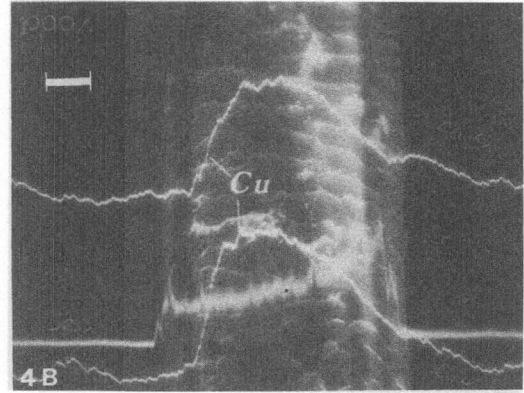

Fig. 4   X-ray line scans across the widths
of a hair immediately above the
root. (The same hair as in Fig.
3). A: S (Kα) intensity; B: Cu
(Kα) intensity. Sec: additional
line scan due to secondary electrons. Scale Factors: 2.5 x 10^2
cps (full scale) for S;10^2 cps
(full scale)for Cu. Bar 10 μm,
20 kV.

Spatial Variations of S and Absorbed Cu

The hint in the previous section has been tested by examining the
line scans due to S (Kα) along length of the same hair segments. The
segments were analyzed without having any prior knowledge as to which end
of a segment was closer to the root. An example of this examination is
shown in Fig. 6. The segment (Fig. 6) was already analyzed in Fig.(s) 5A
and 5B for perpendicular line scans. As the scanning electron beam moved
along the segment (left to right) the line scan of S (Kα) was unvarying

30

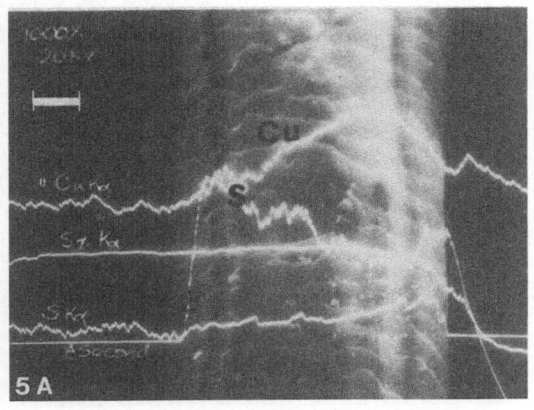

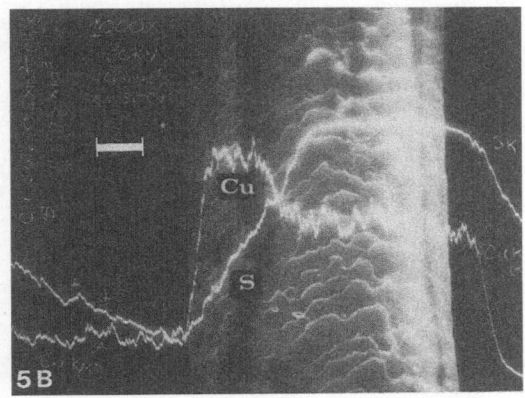

Fig. 5 Superimposed line scans of S ($K\alpha$) and
absorbed Cu ($K\alpha$) intensities. (The
same hair as in Fig. 3). A: 1mm away
from root; B: 2–3mm away from root.
Sec (A): additional line scan due to
secondary electrons. Scale factors:
0.5 x 10^3 cps (full scale) for S (A &
B); 10^2 cps (full scale) for Cu in A
and 0.5 x 10^2 cps (full scale) in B.
Bar 10 μm, 20 kV. (Reproduced with
permission from San Francisco Press,
Inc., Ref. 11).

until a sudden drop was encountered (Fig. 6). The arrow shows this area.
When the segment was reexamined by turning the specimen stub through 180°,
the line scan due to S ($K\alpha$) was reversed. That is, first there was a
trough and then there was a rise. This was followed by an unvarying line
scan due to S ($K\alpha$) along the length of the segment. Simultaneous line
scans of Cu ($K\alpha$) along length of hair have confirmed the varying feature
of absorbed copper ions (Fig. 6, lower scan).

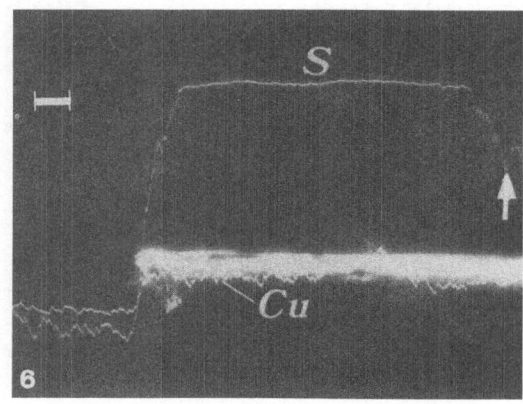

Fig. 6 Line scanning distribution of S
and absorbed Cu in a hair seg--
ment (see text). Scale factors:
0.5×10^3 cps (full scale) for
S; 10^2 cps (full scale) for Cu.
Bar 40 μm, 20 kV. (Reproduced
with permission from San
Francisco Press, Inc., Ref. 11).

A sudden drop, or rise, in the line scans of S ($K\alpha$) have been observed in many segments of hair. The effect was most prominent at the
root of hair. The intensity of S ($K\alpha$) dropped to a minimum when the
scanning electron beam was on top of the bulge ("papilla") of the root of
hair. At this point, studies of control hairs (no Cu++) became more
important.

Characteristics of Sulfur

The combed (telogen) hairs of four individuals and the plucked
(anagen) hairs of fourteen individuals were examined particularly at the
root of hair and up to a distance of about 1.5cm from the proximal end
(root). Because the line scanning variation of S was complemented by the
recorded characteristic spectra, these spectral data will be presented.
For each hair, characteristic x-ray spectra were collected at the bulge
(or tip) of the root and at the various distances away from this area.
For all hairs examined this way, the intensity of S ($K\alpha$) was the lowest at
the bulge of the root (or root tip) (Fig. 7). That is, the intensity of S
($K\alpha$) continued to rise with distance (away) from the bulge of the root
(cf. solid vs dotted spectra in Fig.(s) 7 and 8). This observation was
facilitated with combed (telogen) hairs which seldom contained overlying
sheath cells at the root. The characteristics of two such hairs from two
individuals are shown in Fig.(s) 7A and B. For these hairs (telogen) the
increase in the intensity of S ($K\alpha$) was rather sharp. Such changes
occurred a short distance (50 - 100 μm) away from the bulge of the root.
Despite the irregularity of the hair surfaces at the root, the measured S
($K\alpha$) intensities were comparable on two hairs of the same individual

32

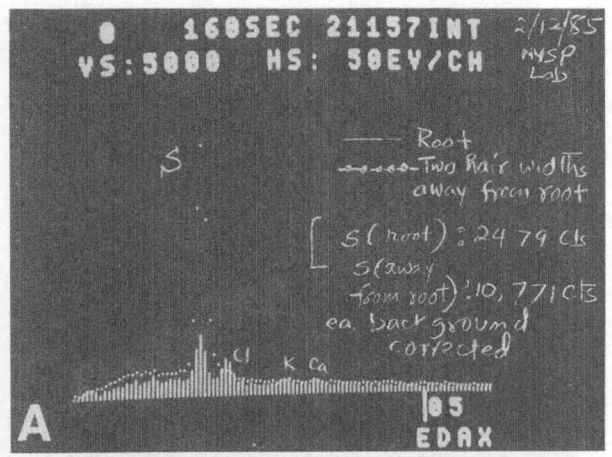

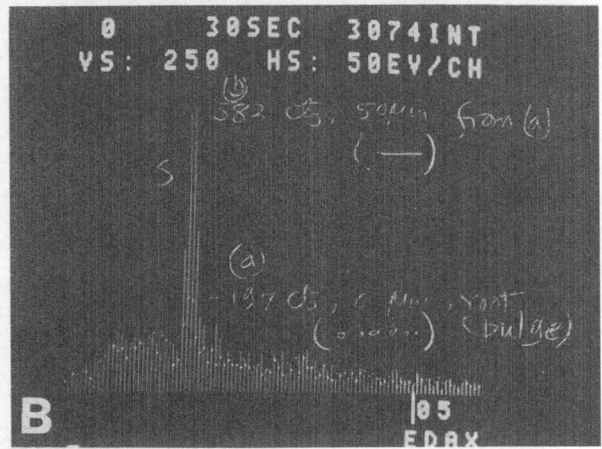

Fig. 7   Energy dispersive x-ray spectra of
combed hairs. A: 1st individual's hair.
Solid Line ____ bulge of root; dotted
line ---- two hair widths (100 μm) away
from the bulge of root. B: 2nd individual's hair. Dotted line ---- bulge of
root, solid line ____ 50 μm away from the
bulge of root. Background-free integral
x-ray cts {S (Kα)} are shown. Time: 160
sec in A and 30 sec in B. 20kV. (Spectra in A reproduced with permission from
San Francisco press, Inc., Ref. 11).

(Fig. 8A vs B).
     Anagen (plucked) hairs containing remnants of attached sheath cells
also exhibited a minimum intensity of S (Kα) at the bulge of the root.
(See these data in the following section). However, the S (Kα) intensities of anagen hairs were always higher than that of combed hairs from the

same head. The combed hairs indicated an early saturation in intensity
{S (K$_\alpha$)} on the hair shaft. This distance measured from the root end of
hair was in the range of 0.5mm to 1cm, depending upon the donor. The
anagen (plucked) hairs of the same individuals (four) and ten other
individuals did not indicate any saturation within 1.5cm from the root of
hair. For these hairs (anagen) the intensity of S (K$_\alpha$) continued to rise
rather slowly within the endodermal portion of the root of hair. The

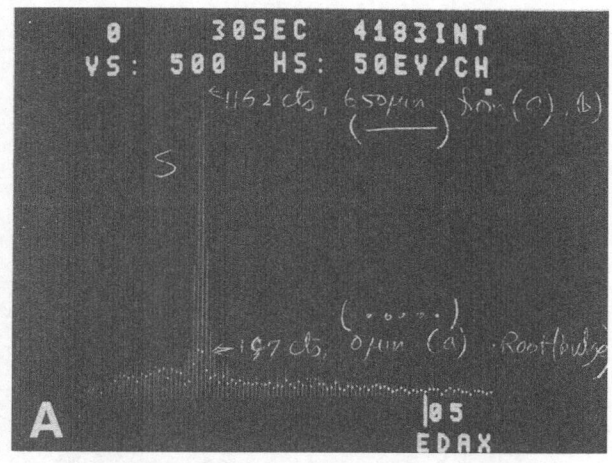

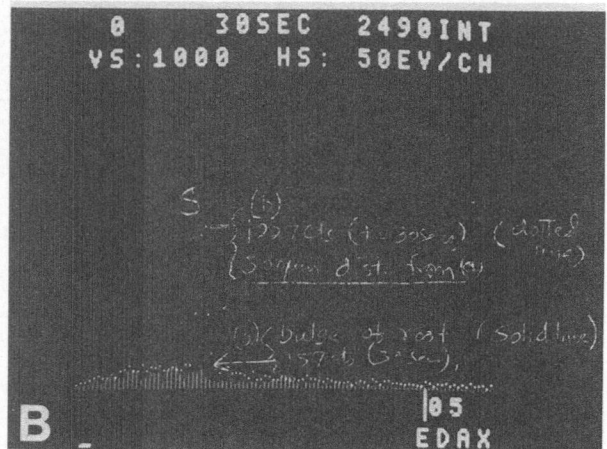

Fig. 8   Energy dispersive x-ray spectra of
         two combed hairs of the same indi-
         vidual. A: 1st hair. Dotted line –
         bulge of root; solid line – 650 µm
         away from bulge of root. B: 2nd hair.
         Solid line – bulge of root; dotted
         line – 500 µm away from bulge of root.
         Background-free integral x-ray cts {S
         (K$_\alpha$)} for 30 secs: at 'bulge' – 157 to
         197 cts; 500 um to 650 um away from
         'bulge' – 1162 to 1227 cts. 20 kV.

increase then continued <u>via</u> jumps along the length of the hair shaft. The
line scans of S (K_α) between successive jumps were not necessarily highly
regular as shown earlier (Fig. 6). Because the saturation values of
sulfur (intensity) on these hairs were unknown the rate of increase in
intensity ($\Delta Is/\Delta x$ at $x_3 \geq 0$) was measured at the root of hair. This rate
was as low as 0.2×10^3 cts/min/cm, or it was as high as 30×10^3
cts/min/cm, depending upon the donor of hairs. The standard deviation
of this rate from combed (telogen) hairs to plucked (anagen) hairs of the
same individual was only 2 to 3%. Because no two individuals indicated
the same rate, this rate will be termed "sulfur index".

Hair Comparison

Because the data on spatial variation of S were promising, the observed characteristic of S was immediately put to a blind matching test.
This test consisted of four hairs in evidence (A, B, C and D) collected
from a rape scene and three control hairs one from each of two suspects
(Sus 1, Sus 2) and a victim (Vi) (Fig. 9). The characteristic x-ray
spectra of these hairs have been shown in Fig. 10 A-G. These x-ray studies were conducted initially in order to avoid dependence on morphologic
properties of these hairs examined in the light microscope.

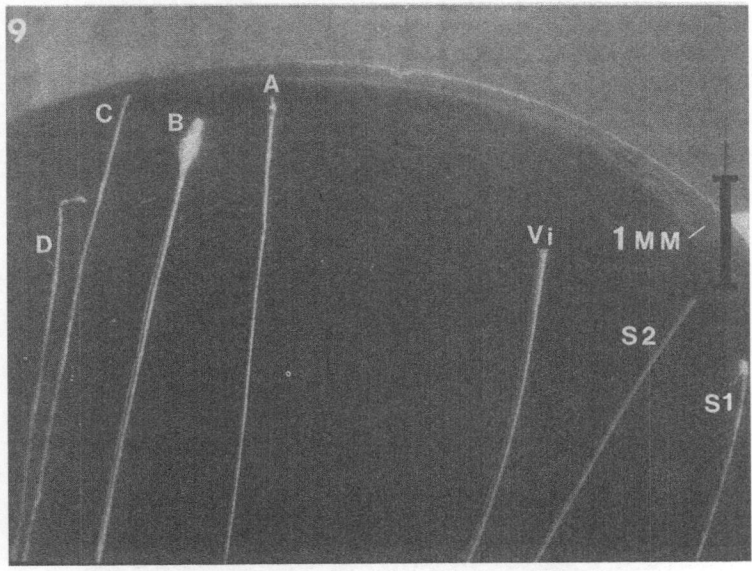

Fig. 9. Seven hairs in a blind matching test.
Evidence hairs: A, B, C and D. Control
hairs: Vi (from victim's head), S1 (from
suspect numbered '1') and S2 (from suspect numbered '2'). Scale Bar: 1mm. 20kV.

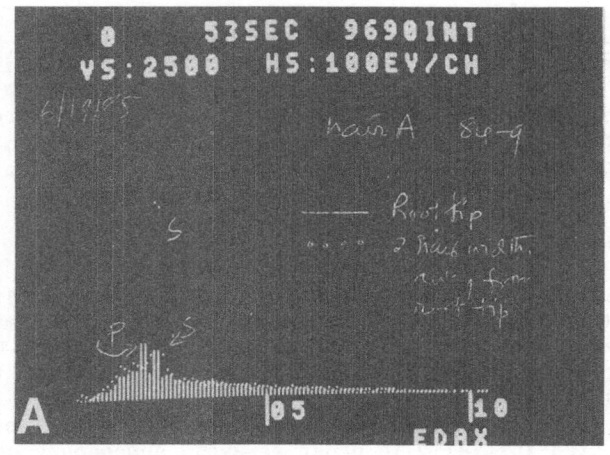

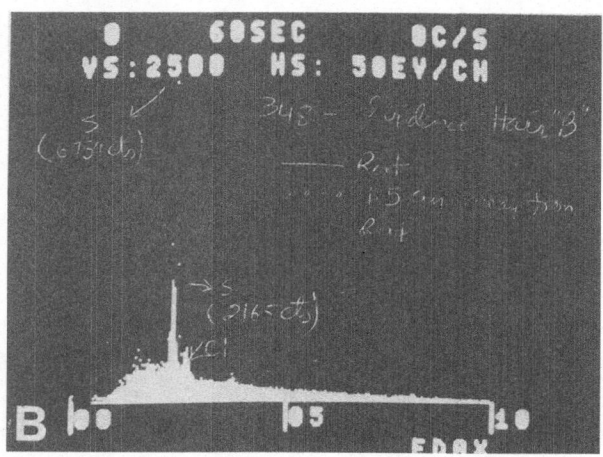

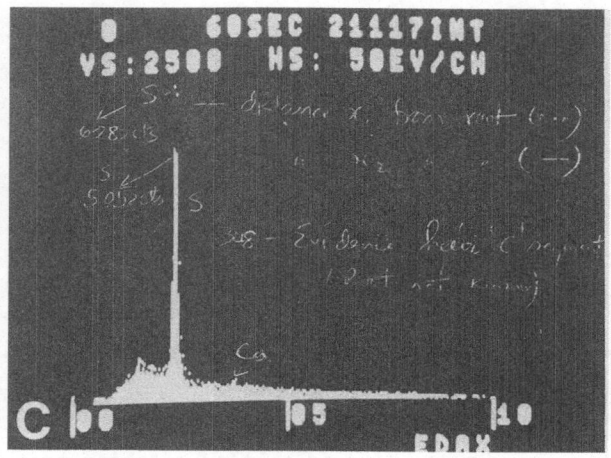

Fig. 10   EDX spectra of seven hairs in a
          matching test.   A-C: hairs or hair
          segments in evidence.   20 kV.

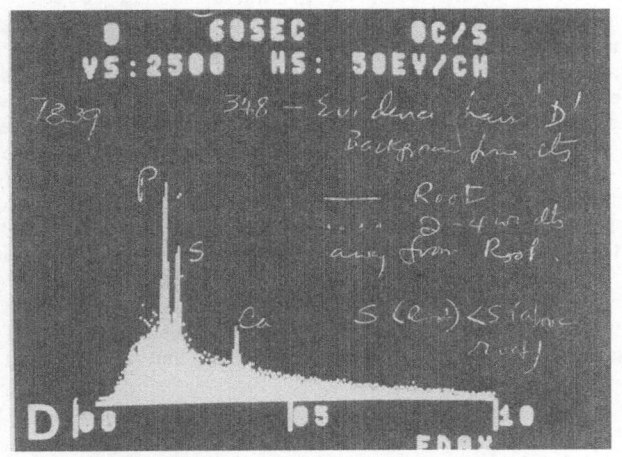

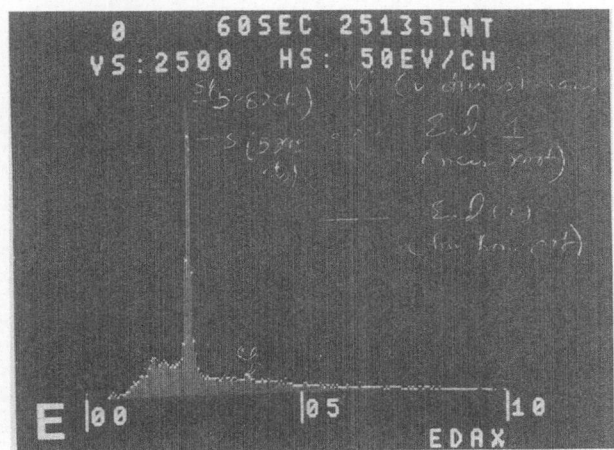

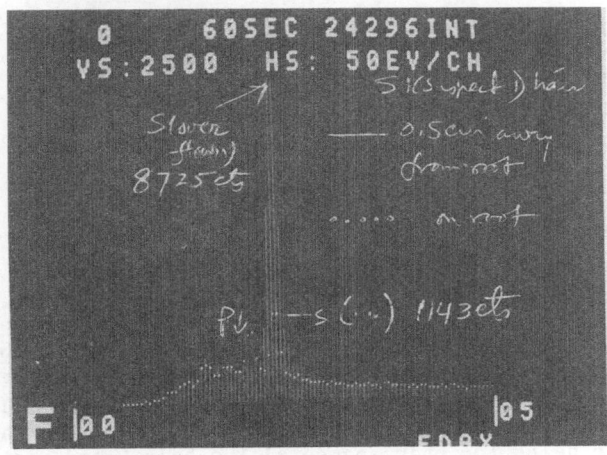

Fig. 10 (Cont.) EDX spectra. D: hair in
evidence, E: control hair from a
victim and F: control hair from the
suspect numbered '1' (Sus 1 or S1).

The hairs marked A, B, D and S1 (Sus 1) had attached roots, which facilitated measurement of sulfur indices of these hairs (Fig. 11). The other hairs were cut segments, each devoid of the root. The immediate observation was that the sulfur indices of the hairs A, D and S1 (Sus 1) were comparable (range: 21×10^3 to 26×10^3 cts/min/cm). The hairs D and S1 (Sus 1) were anagen hairs, whereas the hair A was a telogen hair. Phosphorus (P) was detected to a magnitude (peak height) comparable to that of S only at the root of the hairs A, D and S1 (Sus 1) (Fig.(s) 10A, 10D and 10F). The line scans of P (K_α) matched in shape with the line scans of S (K_α) and therefore the detected phosphorus was endogeneous in origin. The other minor elements of these hairs were calcium and chlorine (hair B). The line scans of these elements indicated that these elements were present as deposits or aggregates. Copper was detected (resolved) in very small amounts in some hairs. Therefore these elements (Cu, Ca and Cl) were disregarded. The hair segments C, Vi and S2 (Sus 2) indicated a

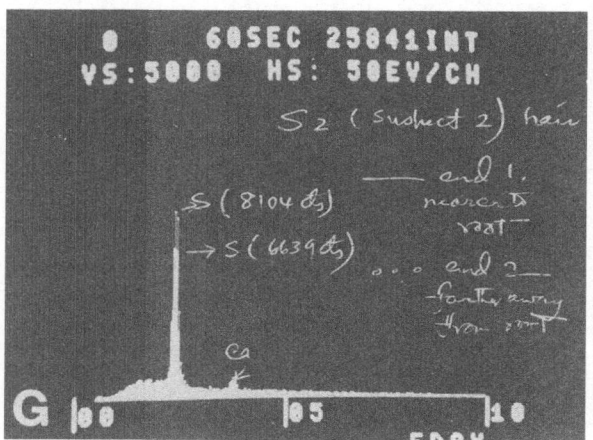

Fig. 10 (Cont.) G: EDX spectra at two ends of the hair segment, S2, from the suspect numbered '2'. Background-free S (K_α) counts: 6639 (solid line, end 1), 8104 (dotted line, end 2), time: 60 sec. 20 kV.

a differential in x-ray counts due to S (K_α) at the two ends. This analysis indicated that the hairs A, D and S1 were quite alike in terms of elemental characteristics. No definite conclusion could be given about the other hairs. Interestingly the suggested match between the hairs A, D and S1 was supported by the light microscopic features of color, pigmentation, medullation, structure and cuticular trait, etc., of these hairs. Unlike the other hairs (or hair segments) the hair B had a strange structure at the root and on the hair shaft.

Incidentally the hairs in the present test were not washed. They were clean and x-ray line scanning studies conducted prior to carbon coating indicated that washing was not needed for these hairs. Evidence has been obtained which suggests that sulfur concentration in scalp hairs remains unaffected by washing. Under the same conditions many elements of hair are variably washed out of hair. The elements which are highly wash-able are sodium (Na), chlorine (Cl), calcium (Ca) and copper (Cu). Ca and Cu were detected to a level of 1% in hairs of two females and these ele-ments were easily removed by washing. Such high levels of Ca and Cu were the result of applying cosmetics to the hair. Certain hair formulas and hair sprays contain heavy metals (e.g. lead in Grecian formula). If the detected element appears to be the result of external application, the

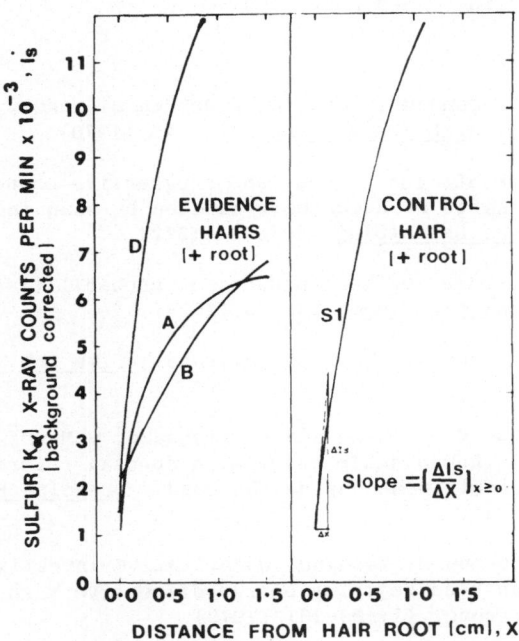

Fig. 11 Increase in sulfur counts (x rays)
with distance from the root end of
hair. Four hairs in a matching
test (See Fig. 10)

question of having it administered *via* food or drug ingestion is ruled out. The x-ray line scanning technique will find applications in this area of forensic toxicology. The sudden rise in intensity of S at the root of hair and the increased trend along the hair shaft suggests that S is continuously incorporated in the keratin of growing cuticles. Since the accumulation of S reaches an immediate saturation with combed (telogen) hairs at a shorter distance from the root, the incorporation "pump" must be at the root of hair.

Regarding the usage of the sulfur index, the technique will be considered for routine hair comparison when more data is obtained on a larger population to make this assessment. Automated line scanning of hair will be considered along this line. Meanwhile an attempt will be made to determine if the increased trend of S away from the root of hair is corroborated by methods of bulk analysis , such as, atomic absorption and neutron activation, etc.

Conclusion

The conventional microscopy test for hair comparison can be significantly aided with the uses of elemental parameters examined in the SEM-EDX.

Acknowledgements

The work has been sponsored by New York State Police at the Headquarters Crime Laboratory. Thanks are due to Robert Miazga for his assistance in photography. The results in Fig.(s) 2, 4-6, 9, 10A and D and 11 will be used for publication of two meeting reports. These are SEM Meetings/1986 May 5-9, New Orleans, SEM, Inc., AMF O'Hare, Chicago and MAS/EMSA - 1986 Joint Annual Mtg., Aug. 10-15, Albuquerque, New Mexico, San Francisco Press, Inc., San Francisco.

REFERENCES

1.   R. Dawber and S. Comaish, Scanning electron microscopy of normal and abnormal hair shafts, Arch. Dermatolog. 101:316 (1970).

2.   E. Bottoms, E. Wyatt and S. Comaish, Progressive changes in cuticular pattern, along the shafts of human hair as seen by scanning electron microscopy, British J. Dermatolog. 86:379 (1972).

3.   B. D. Gaudette, Some further thoughts on probabilities and human hair comparisons, J. Forensic Sciences 23:758 (1978).

4.   B. D. Gaudette, Forensic hair comparisons, Crime Lab Digest 12:44 (1985).

5.   S. Seta, H. Sato, M. Yoshino and S. Miyasaka, SEM-EDX analysis of inorganic elements in human scalp hairs with special reference to the variation with different locations on the head. Scanning Electron Microsc. I:127 (1982).

6.   S. Seta, H. Sato and M. Yoshino, Quantitative investigation of sulfur and chlorine in human head hairs by energy dispersive x-ray analysis, Scanning Electron Microsc. II:193-201 (1979).

7.   S. Seta, "Analysis of biological elements of scalp hairs with SEM-EDX and TEM and its application to the hair comparison," in: "Electron Microscopy in Forensic, Occupational And Environmental Health Sciences," S. Basu and J. R. Millette, eds., Plenum Publishing Corp., New York, NY (in press).

8.   C. Cohen, "Architecture of the α- class of fibrous proteins," in: "Molecular Architecture in Cell Physiology," T. Hayashi and A. G. Szent-Györgyi, eds., Prentice-Hall, Inc., NJ, pp. 169-190 (1966).

9.   E. S. West and W. R. Todd, "Text Book of Biochemistry," Macmillan Co., New York pp. 1190-1192 (1964).

10. J. C. Russ, New methods to obtain and present SEM x-ray line scans, EDAX Editor 9 (No. 2):3 (1979).

11. S. Basu, Spatial variations of sulfur and absorbed copper ions in human hair by x-ray line scanning, in: Proceedings of the 43rd Annual Meeting of the Electron Microscopy Society of America. G. W. Bailey, eds., San Francisco Press, Inc., San Francisco, 124, 125 (1985).

12. L. A. King, R. W. Wigmore and J. M. Twibell, The morphology and occurrence of human hair sheath cells., <u>J. Forensic Sci. Soc.</u>, 22:267 (1982).

13. D. Maes and B. D. Pate, The spatial distribution of copper in individual human hairs, J. Forensic Sciences, 21:127 (1976).

14. L. Reimer, Electron-specimen interactions, <u>Scanning Electron Microsc.</u> II:111 (1979).

15. J. C. Russ, X-ray mapping on irregular surfaces, <u>EDAX Editor</u> 9 (No.2):10 (1979).

16. S. J. B. Reed, "Principles of x ray generation and quantitative analysis with the electron microprobe," <u>in</u>: "Microprobe Analysis", C. A. Anderson, ed., John Wiley & Sons, New York, NY, pp.53-81 (1973).

EVALUATION OF STAPLE IMPRESSIONS IN THE SCANNING ELECTRON MICROSCOPE*

S. Basu

New York State Police Crime Laboratory
Albany, New York 12226

ABSTRACT

When the driving force for stapling is optimal, the desk staplers produce characteristic impressions on the shoe-ends of staples and yield regular bends of the legs of the staple to form a permanent clinch. The impression consists of persistent dents or scratches and parallel striations which appear altogether on a lifted, scraped metal at each shoe-end. The contour of this "metal-lift" is often marked by fractures and its wavy, banded shape reveals bending distorttions. The two composite impressions at the two shoe ends of a staple seldom match. This gives one many useful "individual characteristics" for comparative identification, which are solely due to the shape and the surface peculiarities of the paired slots of the clinching anvil at the base of a stapler, which force the legs of a staple to fold toward one another. The effects of pre-existing marks in the area of impression, different forms of staple and the age of the stapler (anvil) on the resulting impression, are found to be negligible. Hence, the developed method in the scanning electron microscope will be useful for forensic identification of the source of a staple on questioned documents. These studies have also laid down the foundation of a new hypothesis that attempts to explain the disappearance or obliteration of the expectant tool marks involved with bending or rotation of the driving force.

INTRODUCTION

The scanning electron microscope (SEM) was introduced into forensic work immediately after its commercial debut in 1965,[1,2] and has achieved since then a position of eminence in the area of topographic comparison, identification and interpretation of physical evidence, which could not otherwise be properly addressed by either light microscopy or transmission electron microscopy (See Table 1 in Ref. 3 for a comparison).

The principal mode for image formation in the SEM is due to secondary electron emissions from the specimen surface.[4] While the contrast due to secondary electrons is strongly dependent on topographic details, the effective area of their emission corresponds very closely to the diameter of the incident electron probe. The result of this is the combination of remarkable depth of field (300-500 times that of a light microscope) and high resolution (10 nm or better) at wide ranges of continuously variable magnifications.

---

*Presented at the Joint Annual Mtg. of EMSA/MAS, Louisville, Kentucky, 5-9 August, 1985.

43

Quite aware of these advantages, forensic investigators have used the SEM for the examination of tool marks, particularly firearm markings, including firing pin impressions,[3,5-8] extractor and ejector impressions on cartridge cases,[3,5] striations and microstriations along the shoulder of the land impressions of copper clad lead bullets,[3,9] inscriptions on coins,[10] numeral impressions of hand stamp dies,[11] and impressions of cutting tools on metallic wires, etc.[3,12-14]

Staplers are unique office tools used for stapling documents, and have been used to date old documents.[15] The problem of identifying a fastened staple on documents as having originated from a particular stapler is not a new issue in questioned documents,[16-18] but has remained to be solved because researches in this area were routed in directions that produced no reliable standard. This has been partly due to limitation of light microscopy and partly due to lack of emphasis given to random characteristics which are essential for individualizing a stapler. The shoe-ends of a staple are prone to yield random characteristics of the shear induced impressions associated with bending of the staple legs in the paired concavities (slots) of the clinching anvil (Fig. 1). When the stapler's ram (driver) pushes down on the staple, it also produces marks on the back of the staple. If the stapler's ram possesses working tips (sharp points), these also produce marks (class characteristics) that can be used to identify the make or model of the stapler. However, those marks on the driver side and at the bends of a staple possess fewer individual characteristics of a stapler. Only the impressions at the shoe-ends possess a multitude of characteristics suitable for comparative identification in the scanning electron microscope.

In order to establish a reliable basis for comparison of staples and to identify the stapler in question, eleven staplers of different makes and models were used. The effects of pre-existing marks and different forms of the staple ends on the resulting impression were examined by using different brands of staples in these staplers. This experience was applied to a blind matching test consisting of seven staples and seven staplers. Since these studies were conducted (1981), exhibits*,[19] (poster) of staple impressions have been presented on two occassions, including the EMSA/MAS Meeting, the proceedings of which included an extended abstract.[19] Although the present report is based upon this published abstract, it contains additional studies conducted recently (1985). Because the impressions at the shoe-ends of a staple are dependent on the shape and the defects of an anvil, the effect of the natural wear of the anvil on the resulting impression has been examined, using the same staplers that were used beforehand (1981).[19]

Hypothesis of Uncharacteristic Tool Marks

The studies of staple impression had another premise which applies to other tool marks, particularly those that are strangely uncharacteristic, such as, the firing pin impressions of selected shotguns and rifles.[8] A hypothesis is proposed that is applicable to only deformable (or malleable) materials that are susceptible to tool marking. It suggests a polishing effect, due to sudden breakdown of advancing forces of friction, or shear fields, toward and surrounding the developed impressions that had taken place earlier. This effect acts by effacing or filling the pre-formed impressions partially or fully. Tool marks in which the driving force is involved with bending (staples) or with revolution (e.g. drills, firing pins of selected shotguns and rifles) are most affected. When the driving force is sub-optimal, the expectant defects may not appear. When the driving force is excessive, the expectant defects will be partially or fully obliterated. Staplers were the chosen systems for testing this hypothesis because it was the unusual shape of a staple on fastened papers that gave the clue that the driving force was either inadequate or excessive.

---

*Poster presented (but not abstracted) at a workshop of the "Forensic Microscopy Symposium", NY Microscopical Society, John Jay College, New York, April 27, 1984.

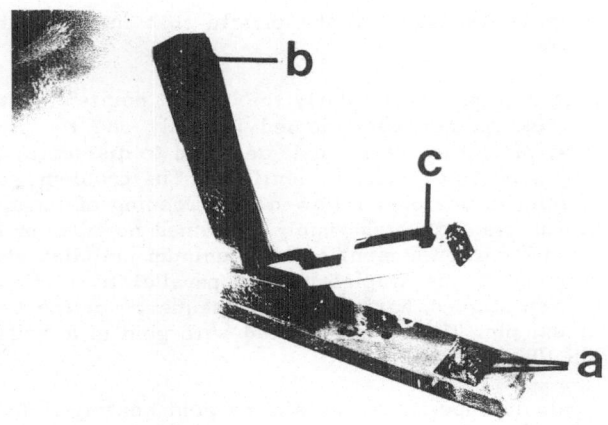

FIG. 1. Photograph of a standard stapler (Swingline® 747, #8 per list in <u>Materials</u> used for stapling experiments in Fig.(s) 2-10. Arrows: a – anvil (represented by two concavities), b – ram (driver), c – place for cohered staples as a stick of 1/2 in. (13 mm) crown and 1/4 in. (6.5 mm) legs.

MATERIALS

Eleven desk staplers, each standardized to a staple size of 0.5025 ± 0.0025 in. (13 mm) crown and 0.258 ± 0.008 in. (6.5 mm) legs, were used. These included the seven staplers of a blind matching test. The makes and models of these seven staplers were Swingline®27 (#1,2,3 and 4), ASCO®40 (#5), ASCO®40 (#6) and Swingline®27 (#7). The remaining staplers were Swingline®747 (#8, 9), Swingline®27 (#10) and ACE®102 (#11). These staplers will be identified in the results section by their serials.

Standard staples of four different brands were used in each stapler. The brands were Swingline® Standard No. S.F.-1 ("100% round with sharp point"), Swingline® R.W.-35 ("100% round"), Bostitch® Standard SBS 19 ("chisel points") and ACE®2025 ("chisel points"). The different forms of staple legs (or shoe-ends) have been identified by the labels within parenthesis following each brand. These were all cohered staples of 1/2 in. crown and 1/4 in. legs. They were made from 19 gauge wire (diameter 19/1000 in. or 0.482 mm) of an alloy of iron (Fe) and zinc (Zn) as determined by energy dispersive X-rays (EDX) in the scanning electron microscope.

METHODS

Three pages of bond paper were used for each stapling experiment with a particular stapler and staple brand. The staples were tacked in three different rows, using different hand pressure, namely, (a) an optimal pressure to produce regular bends with a clinch, (b) a sub-optimal or inadequate pressure to yield bends without a clinch (i.e. the ends remaining open), and (c) an excessive pressure, or on impact, to create irregular bends and/or loops on the legs. When the expectant bends were not consistent with the rest in the same row, that staple was rejected. With a single-edge razor blade, the papers around the selected staples were scored in small rectangles. The residual papers, fastened to each staple, were removed with fingers, and the legs were straightened out with fingernails to give a U-shape to each end of the staple. Irregular bends or loops required extra caution and careful handling of the shoe-ends. The screened staples of the three categories (a-c) were collected in three small beakers, and then cleaned ultrasonically in methanol (10 min) and acetone (10 min). Acetone is especially needed to remove the glue or the plastic that manufacturers apply to hold the staples together.

The pointed ends of a staple, particularly the "chisel points", capture paper fibers during stapling. These points were flamed for only one to two seconds, or a solvent (10 M Urea at pH 3.5 in acetic acid) was used to dissociate the fibers. These fibers are annoying sources of charging artifacts. The problem was avoided by further sonication (5 min) in acetone, followed by screening of cleaned staples in a stereo microscope. The staples which visibly contained no fiber or less fibers were glued side-by-side with carbon cement on an aluminum pin (diameter 13mm), while keeping the shoe-ends of the staples up and parallel to each other (Fig. 2). A total of four to seven staples, having a known sequence of the bends, were mounted on each aluminum pin. These were coated with gold in a sputter coater (E. F. Fullman, Inc., EMS 40).

There was a net gain in topographic details by gold coating. (The extended abstract in the proceedings of EMSA presents micrographs of uncoated staples.) Secondary electron emission increases systematically with atomic number,[20] and is certainly higher with gold (Au) than with iron (Fe) and zinc (Zn) of the specimen staples. These were examined in an AMRay model 1000 SEM at 20 kV with no tilt of the shoe-ends (e.g. Fig. 2).

RESULTS

Regular Bends

Composite impression. Fig. 2 represents secondary electron images of an intact staple and three other staples of the same brand, each of which had regular bends at the legs, as they were tacked on papers with the shown Swingline® stapler in Fig. 1. Usually, as the staple legs are bent over the anvil (Fig. 1), the metal at each shoe-end is scraped, or the metal is lifted from the base. The impressions appear on this metal-lift and possess a composite pattern, consisting of an outer band, due to the metal lift and underlying dents. The latter rise vertically from the baseline of the shoe-end. Each dent contains numerous striations of various thicknesses. These results are shown in Fig.(s) 3 and 4.

These impressions are the results of moving frictions between a shoe-end and the corresponding concavity (slot) of the anvil, which the staple leg had to traverse to form a clinch. The surface features of the two concavities of an anvil are rarely alike. Therefore, the impressions at the two shoe-ends of a staple seldom match (Fig. 3 vs Fig. 4).

The metal-lift (Fig. 3, 'ml') is wavy, thus indicating a multiple number of peaks (p_1, p_2, p_3, etc.) and troughs of a band (Fig. 3). It may also represent a single broad band (Fig. 4). Like the petals of a flower, more than one metal-lift may occur in superimposition on the same shoe-end (Fig. 3, 'Sml'). The contour of a metal-lift is often perturbed by fractures. With successive stapling, the fractures occur in the same general area of the metal-lift (Fig. 3, 'fr'). During the motion of a shoe-end, some dents occur earlier ('d earl') and these are obstructed from view by the dents occurring later ('d late') (Fig. 4). The apparent distance between these two groups of dents, as imprinted in the final impression, can vary with successive staples because the shoe-ends, under bending motion, do not necessarily retrace the same path. Because the anvil is slightly wider than the thickness of the wire, the anvil allows the staple leg to move sideways. Despite these inconsistencies, the stated defects persisted in numerous tackings with a given stapler and a particular form of the legs (e.g. "100% round with sharp point").

Infact, each of eleven enlisted staplers (Materials Section) produced characteristic impression. The impressions obtained with two staplers of the same make and model (e.g. Swingline®747, ASCO®40) never matched. The impressions obtained with six staplers of the same make and model (Swingline®27) were all different,

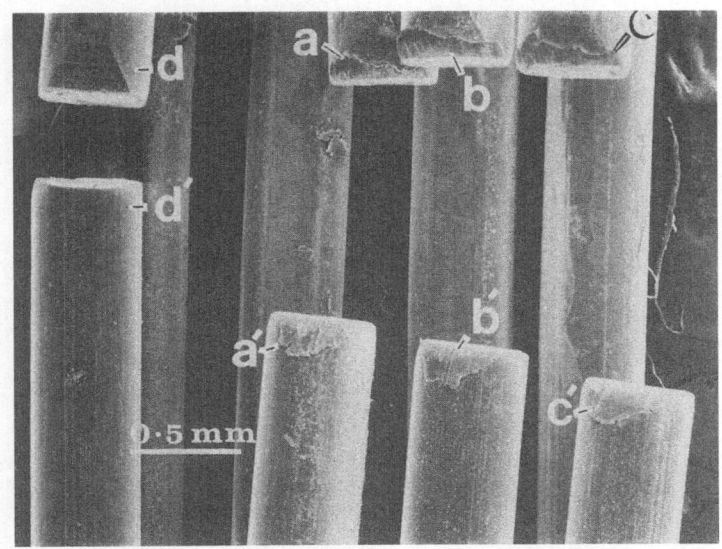

FIG. 2. An SEM mount (13 mm diameter aluminum pin) containing three tacked staples (aa', bb', cc') and an intact staple (dd') of the same brand. a,b,c: shoe -ends on top having matching impression; a',b',c': shoe-ends at bottom
having matching impression. Stapler shown in Fig. 1.
Staple brand: Swingline® Standard No. S.F.-1, "100%
round with sharp point". Original magnification:35X.

due to random characteristics of the persisting defects. Therefore, the composite
pattern of staple impression had no class characteristic that would identify the
make or model of the stapler. Only two staplers (e.g. ASCO®40 and Swingline®27)
produced similar looking metal-lifts (outer bands) at only one shoe-end, each
having two peaks at the edges and a trough in between; but the other characteristics of these metal-lifts did not match. Also, the metal-lifts at the remaining
shoe-end(s) had different shapes. The dents containing a multitude of striations
were much stronger characteristics of a stapler (anvil). This observation has
been stressed in another section on microstriations.

Effect of pre-existing marks. Intact "100% round with sharp point" staples possess
a sharper end (Fig. 5, top), cut to the shape of a trapezoid. The other shoe-end
(Fig. 5, bottom) is roundish. Actually, this round end and the staple wire itself
possess a closed "c" shape. These staples normally exhibit parallel striations
on the shoe-ends. These may be vertical or horizontal to the base-line (arrows,
Fig. 5). These striations which pre-exist in the area of impression, were smeared
out, making allowances for the composite pattern to occur. Therefore, the staple
impressions (Fig.(s) 3 and 4 etc) were not influenced by the marks on the legs
of a staple.

Effect of staple forms and age of the anvil. It is interesting to note that, with
different standard forms of staplers ("100% round", "100% round with a sharp
point" and "chisel points") and with natural wear of the anvil, the composite
impressions at the shoe-ends of staples do not change significantly. An example
of this study is given in Fig.(s) 6 and 7. The shown staples had chisel points, and
these were recently tacked on papers with the same stapler (Fig. 1) that had
been used four years ago to obtain the results shown in Fig.(s) 2-4. During these
four years, this stapler (Fig. 1) and two other staplers were used for routine office
work. These three staplers have been used recently to examine the effect of
natural wear of the anvil on the resulting impression.

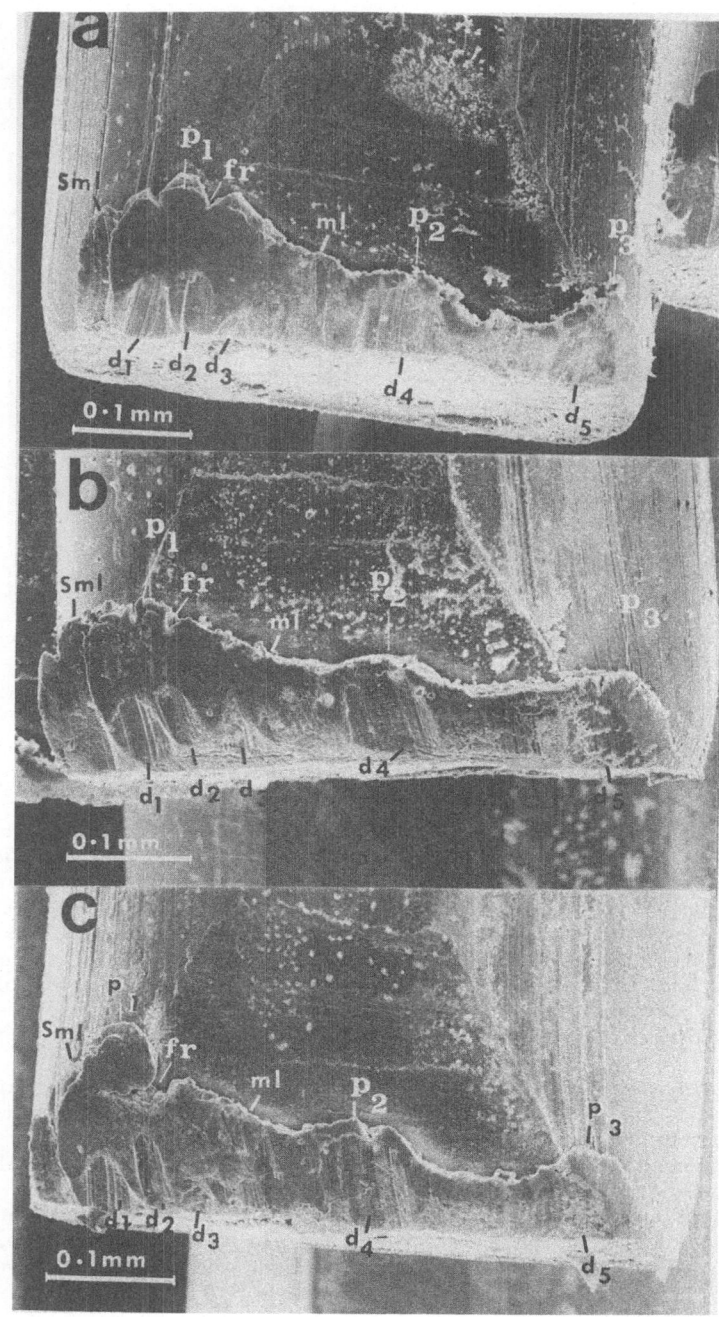

FIG. 3. Magnified images of matching impressions on the
top shoe-ends (a,b,c) of tacked staples in Fig. 2.
This arrangement in which the metal-lifts (ml) are
directed upward facilitates vertical comparison of
tool-marks. But the shown five dents ($d_1, d_2, d_3,$
etc) are not convex. They are actually concave.
Remaining notations: p_1, p_2, p_3 – Peaks; Sml-
superimposed metal-lifts; fr - fractures. Original
magnifications:190X (a) and 200X (b,c).

48

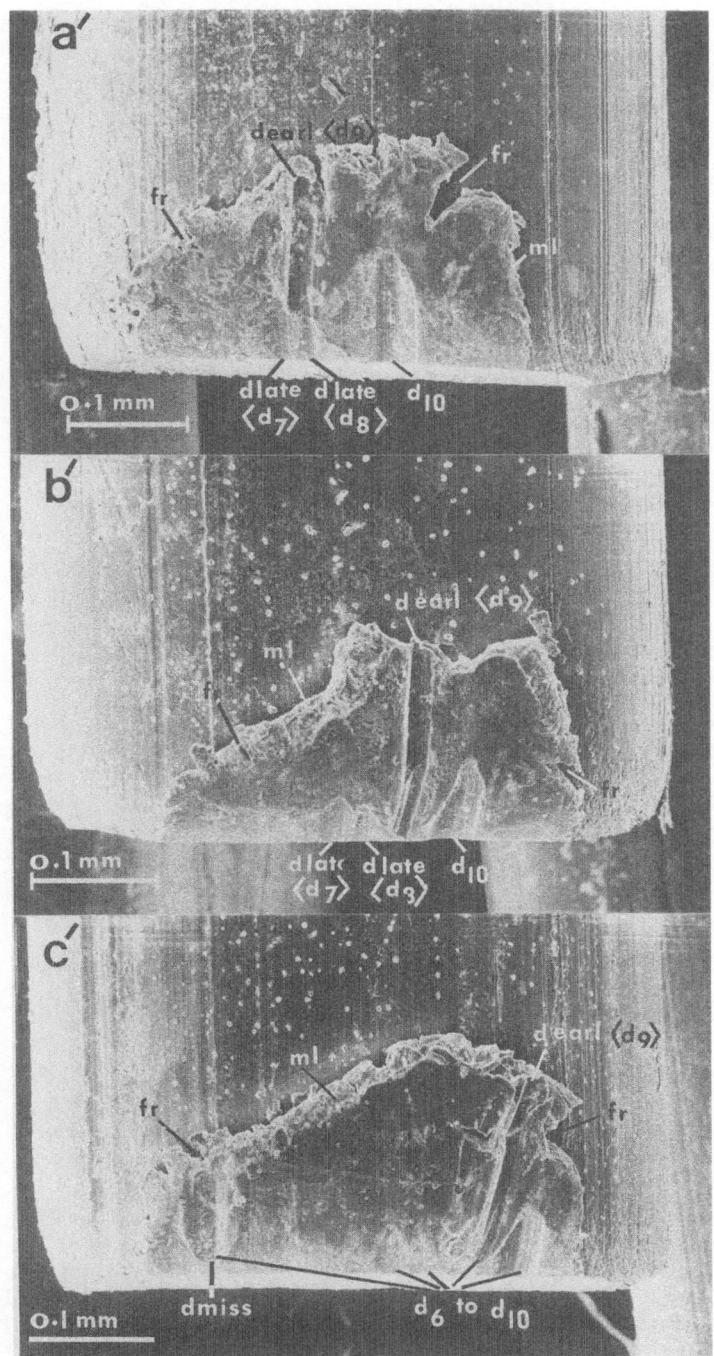

FIG. 4. Magnified images of matching impressions on the bottom shoe-ends (a',b',c') of the three tacked staples in Fig. 2. The dent, d earl, in a',b' has been masked at the base line by the group of dents, d late. The dent, d miss, in c' is missing from a',b' because the staple leg in c' moved more to the left on anvil. Remaining notations: d_6 to d_{10}: five dents; ml - metal-lift, fr-fractures. Original magnifications:190X (a) and 200X (b,c).

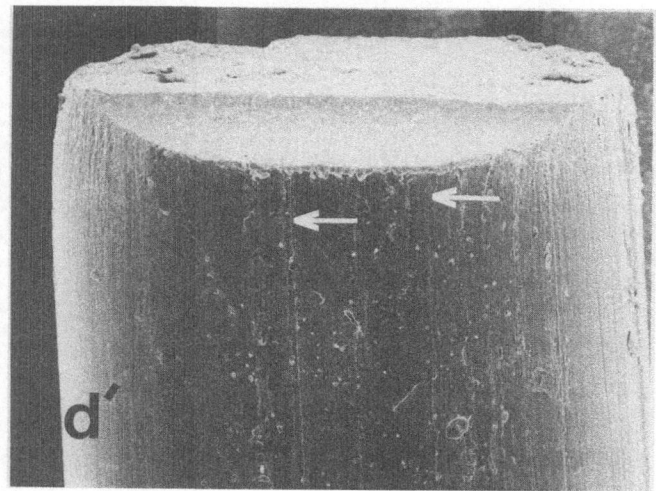

FIG. 5. The top (d) and bottom (d') shoe-ends of the intact
staple in Fig. 2 at a higher original magnification
(200X). Arrows: pre-existing striations in the area
of impression, which are smeared out on stapling.
Scale bar: same for both panels d and d'.

The main differences between the impressions shown in Fig.(s) 3 and 4
and those in Fig.(s) 6 and 7 are underlined in the thicknesses of the respective
metal-lifts and the lengths of dents. The age of the anvil had nothing to do with
these differences because, the impressions produced recently with "100% round
with sharp point" staples had an exact match with the ones shown in Fig.(s) 3
and 4. The metal-lift is always thinner with "chisel points", but thicker with
"100% round with sharp point". Obviously, the peaks of metal-lifts and the dents
were sharper with the "chisel points" than with the other (round) form (cf. Fig.
6 vs Fig. 3; Fig. 7 vs Fig. 4). Otherwise, these impressions were equivalent in
terms of the total number of peaks (three) and dents (ten), which were matchable.
This suggests that the degree of sharpness of the shoe-ends may influence the
overall appearance of an impression, but the main characteristics (i.e. form)
of the impression are strictly dependent on the persistent surface features of
the anvil.

50

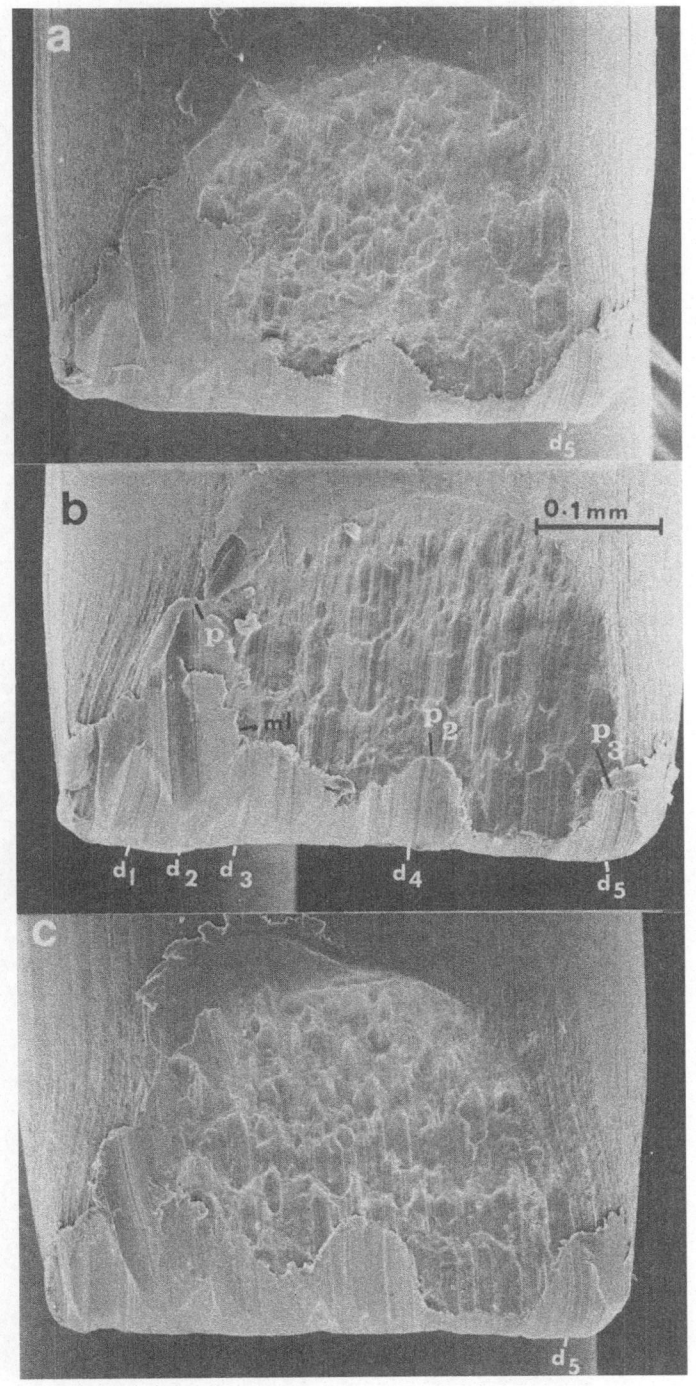

FIG. 6. Matching impressions on top shoe-ends (a–c) of three "chisel point" staples. Staple brand: Bostitch®Standard No. SBS-19, "chisel points". Stapler shown in Fig. 1. Notations: Same as Fig. 3. Original magnifications: 200X. The notations and the scale bar shown in b apply also to a and c. Compare vertically.

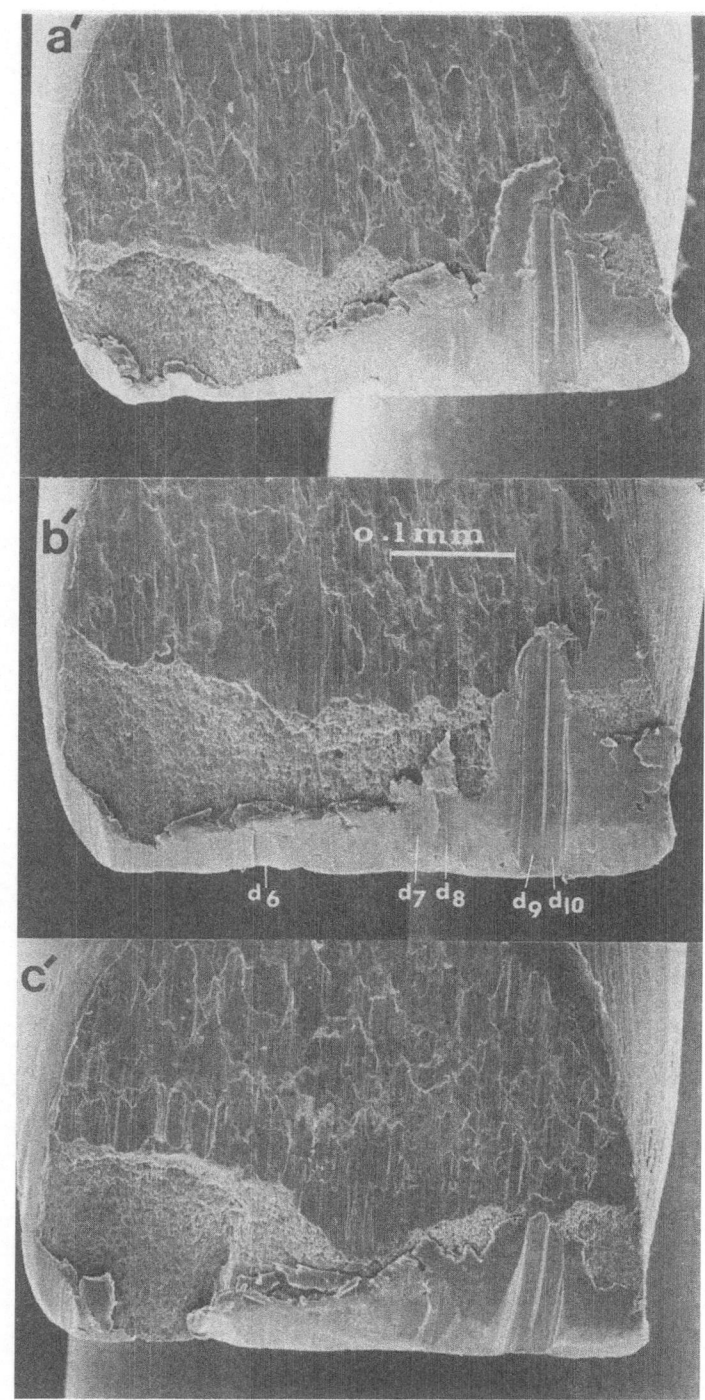

FIG. 7. Matching impressions on bottom shoe-ends (a'-c') of
three "chisel point" staples (The top shoe-ends of
these staples were shown in Fig.6). The staple brand
and stapler: same as Fig. 6. Notations: same as Fig.
4; Original magnification:200X. The notations and
the scale bar shown in b' apply also to a' and c'.

Microstriations. The ability to match two staple impressions has been confirmed by a vertical comparison between specific dents, in terms of submicroscopic striations in them. Fig.(s) 8 and 9 present two examples. The microstriations could be straight up or curved, but are always parallel to each other, due to uniform motion of the legs at the shoe-ends. The comparative value of a staple impression is thus enhanced by the increased frequency of matchable microstriations in a dent.

Irregular Bends and Forked Loops

Effects of excess tacking pressure. When the ram (Fig. 1) of a stapler was pressed hard, and then pressed slightly harder, in consecutive staplings, the following was observed: (1) The staples still possessed regular-looking bends with a clinch; but the dents and the striations in them, as well as the metal-lifts (chisel points), were all slanted to the right or to the left, depending upon the inclination of the driving force, which caused the legs to bend slightly outward or inward (Fig. 10aa'). (2) The staple legs bent too much, to form forked loops and the expectant defects started to be obliterated (Fig. 10bb'). (3) Irregular bends and incomplete bends were produced when the ram (Fig. 1) was struck with an impactsome force. The progression of a polishing effect on the resultant impression was clear with this type of tacking habit. Effacing of the impression takes place from the top (Fig. 11), or from the bottom of the metal-lifts (Fig. 12a). (4) When the driving force was excessive, but non-uniform, quite unmatchable dents and striations occurred in two to three horizontal layers on the metal-lifts. These defects started to be smeared out from the top of the metal-lifts (not shown). This makes it evident that the resultant impressions are rubbed out by the moving forces of friction.

However, the metal-lift may also be featureless, when the stapling is performed so gently as to not result in a clinch (Fig. 12b). The bends would remain significantly open at the ends, quite similar to the "straight bend" configuration (U-shapes) shown in Fig. 2. The shoe-end impressions of these staples were indicative of the legs having simply glided over the clinching anvil.

Those results (cf. Fig. 12a vs b) are also consistent with the metal-lift having occurred concomitantly with the bending of the legs, which precedes the occurrence of dents and striations.

Analysis

Even under controlled conditions of the driving force, not all staples of a stick produce identical bends with a clinch. The shapes of shoe-ends, the thickness of documents, the uncohered state of staples in a stick and the condition of the ram (driver) all influence variably, and may determine whether a shoe-end will take the same trajectory on the anvil surface in successive staplings. Changes in this trajectory are concomitant with the changes in the driving force, which in turn determines the success rate in obtaining all expectant defects in an impression.

A summary of success rates, based upon staplings on three pages of bond papers, with eleven staplers will be given. The success rates have been averaged over seven staples of each form ("100% round", "100% round with sharp point" and "chisel points") used in each stapler.

Regular bends. About five out of seven staples per stapler possessed all persisting defects. In other words, the success rate with regular bends was about 70%. The remaining 15% to 30% of the staples (e.g. two out of seven) had about 2/3 of all expectant defects.

Irregular bends and forked loops. Not more than two out of seven staples per stapler possessed 2/3 or 67% of all persistent defects. Not even one out of seven staples had all expectant defects. Therefore, the success rate with irregular bends and forked loops varied from 0 to 30%.

53

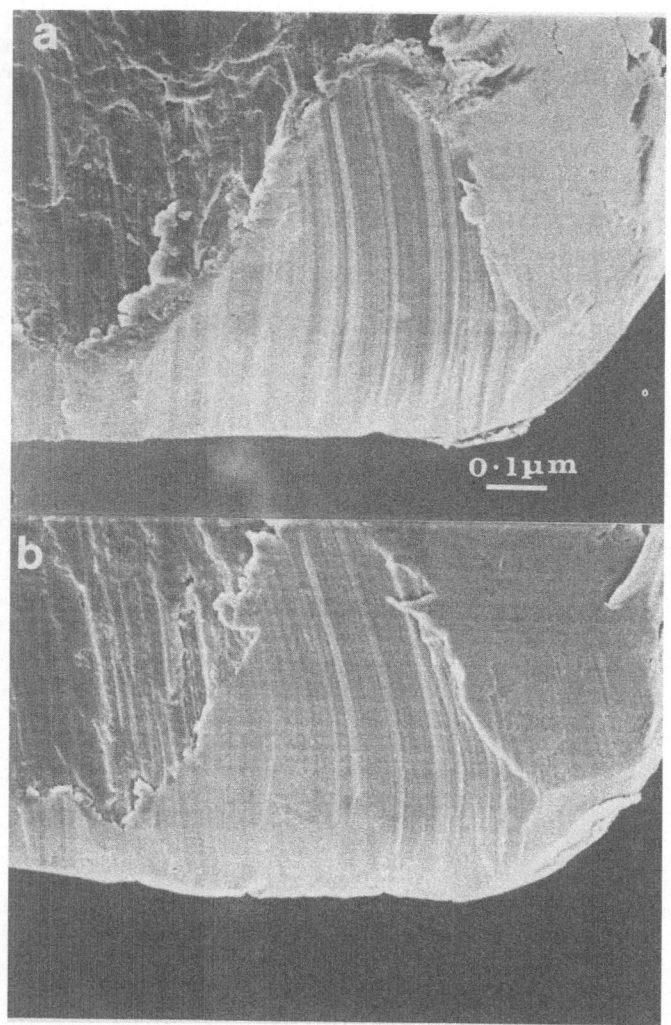

FIG. 8. Matching comparison of the dents, d_5's in Fig.(s) 6a and
6c in terms of microstriations. Original magnification:
1000X. Scale bar: same for both panels a and b.

A blind matching test. The analysis of staple impression was concluded with
a blind matching test, involving seven staples and seven staplers. The test was
solved using about 2/3 (or 67%) of all characteristics due to metal-lift, dents
and striations of a staple impression. Three ruled papers, containing seven tacked
staples, each marked with a serial, were received along with the list of seven
staplers. All seven staples had regular bends with a clinch. An examination in
a stereo microscope revealed that only one (#2) staple was "100% round with
sharp point", three (#3,5,6) were "100% round" and the rest were "chisel points"
(#1,4,7). These three forms of staples were used in seven staplers to derive the
individual characteristics of impression due each stapler. Altogether, the weight
of these characteristics was at least 2/3 or 67% of all possible defects. This
evaluation was subjective, as it took into consideration equivalent weight (33%),
due to each category of defects, namely, metal-lift, dents (or scratches) and
microstriations. The impressions on the evidence staples #3,5 and 6 were identified
as having originated from a Swingline® 27 stapler (#2 in the list). The matching
correspondances between the other evidence staples and the staplers indicated

54

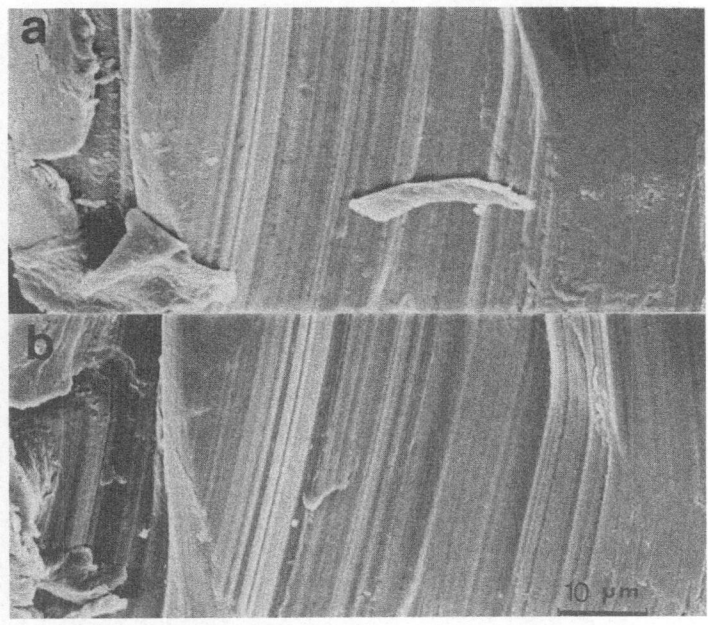

FIG. 9. Matching comparison of the dents, d_1's in Fig.(s) 6a and 6b in terms of microstriations in them. Original magnification:1350X. Scale bar: same for both panels a and b.

that only five out of the submitted seven staplers were infact used in this test (#1 staple, #3 stapler; #2 staple, #5 stapler; #4 staple, #7 stapler, and #7 staple, #6 stapler).

DISCUSSION

As introduced earlier, impressions due to a stapler may occur on various areas (crown, legs and shoe-ends) of a staple. Among these, the lifts at the shoe-ends possess the most random characteristics of bending distortions. Staples labelled "100% round" and "100% round with sharp point" show impression on the ground (base) of only one leg (Fig. 13). These impressions are useless, as the "chisel points" have little grounds (or bases) to have marks. Most staplers have a two position clinching anvil on the base that either forces the legs of the staple to fold toward one another to form a permanent clinch, or spread outward to form a temporary clinch. The present studies were concerned with only the previous group. With the latter, impressions occur on the inside ends, which bend outward.

The best tilt for comparative studies of staple impression is zero degree (Fig.(s) 2-12). Higher tilt is useful for perception of an impression in three dimensions. Topographic details are sacrificed in part by not tilting the specimen, but they are reinstated by gold coating (cf. Fig.(s) 3-11 vs Fig.(s) 12 and 13).

The plastic glue, which holds the staples together as a stick, is usually applied by a flow method, to the outside or inside only.[21] This glue and the paper fibers which cling to shoe-ends must be removed.

Hypothesis. The comparability of staple impression at the shoe-ends results from the specificity of the composite impression, the constituents of which (metal-

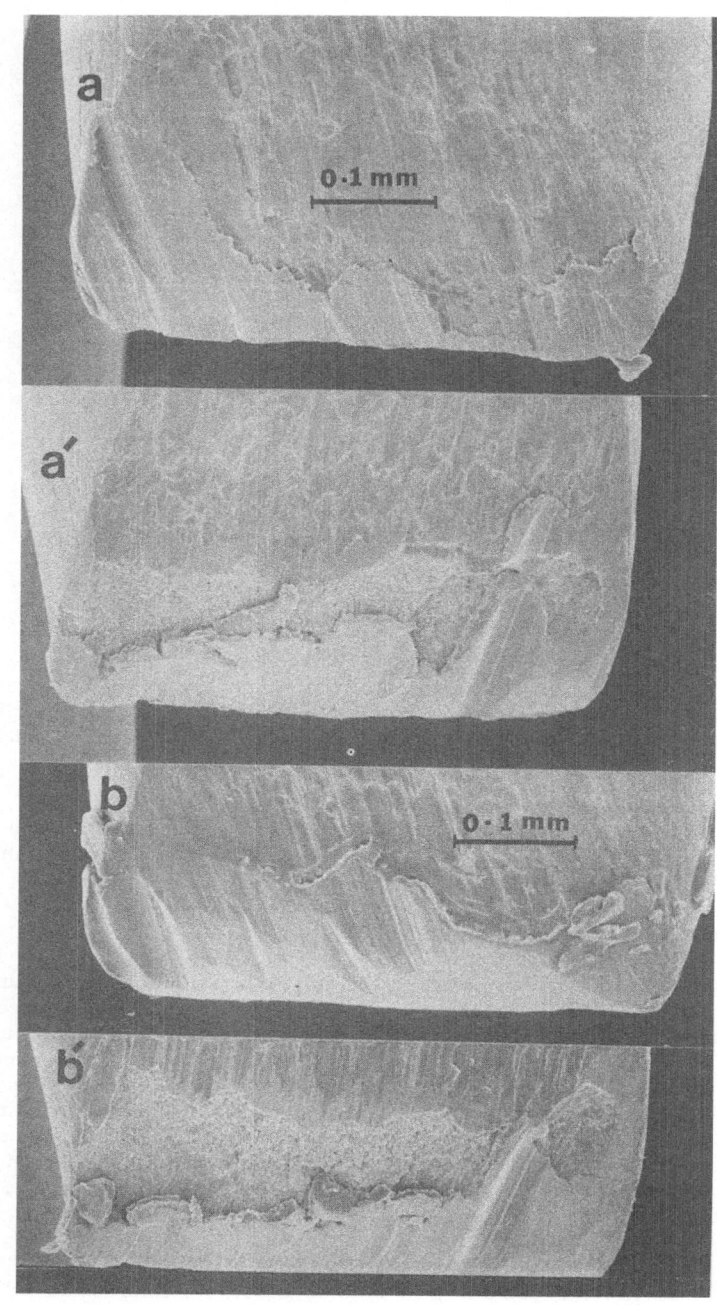

FIG. 10. <u>Effects of excessive tacking pressure.</u> a,a': slanted
impressions of a first staple which was tacked hard;
b, b': slanted impression <u>plus</u> obliterated dents (d
obl) of the second staple which was tacked harder.
Staple brand and stapler: same as <u>Fig. 6.</u> Original
magnification: 200X (approx) Scale bars: same for
a and a'; b and b'.

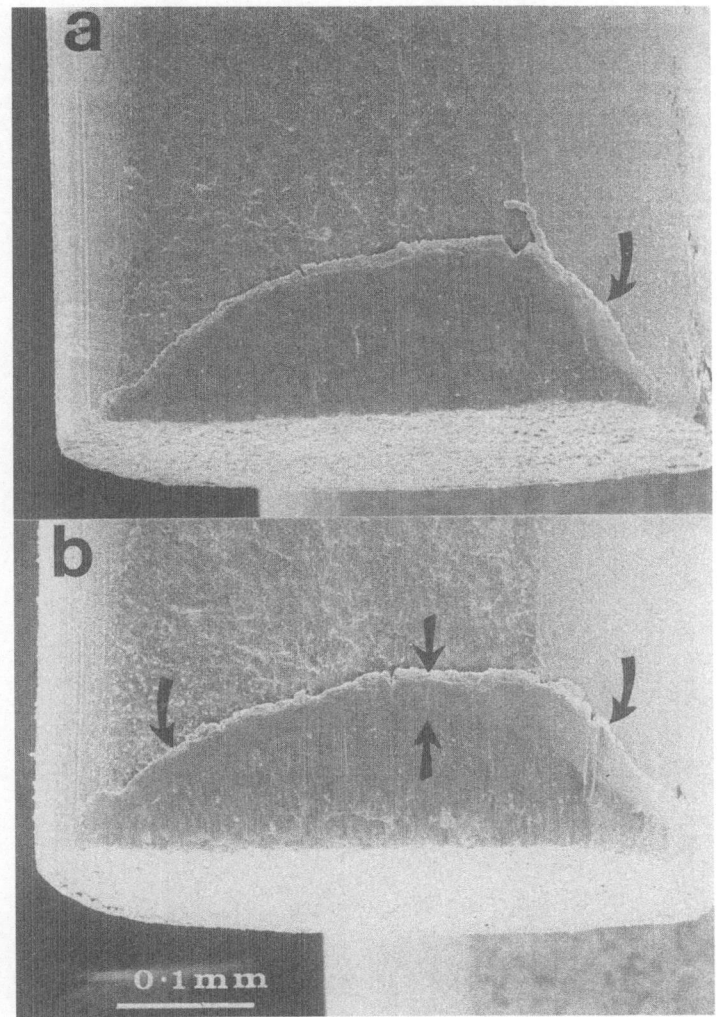

Fig. 11. <u>Effaced staple impression</u>. The ram of a stapler was
slammed in two staplings (a & b). Only top shoe -ends
are shown. The dents and striations started to be effaced from the right hand top corner of the metal-lift
(a) (arrow). This polishing effect progressed through
about one-third of the metal-lift in (b). See arrows in
b . Stapler and staple brand: Swingline®27, #2 as per
list in <u>Materials</u> and Swingline®R.W.-35, "100% round".
Original magnification:220X. Scale bar: same for both
panels a and b .

lift, dents or scratches and striations) are solely dependent on the anvil surface.
The harder anvil surface can be imagined to possess valleys, hills and protrusions
in the scale of microns (one to one hundred). The driving force determines the
reproducibility of impressions due to these structures. As the legs start to bend,
the driving force is counteracted by the frictional resistance, which increases
the area of contact and so reduces the angle of contact. The lifting of metal
at the shoe-ends is the result of this.

As the legs continue to move, the counteracting forces of friction are stream

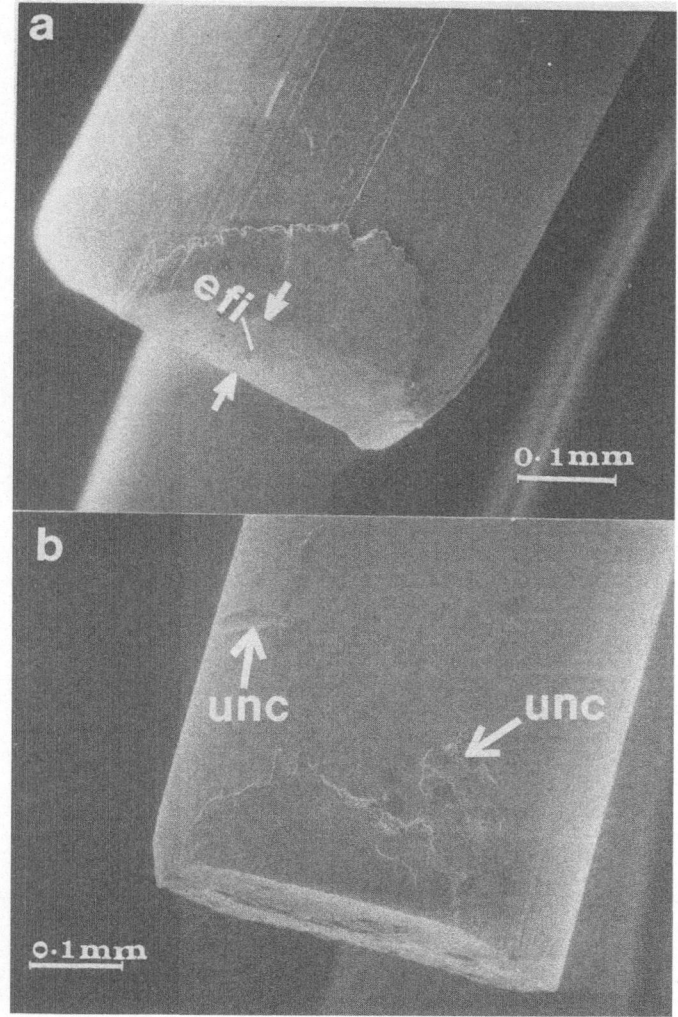

FIG. 12. Effaced staple impressions due to smearing from bottom of a metal-lift (a) vs unfeatured impression with gentle tacking (b). Notations: efi: effaced impression in (a) shown between blunt arrows; unc: uncharacteristic defects. Stapler and staple brands: Swingline®747 (Fig. 1) and Swingline®S.F. No.-1, "100% round with sharp point" (cf. Fig. 4). Original magnification:165X and 155X (b); uncoated specimens.

lined at an optimal driving force, which allows the impressions due to so called "hills" and "protrusions" of the anvil surface to prevail. If the driving force is enhanced, the stream lines of friction breakdown and take random directions. This phenomenon is known as turbulence in fluid viscosity[22] and should occur when a deformable solid surface is involved in forced, tangential motion on a harder surface. The resultant is a polishing force field, which acts like mud slides by rubbing out or by filling the pre-formed impressions partially of fully. When the driving force is sub-optimal, the forces of friction are not laminated. This allows the shoe-ends to glide over the anvil, and the impression contains nothing else but the metal-lift. Therefore, tool marks involved with bending, revolution

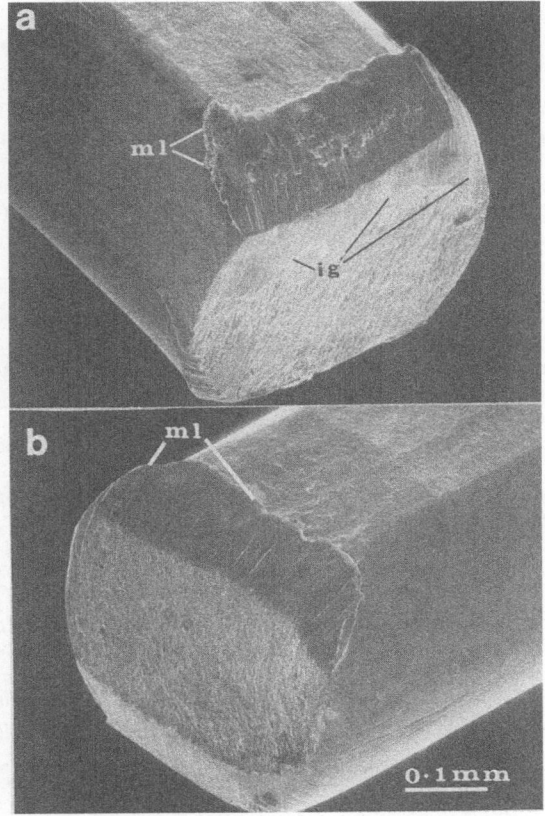

FIG. 13. Two ends of a control, uncoated staple ("100% round, Swingline®R.W.-35) at a 60° tilt. a. Top shoe -end, b. Bottom shoe -end. Arrows: ml- metal-lifts, ig - impression on ground. The shoe -end (b) has no impression on ground. Stapler: Swingline®27 (#1 stapler in the "Blind Matching Test"). Staple: Swingline®R.W.-35. Scale bar: same for both panels a and b.

and translational motion are likely to be obliterated when the forces of impression are excessive or irregular (cf. firing pin impressions of selected shotguns and rifles, drilling impressions, etc.)

CONCLUSIONS

1. The composite pattern of staple impression is a characteristic of the stapler. Since the staple form can be easily identified, matching impressions can be obtained with the stapler and the staple combination. The technique in the SEM would be useful for forensic identification of the source of a staple on questioned documents.

2. The studies of staple impression have explained the obliteration of tool marks.

Acknowledgement

Thanks are due to the New York State Police for sponsoring this work, to Mary Beth Larmour for editing and preparing the typescript and to Robert

Miazga for photography. Thanks are also due to Leonard Kirsch of Capitol Staple Co. Inc., Albany, N.Y. for technical information.

REFERENCES

1. P. R. Thornton, "Scanning Electron Microscopy", Chapman and Hall, London (1968).
2. J. P. Williams, Symposium on Forensic Chemistry , American Chemical Society, 156th National Meeting, Atlantic City, September 11 (1968).
3. E. J. Korda, H. L. MacDonell, and J. P. Williams, Forensic applications of the scanning electron microscope , J. Crim. Law, Criminol and Police Sci., 61:453 (1970).
4. J. I. Goldstein, J. J. Hren, and D. C. Joy, "Introduction to Analytical Electron Microscopy", Plenum Publishing Corp., New York (1979).
5. H. L. MacDonell and L. H. Pouden, "Application of the scanning electron microscope to the examination of firearms markings", Scanning Electron Microsc., II:569 (1971).
6. C. A. Grove, G. Judd, and R. W. Horn, Examination of firing pin impressions by scanning electron microscopy , J. Forensic Sci., 17:645 (1972).
7. G. Judd, R. Wilson, and H. Weiss, A topographical comparison imaging system for SEM applications , Scanning Electron Microsc., I:167 (1973).
8. C. A. Grove, G. Judd, and R. W. Horn, Evaluation of SEM potential in the examination of shotgun and rifle firing pin impressions , J. Forensic Sci., 19:441 (1974).
9. G. Judd, J. Sabo, W. J. Hamilton, Jr., S. Ferriss, and R. W. Horn, SEM microstriation characterization of bullets and contaminant particle identification , J. Forensic Sci., 19:798 (1974).
10. P. G. Rodgers, J. R. G. Jacob, P. Blais, and D. C. Harris, The scanning electron microscopy of counterfeit coins , Scanning Electron Microsc., II:585 (1971).
11. R. F. Dragen, A. E. Wilimovsky, R. Diederichs, M. J. Camp, and M. A. Haas, The application of the scanning electron microscope to the examination of tool marks , Assoc. Firearm and Tool Mark Examiners Journal, 6:1 (1974).
12. M. E. Taylor, Forensic applications of scanning electron microscopy , Scanning Electron Microsc., II:537 (1971).
13. J. R. Devaney and L. W. Bradford, Applications of scanning electron microscopy to forensic science at JET Propulsion Laboratory , 1969-70, Scanning Electron Microsc., II:561 (1971).
14. R. P. Singh and H. R. Aggarwal, Identification of wires and the cutting tool by scanning electron microscope , Forensic Sci International, 26:115 (1984).
15. A. Osborn, "Questioned Documents", Boyd, Albany, 2nd ed., p. 579 (1929).
16. O. Hilton, "Scientific Examination of Questioned Documents", Elsevier, New York, p. 93 (1982).
17. N. Duxbury and A. Leslie, The identification of staples with stapler , Paper presented at the American Society of Questioned Document Examiners Meeting (1965).
18. J. Murdock, The individuality of tool marks produced by desk stapler , Paper presented at California Association of Criminalistics Meeting, May (1972).
19. S. Basu, Evaluation of staple impressions in the scanning electron microscope , in Proceedings of the 43rd Annual Meeting of the Electron Microscopy Society of America, G. W. Bailey, ed., San Francisco Press, Inc., San Francisco, pp. 514,515 (1985).
20. D. A. Moncrieff and P. R. Barker, Secondary electron emission in the scanning electron microscope , Scanning, 1:195 (1978).
21. J. J. Horan and G. J. Horan, The staple , paper presented at the Joint Meeting of the American Society of Questioned Document Examiners and the Canadian Society of Forensic Science, Montreal, Quebec, Canada, September 23-27 (1985).
22. G. Barr, "A Monograph of Viscometry", Oxford University Press, London, (1931).

FORENSIC USES OF DEFLECTION (Y) MODULATION AND X-RAY DOT MAPPING[1]

Samarendra Basu

New York State Police Crime Laboratory
Albany, New York 12226

ABSTRACT

The image processing in the scanning electron microscope by deflection (Y) modulation allows augmentation of depths and topographic features of line crossings that naturally occur in many kinds of handwritten documents. A critical evaluation of this method, in comparison with the secondary and the backscattered electron imaging methods, has been made in order to establish that deflection modulation is advantageous for sequencing of intersecting lines of pencils and ball-point pens which leave relatively deep and equal impressions in the paper. The most reliable features of sequencing were less frequent with plastic-point pens. These results have indicated that the sample condition of line crossings is improved by the hardness of the writing tip, by increased writing pressure, and by thick or non-dispersive deposition of ink. The effectiveness of the method is determined by the selected signal (secondary or backscattered), by the accelerating potential of the incident electron beam, and by the specimen-to-electron detector geometry. Knowing these limitations is crucial to successful applications of deflection modulation, since this method is sensitive to the flaws contained in the original signal. The method qualifies for re-examination of the apparently uncharacteristic firing pin impressions of shotguns. The x-ray dot mapping method is well-suited for elemental matching comparisons of multilayered paint chips. Despite the time involved, this method and the backscattered electron imaging method are suitable for examination of fingerprints of magnetic inks.

INTRODUCTION

The imaging capabilities of the scanning electron microscope (SEM) in conjunction with an energy dispersive x-ray spectrometer (EDS) allow both topographical and compositional displays that can be readily interpreted. Illustrative evidence of this type would be valuable in forensic determinations, since the associated techniques of image formation, chemical analysis and specimen preparation are well

---

[1]Presented at the 43rd Annual Meeting of the Electron Microscopy Society of America held jointly with the Microbeam Analysis Society, Louisville, Kentucky, 5-9 August, 1985.

understood and documented. Various methods of signal processing are also available, which allow intuitive, stylistic and synthetic interpretation of the image. Forensic applications of three such methods will be shown in this report. These are defection modulation (DM) or "Y-modulation"[2], x-ray dot mapping[3] and backscattered electron imaging.

One of the most intriguing problems encountered in questioned document examination is that of determining which of two intersecting lines was drawn last. Because this line is upper, one would expect that this line is continuous and may represent definite features of 'edge dominance' over the line written first (lower). This problem can be approached by augmenting depths and topographic features on a background the gray levels of which covers a little less than the full range of signal intensity. In fact this is done by DM.[2] In this method the time varying scan signal is mixed with the original video signal (raster) to modulate the Y axis of the line on the CRT (cathode ray tube) screen. When the DM gain is adequate, if not at the maximum level, significant surface reliefs (Y-shift) can be achieved. These images contain valuable information supplementing normal mode images using either secondary electrons (SE) or backscattered electrons (BSE). Another potential application of DM is the studies of firing pin impressions of certain shotguns, which are apparently uncharacteristic.[4] The SEM images of such cavities, grooves and faceted objects suffer from shadow effects depending upon the shape and orientations, towards or away from the Everhart-Thornley[5] detector. If the cartridge case contains a magnetic element, the magnetic field interferes with the focusing of the electrons. The topographic details at low magnifications may also remain buried under significant variations in image contrast due to different surface potentials (voltage contrast). All three images (SE, BSE and DM) recorded with the the Everhart-Thornley[5] detector can be critically affected by the specimen condition, operating voltage of the SEM and by the 'specimen-to-detector geometry'. The latter takes into account the angle of incidence (tilt), the angle of emission (i.e., the take-off angle or the angle of elevation of the detector above the specimen tilt plane) and the orientation of the specimen on the plain of tilt. Therefore this report provides a critical evaluation of the DM images, in comparison with the corresponding SE or BSE images. Oron and Tamir[6] first examined cross-overs of lines in questioned documents by a slight modification of DM[2] and a derivative mixing process.[7] Their work has been duplicated here, using only the DM method.

The x-ray mapping capability of the SEM-EDS has been used earlier to decipher the chemical makeups of particulates, such as, gunshot residues by studying them in cross-sections.[8] Such determinations as, whether the spectral (x-ray) elements are distributed in multilayers (e.g. paint chips), or they are highly heterogeneous, can be made by x-ray dot mapping. This method and the BSE imaging method have been used for examination of fingerprints of magnetic inks.

MATERIALS AND METHODS

Specimen Preparation

DM: orientation dependence. A phonograph record chip (1cm x 1cm) containing oriented grooves was coated with gold in a sputter coater (E.F. Fullam, Inc.®, EMS40). The record chip was mounted with silver paint on a 1/2-inch diameter aluminum disk and then examined.

Trial line crossings. The trial line crossings of known sequences (line 1 drawn first; line 2 drawn later) were prepared using combinations of writing instruments and paper types. The writing instruments used were: #2 and #3 pencils of Winthrop ® (Bondexed lead, 96), Reliance ® (Durolead, Templar 777), Berol ® (Mirado 174), Blackwell ® (Sundance) and Empire ® (Choice); black medium and fine ball-point pens of Bic ® (Biro), Longlife ®, Faber Castell ® (Spirit, M-Blk) and IBM ®; and plastic-point pens (extrafine) of Papermate ®, all made in the United States. The papers used were smooth photocopy papers of International Paper Company ®, smooth blue-ruled writing pads of Mead ® and coarser bonds (25% cotton fiber, heavy weight) of Four Star ®. In most line crossings examined in this report the lines 1 and 2 were made one immediately following the other. The directions of individual strokes were marked on the lines with arrow heads. The papers containing these trial intersections were enclosed in envelopes and then stored at least 3 months in vacuum prior to mounting and examination in the SEM. As an exception to the previous samples, a time lapse of about two weeks (14 days) was allowed between the drawing of the lines 1 and 2 in a number of intersections and these were also examined 3 to 4 months later. A few intersections were also prepared and examined on the same day.

Because the courts might be reluctant to allow large scale sampling and coating of documents, the line-crossing samples were cut to smallest possible sizes (~ 2mm x 2mm). These were mounted on a relatively larger diameter aluminum pin (diameter 1-inch) using a spread, thin layer of diluted carbon cement or rubber cement. Selected samples were coated with carbon by vacuum evaporation in a Ladd ® evaporator (See Results).

Firing pin impressions. Test-firings (3 shots per firearm) were conducted with selected 12 and 20 gauge shotguns (Ithaca ®) in the Laboratory range. The ammunition used were Remington ® Skeet Shot shells. The plastic wading was removed from each case and the collected cases were cleaned ultrasonically in methanol (20 min.) and acetone (20 min.). Because the primer (housing) surface bearing the impression of the firing pin remains separated from the surrounding case by a narrow circular groove, this groove was filled with carbon cement. This improved conduction within the specimen.

Paint chips. Under a stereomicroscope of low power (1X-70X) the paint chips were given smooth cuts along two parallel edges. Each chip was held at one edge in a tiny droplet of diluted (1:4) rubber cement (Carter ®) on a polished carbon planchet (diameter 1/2 inch), or a carbon pin mount (diameter 1/2 inch). For delivery of the pin mounts a metal block (1 3/4 in. x 1 3/4 in.) was used which contained four pin holes. The block containing the sample mount was kept for 30 min. in a vacuum oven at 90°C. The dried droplets of rubber cement usually provide good anchorage to the paint chips held on edge. Depending upon the case and the sample size, the supplied paint chips have been used for other tests which followed after the examination in the SEM-EDX. Therefore an initial examination was performed in the SEM in order to decide whether the paint chips were to be made conductive by carbon evaporation.

Fingerprints. Aluminum foils containing several fingerprints of a magnetic ink (Flint Ink Co. ®) were cut to 3/4 inch x 3/4 inch segments. Each segment containing a fingerprint was mounted on a 1-inch diameter aluminum pin. No carbon coating was necessary for these specimens.

63

SEM and signals. The specimens were examined in an AMR 1000 SEM using a grounded specimen stage that increased conduction within the specimen. The SEM was equipped with a deflection modulation device, an energy dispersive x-ray spectrometer and a multichannel analyzer (EDAX 707A), and a solid-state backscatter detector ("quad") system mounted in the specimen chamber around the exit of the final lens. A standard Everhart-Thornley[5] detector was used to produce a secondary electron (SE) signal at a +300 V (volts) bias of the collector screen while a +11kV potential was applied on the scintillator of the detector to accelerate the collected electrons. The same detector was used to form backscattered electron (BSE) images, for which a negative bias (-100 V) was applied to the collector screen to reject the secondary electrons. Only these two signals (SE and BSE) were used for deflection modulation (DM) and the modulated images will be called DMSE and DMBS (not DMBSE) images, respectively. The images in the normal mode (SE or BSE) and in the DM mode were recorded in succession. No altering of the brightness and contrast levels of the CRT was necessary for the DM images. The SEM used did not have the accessory required to rotate the direction of scan and to compensate for the apparent distortion introduced by the angle of tilt of the specimen disk.

The solid-state ("quad") detector for higher energy (> 10 KeV) backscattered electrons was used to separate atomic contrast (A + B) from topographic contrast (A - B).[9] Because this method was corroborative, it has been used only to check the line-crossing sequences observed with the Everhart-Thornley detection.[5] These results will be used elsewhere.

Detector geometry. The specimen-to-detector geometry is a relevant point in the studies of line crossings. The Everhart-Thornley detector[5] (fixed) used was a horizontal, low take-off angle detector. With normal incidence (no tilt) when the working distance (distance from the final lens aperture) was about 11 to 12mm the focused electron spot on the specimen was coincident with the line of detection. Therefore higher take-off angles for backscattered and secondary electrons were achieved by tilting the specimen by about 45° (degree) toward the detector (on right) and by taking advantages of larger working distances (15-20mm) since the required magnifications (15X-200X) were low. The horizontal angle between the detector and the specimen stage on the chamber door was about 72° (degree). Because this angle was less than desirable (i.e. 90° degrees) the goniometer stage containing the sample disk was turned clockwise through about 13° (degree). This made about 85° (degree) angle between the detector axis and the axis (horizontal) of tilt of the specimen. Consequently the viewing direction of the detector was not exactly from the top to the bottom of the SEM images: it was still off by about 5° (degree) to the left of the vertical axis (Y) of the SEM images or the display CRT. The horizontal or the longer axis (X) of the SEM images (or CRT) corresponded to horizontal motion(X) in the sample chamber across the detector. The Y-motion was towards or away from the detector.

SEM operating conditions. The instrumental parameters were apparently dependent on the sample conditions of line crossings. The line crossings indicative of relatively deep and equivalent impressions were suitable for the BSE imaging at 20kV electron potential (cf. Koons[10]). However, the attenuation of oriented lines and grooves along the direction of Y-modulation (DM) was associated with the use of this higher accelerating voltage (20kV). The solution to this was the use of SE at a much lower accelerating voltage (5kV) (cf. Fig.(s) 1,2 vs. Fig. 3) (cf. Oron and Tamir[6]). The best orientation of the line crossings in the plain of tilt was the one in which the lines (1 and 2) were 'X'-like diagonals on the CRT screen or the SEM images. Because this orientation is dependent on the detector geometry, it can be assessed for other SEM's using the method in Fig. 1. This criterion has been used in other reports but was not explained.[6, 10] Besides the orientation in the plain of tilt there are other factors that determine take-off angles (e.g. tilt, working distance, Y-motion, etc.) of SE and BSE. These were optimized for line crossings of the same kind. The firing pin impressions were examined at 20kV. The best tilt for DM images of firing pin impressions was 45° (degree). The method used for x-ray (EDS) dot mapping of paint chips and magnetic fingerprints is described elsewhere.[8]

The avoidance of charging of the line-crossing samples was a dominating problem. These samples were irradiated with 1-5kV electrons before recording the image with 20kV electrons at the fastest possible scan rate (20 secs per frame of 1000 lines).[11] The emission currents with a hair pin filament were 75µa at 20kV and 20µa at 5kV. A 100 µm final aperture was used for the SEM images and for the dot maps.

RESULTS AND DISCUSSION

DM of Oriented Grooves

In the SEM the image formation process suppresses some of the flaws contained in the original signal. This information becomes available upon signal manipulation by deflection modulation (DM). The most relevant information to be reported here is that the SEM images of highly oriented grooves and indentations in paper can be critically affected by the 'sample-to-electron detector' geometry. The effect has been analyzed at various operating conditions of the SEM with a view to suppress the problem (following section).

The experiment was performed with a conductive (i.e gold-coated), broken chip of phonograph record containing parallel grooves which had zigzag and straight local orientations depending upon the sound track (Fig. 1). The sample disk (Al pin, 1/2 inch diameter) was rotated clockwise through 90° (degree) to obtain three orientations on the plain of tilt (45° degree) (Fig. (s) 1A & B, C & D and E & F). The BSE and the corresponding DMBS images were recorded for each orientation. The position of the Everhart-Thornley[5] detector was, as if, along-side the top of these images while the viewing direction of the detector was downward (~ 85° degrees to the longer axis of the images). The shown images on top (Fig. 1A, C and E) are like classic illustrations of how surface topography contrast arises in the SEM images[12]; whereas the bottom panels (B, D and F) illustrate the flaws in those original images (Fig. 1).

The images on top (Fig. 1A, C and E) appear to be 'shadowed' in one direction; that is, only one side of the grooves indicates the shadow (Fig.(s) 1A and C). This effect is such as would be seen if the reader looked down onto the specimen along the incident electron beam while the Everhart-Thornley[5] detector was shining light onto the specimen. However, in the DMBS image (Fig. 1D) shadows occur on either sides of the grooves. The shadow lengths measured across the grooves have no correspondence between the original image and the processed image (cf. Fig.(s) 1A vs B). The original BSE signal utilized the shadow effect to generate topographic contrast. This effect appears to be maximum in Fig. 1E. In the corresponding DMBS image, topography persisted only with the zigzag or convoluted grooves while the straight grooves disappeared (See arrow in Fig. 1F). The attenuation of the straight

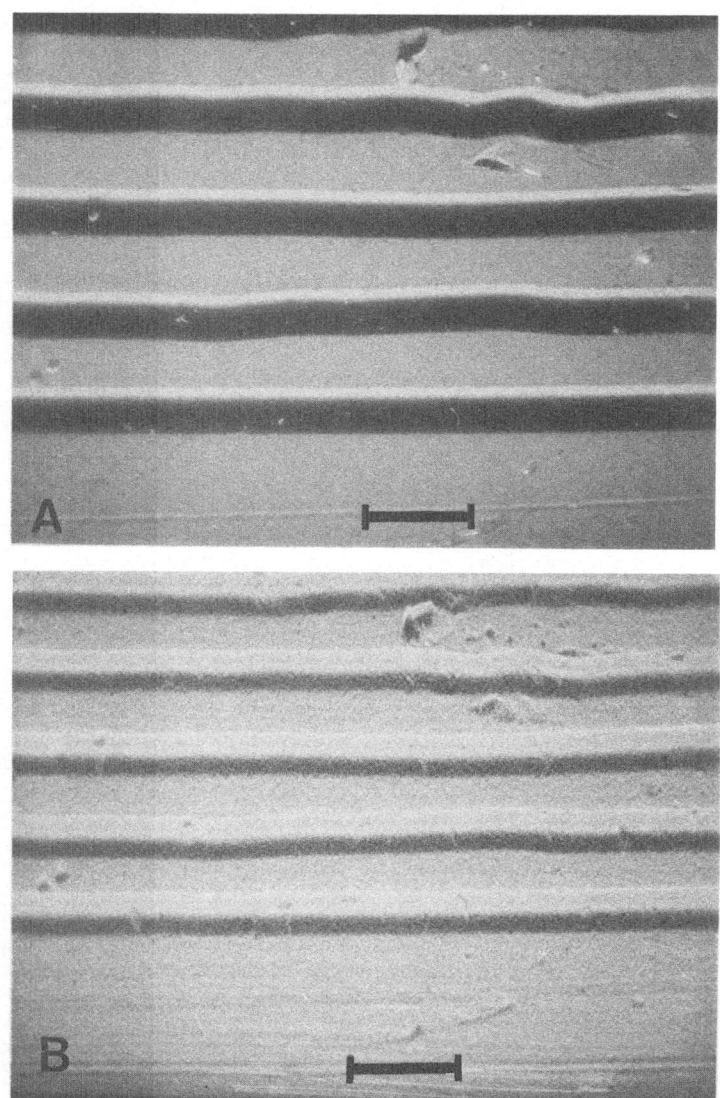

Fig. 1 BSE (A) and DMBS (B) images of parallel grooves
         in phonograph record. Grooves oriented
         parallel to the axis (X) of tilt. Tilt:
         45°, accelerating voltage: 20kV, working
         distance: 18mm and magnification: 160 X.
         Specimen gold coated. Bar: 0.1 mm.

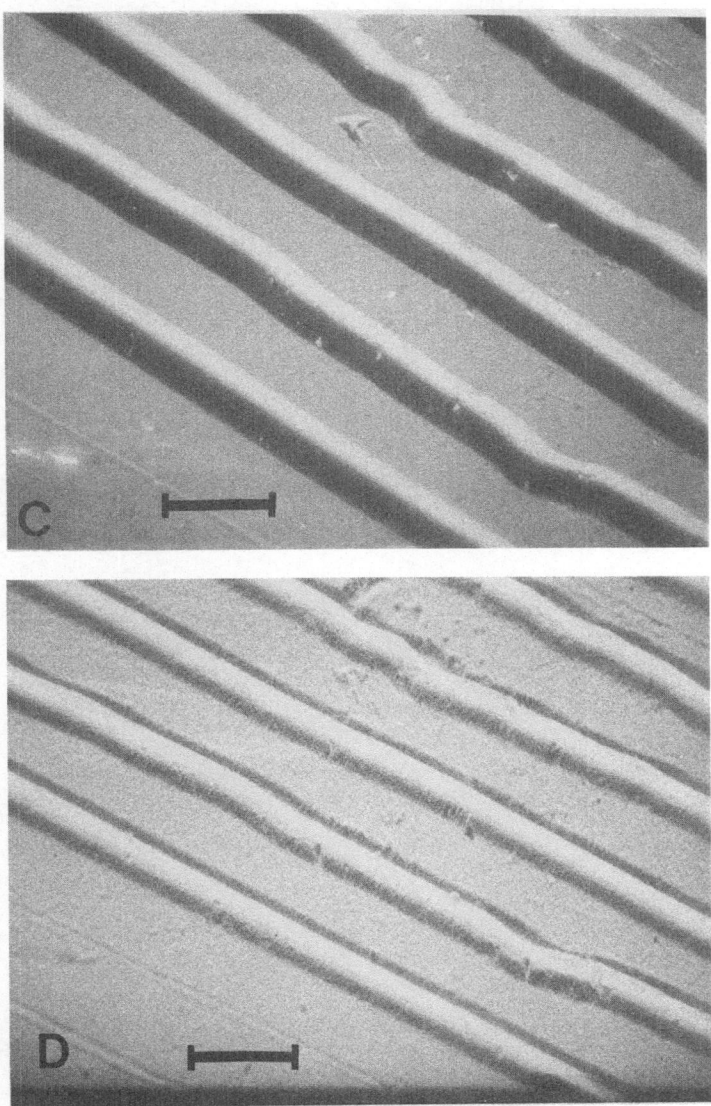

Fig. 1 (Contd.) BSE (C) and DMBS (D) images of
        parallel grooves in phonograph record.
        Grooves inclined to the axis (X) of tilt.
        All other conditions same as Fig. 1A and
        B. Bar: 0.1 mm.

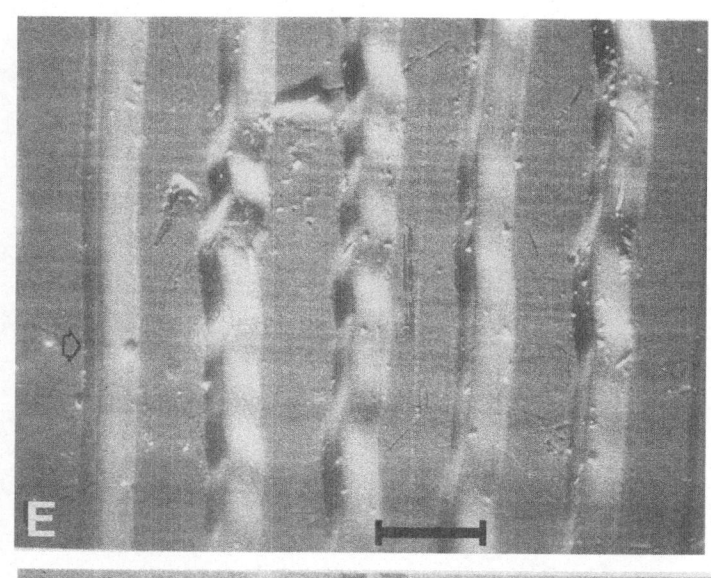

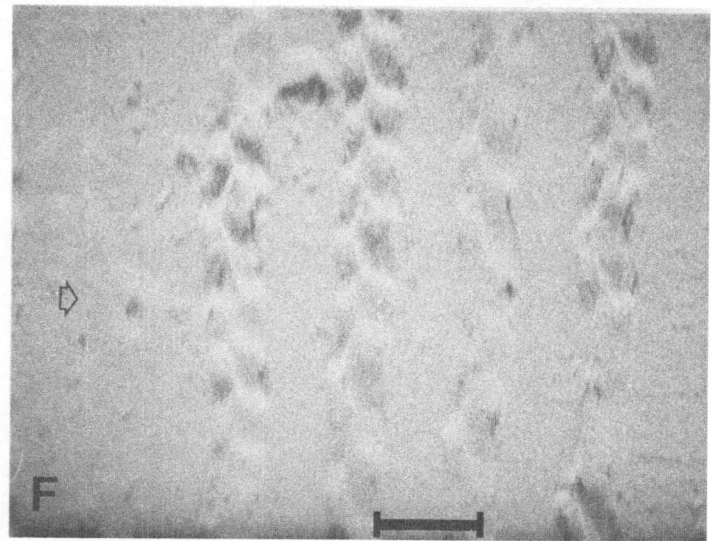

Fig. 1 (Contd.) BSE (E) and DMBS (F) images of
parallel grooves in phonograph record.
Grooves orientation perpendicular to the
axis (X) of tilt. (i.e along the Y axis
of CRT). All other conditions same as in
Fig. 1A and B. Arrow shows the attenuated
straight groove. Bar: 0.1 mm.

grooves persisted at any detector take-off angle (tilt 0°-45°, working distance 12-20 mm).

Sequence of Intersecting Lines

The observation in the previous section (Fig. 1) infers that line impressions by writing instruments will be attenuated in the DMBS images if the image lines are vertical, i.e. parallel to the Y axis of CRT. The experiment was performed with carbon-coated trial samples of line crossings made with pencils (Fig. 2) and ball-point pens (not shown). These samples were rotated in the plain of tilt (45° degree). The shown lines in Fig. 2 were drawn with medium-to-heavy pressure with the sharpened lead of a soft pencil (Reliance ®). As expected, line 2 (line drawn last) in Fig. 2B has been attenuated at the top part of its DMBS image. This portion of line 2 was parallel to the Y axis, while line 1 was along the X axis. Line 2 was upper and should show 'edge dominance' over line 1 . In fact this information is evident in the images in Fig.(s) 2C and D in which both lines 1 and 2 are inclined to the axes (X, Y) of the SEM images (or CRT). The orientation geometry of the lines (1 & 2) in Fig.(s) 2A and B was clearly unfavorable to sequence determination.

The attenuation of straight grooves (Fig. 1F) and lines (Fig. 2B) was due to lack of edge contrast. In other words 'shadowing' was minimal at the edges (two) of these structures. Since the 'shadowing' effect results from varying take-off angles of the BSE emissions [12] and because the tilting of the specimen toward the detector causes the emission maximum to move toward the detector,[13] the attenuation effect is indicative of 'optical' reflection of the primary electrons.[14] If so, the energy of incident electrons would be of more concern than the detector. Incidentally the results in Fig.(s) 1 and 2 were obtained at 20kV accelerating voltage of the incident electron beam. The attenuation effect was also observed with a solid-state ("quad") detector for higher energy (> 10 KeV) backscattered electrons with normal incidence (tilt 0° degree) of a 20kV electron beam.[9] Furthermore the effect persisted with the DMSE images of lines and grooves recorded at the same operating conditions of the SEM. The effect, however, was significantly less when the incident electrons were less energetic (e.g. 5kV or 1kV) (cf. Fig. 3).

The SE coefficient increases several-folds with the decrease of the electron beam energy from 20kV to 1kV.[15] The charging effects are also reduced by the decrease of the beam voltage (cf. Fig. 4 in Reimer[16]). Therefore the line-crossing samples were examined with the secondary electron signal at 5kV or 1kV. An example of this examination has been shown in Fig. 3A and B which are the DMSE and the SE images of a cross-over of two lines. Line 1 was a soft pencil line (Reliance ®) drawn first; line 2 was a ball-point pen line drawn next. Both lines were drawn with normal (i.e. medium) pressure. Although line 1 (pencil line) is oriented along the Y axis of the CRT screen (image), it has not been attenuated in the DMSE image (Fig. 3A). Line 2 is associated with a bridging material (arrow 's') which runs across line 1 . The 'bridge' can be reviewed from the top and the bottom of the image and from either sides (left or right) of line 2 . The bridge is a reliable feature of 'edge dominance' of line 2 over line 1 . The SE image (Fig. 3B) of the same line-crossing did not have this indication. However such high contrast images also cause optical illusion, that is line 1 appears to be in front of line 2 if the images are turned through 180° (degree) and viewed along line 1 . Oron and Tamir[6] had a similar experience with the relatively low voltages (2-5kV) they used for modified DM images of line crossings. This effect

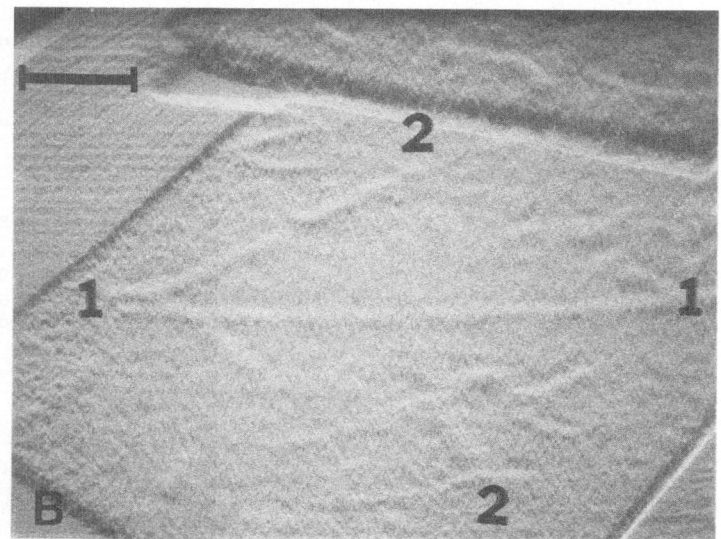

Fig. 2 BSE (A) and DMBS (B) images of a line crossing made with a soft pencil (Reliance ® #2). Line 1 : drawn first, oriented along the X axis (axis of tilt). Line 2 : drawn last, parallel to the Y axis at the top part. Arrow heads show the directions of the strokes. Tilt: 45°, accelerating voltage: 20kV; working distance 17mm and magnification 17X. Specimen carbon coated. Bar: 1 mm.

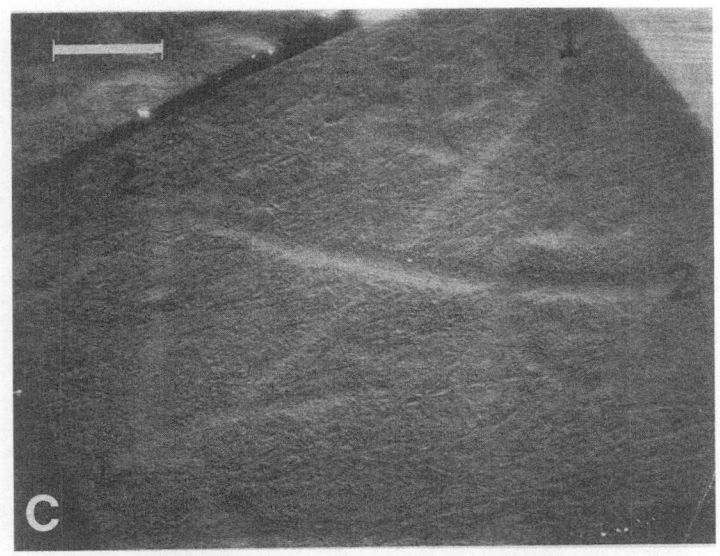

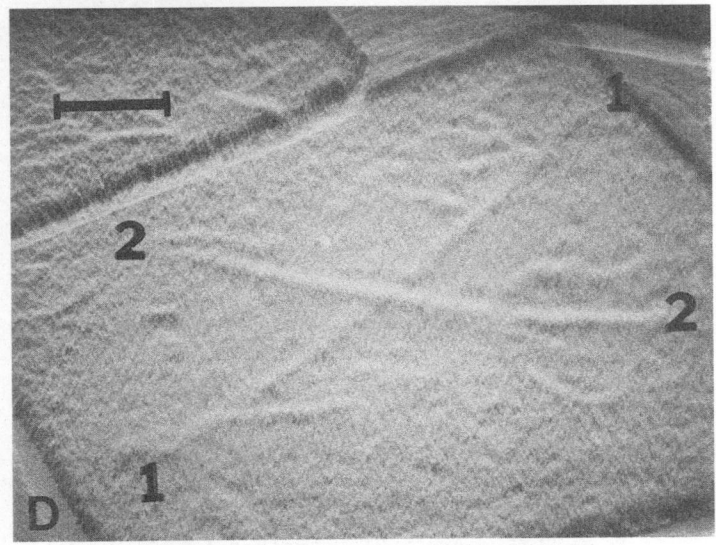

Fig. 2   BSE (C) and DMBS (D) images of a line crossing
made with a soft pencil (Reliance ®, #2).   The
specimen in Fig. 2A and B was turned
counter-clockwise until the lines 1 and 2
were inclined to the X, Y axes, as shown.   See
Fig. 2A and B for all other conditions.   Bar:
1.0mm

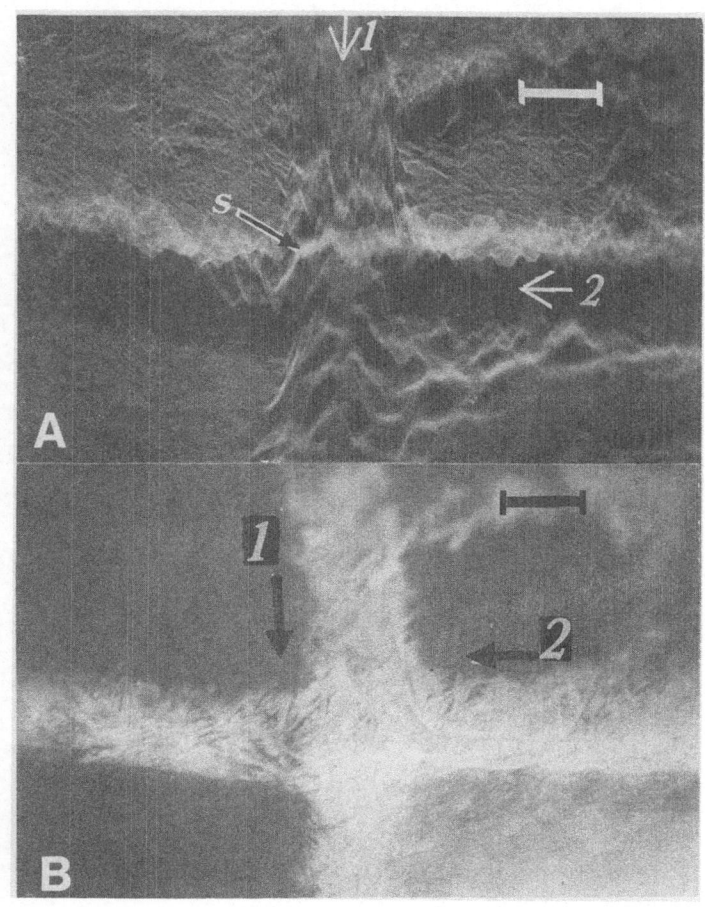

Fig. 3 DMSE (A) and SE (B) images of a cross-over of
two lines. Line #1 (arrow): a soft
pencil line (Reliance ®) drawn first.
Line #2 (arrow): a ball-point pen
(Longlife ® medium black) line drawn
next. Arrow S: streaked (or bridged)
material across #1 line. Tilt: 45°
(degree); accelerating voltage: 5kV,
working distance: 20mm and magnification:
56.3X. Specimen uncoated. Bar: 0.2mm.
(Reproduced with permission from ref. 1)

does not arise with the BSE signal. The optical illusion in the DMSE images of line crossings is perhaps avoidable by taking stereopairs of the image. This work would be possible in due course.

The ability of the BSE signal to determine line sequence has been subject to critical tests involving writing instruments for which inking takes place by a flow mechanism. In these tests the best orientation of the line crossings was the one in which the lines 1 and 2 were 'X'-like diagonals on the CRT screen (cf. Oron and Tamir[6]). No key features were observed with the crossings made with felt tip pens. Extra fine plastic point pens (Papermate ®) occasionally produced distinctive features when the lines were made with heavy pressure (Fig.(s) 4A and B). The impressions made on paper by these writing instruments were much shallower and smoother than those obtained with pencils and ball-point pens. The key features of sequencing appeared only when the ink was non-dispersive. As Fig. 4 shows, the ink deposit at the crossing was streaked along the edges (two) of line 2 . The deposit gathered like 'snow-banks' across line 1 . This feature resembled the 'bridge' found with the pencil vs ball-point pen crossings (cf. Fig. 3). The consistency of these materials ('bridges') was assessed by a rotation of the sample disk containing the trial line crossing (Fig. 4A and B).

These exhibits of DM (Fig.(s) 1-4) should not give the idea that the original signals (BSE and SE) are not useful at all. The quality of impression and the quality of ink improves the sample condition of line crossings.[10,17,18] The quality of ink refers to thickness (or amount) of the ink deposit and to non-dispersive deposition of the ink. When these conditions prevail the conventional methods (BSE and SE) are sufficient to provide elegant demonstration of the sequences of

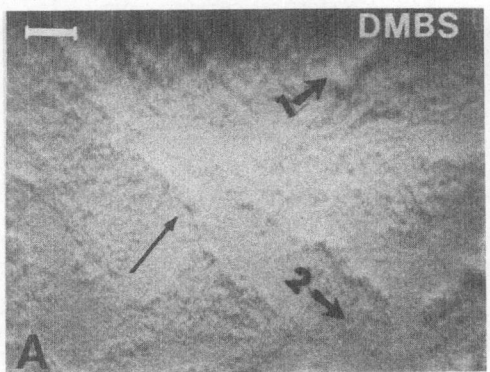 

Fig. 4 DMBS images of a line crossing made with heavy pressure with an extrafine plastic point pen (Papermate ®). Line 1: drawn first; Line 2: drawn last. Arrow shows streaked materials which gather like 'snow-banks' across line 1 (cf. 'bridge' in Fig. 3). A: image recorded first; B: image recorded after rotating the cross-over clockwise through 270° (degree). Tilt: 0° (degree); accelerating voltage: 20kV; working distance: 20mm and magnification 37.5X. Specimen uncoated. Bar: 0.2mm.

intersecting lines. The line crossing in Fig. 2C had this potential and so this sample has been recorded at a higher magnification, using the BSE signal (Fig. 5). The continuity of line 2 , or the 'edge dominance' of this line over line 1 , has been verified by turning the specimen (disk) through 180° (degree) (cf. Fig. 5A vs 5B). The pencil line crossings do not reveal 'bridges' across the lower line (i.e. line 1). Pencil deposits consist of very fine grains. Laminar streaks of these grains are made possible with medium-to-heavy writing pressure and are discernible at the cross-over (Fig. 5).

Unlike the pencil deposits, the ink deposits of ball-point pens are lumpy and semi-solid, and therefore the ball-point pen inks are streaked along the drawn lines. The effect is enhanced at the crossover where the overlapping deposits from two writing instruments are subject to final streaking along line 2 . This effect is easily recognized in the SE (Fig. 6A) and the BSE (Fig. 6B) images if the sample document is coated with evaporated carbon. The excess of ink deposits at the cross-over occurs as a 'snow-bank' across line 1 (See arrow heads, in Fig. 6A and B). This confirms the continuity, or the 'edge dominance', of line 2.

In Fig. 6 both the lines (1 and 2) were drawn with light (line 1) to heavy (line 2) pressure with two ball-point pens, one immediately following the other. This pressure differential was consistent with the different widths of the lines, line 1 having been narrower than line 2. If the condition was reverse, i.e., line 2 was much narrower than line 1 , the sequencing would have been difficult, if not impossible. Such difficult line crossings require much higher magnifications than used in Fig. 6. The criteria sought for these examinations is the stacking sequence of ink deposits in the vertical direction of the smaple substrate.[10,17-20] If the ink deposits consists of different compounds, identification with respect to localization of a particular ink deposit is possible by elemental analysis with energy dispersive x rays and x-ray induced fluorescence (XRF).[19] Alternatively the SE signal is used for comparative high resolution studies of the ink layers.[10,17-20] Recently the cathodoluminescence mode in the SEM has been successfully used for establishing the sequence between intersecting felt pen writing and textile ribbon typescript.[21]

A total of 110 line crossings were examined for DM imaging. The success rate was very poor with plastic point pens (10%) and soft pencils (#1) (30%). A high success rate (80%) was achieved with medium (#2) or hard (#3) pencils and ball-point pens when the writing pressure of line 2 was matched with the writing pressure of line 1 . The line crossings involving two different writing media also produced good success rate (67%) under the above condition and also when line 2 was drawn with heavy pressure. The most tempting examples were obtained with pencils and ball-point pens which leave relatively deep and equal impressions in the paper.

The SEM images of indentations in paper caused by writing intruments can be critically affected by the 'sample-to-electon detector' geometry. This experience could be disturbing. In fact Nolan etal[19] have stated that the appearance of a crossover can be completely reversed by rotating the sample, or by altering other parameters of the SEM. A directionally sensitive detector will worsen the conditions required for sequencing of lines. The attenuation of straight lines and grooves is physically existent even at 1-5kV: the effect is only suppressed by the higher yield of secondary electrons (Fig. 3). It is hoped that these studies have put the above problem into its proper perspective and Oron

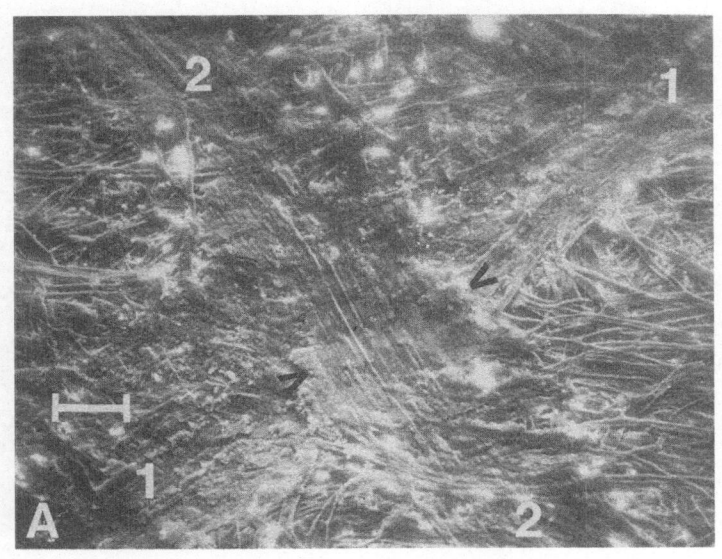

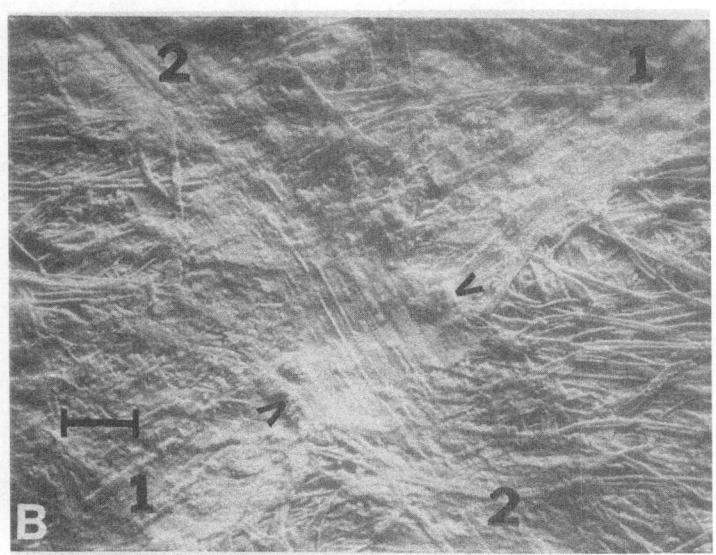

Fig. 5 BSE images of the pencil line crossing
in Fig. 2 at higher magnifications. A:
image recorded first. B: image recorded
after rotating the cross-over clockwise
through 180° (degree). Tilt: 45°
(degree); accelerating potential: 20kV,
working distance: 12mm, and
magnifications: 80X (A), 60X (B).
Specimen carbon coated. Bar: 0.2mm.

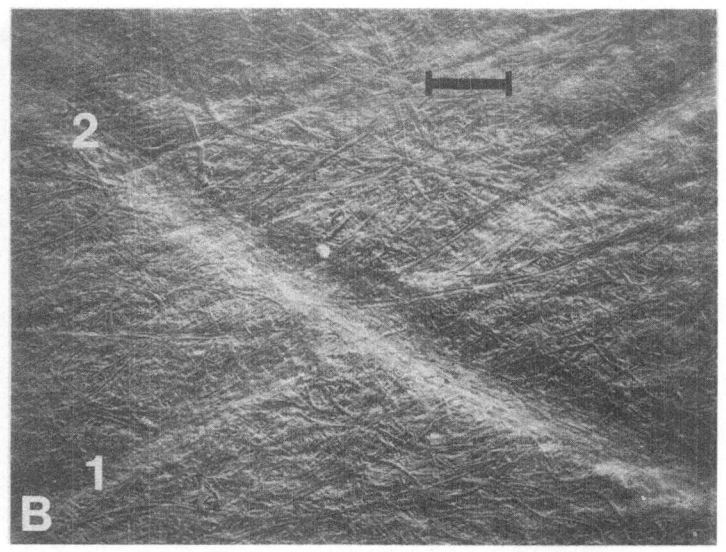

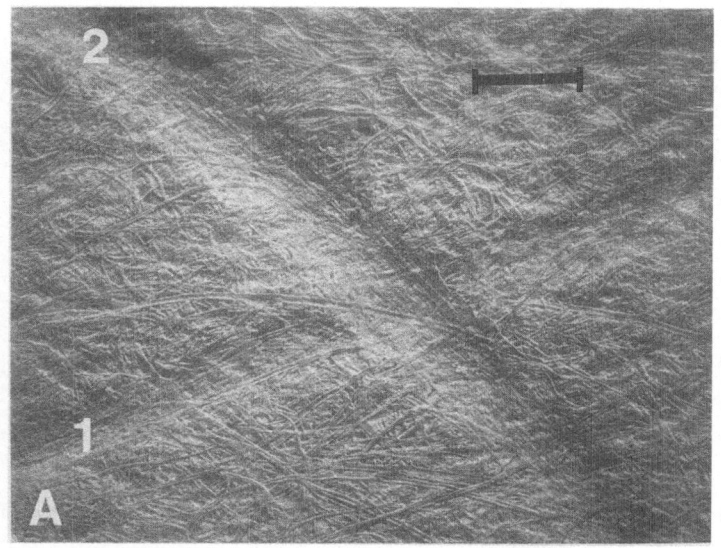

Fig. 6 SE (A) and BSE (B) images of a cross-
over of two ball-point pen lines. #1:
line drawn first with a Bic Biro®
(medium, black) ballpoint pen; #2: line
drawn last with a Longlife ® (medium,
black) ball-point. Arrow heads show 'snow
bank'-like materials across line 1. Tilt:
45°, accelerating voltage 20kV; working
distance 11mm, and magnification 105X.
Specimen carbon coated. Bar: 0.1mm.

and Tamir[6] should be credited for the work they initiated.

Firing Pin Impressions

The DM method has been applied to firing pin impressions of 'difficult' shotguns. The 'difficult' shotguns were those which did not produce characteristic features of impresssion on the cartridge cases. The latter were examined in the comparison microscope, or in the SEM using the SE signal. The application of DM to these cartridge cases has indicated that the method is able to recover characteristic features of indentation from the SE image. The searched identifying features do not require a magnification more than 45-to-50 times, that is normally used.[4] An example is shown in Fig.(s) 7A and B. The characteristic features are found independent of orientation of the firing pin impression (Fig.(s) 7A and 7B, arrows e, f, g and h). Certain ambiguous features (questioned marks in Fig. 7C) in the secondary electron images are also avoided by the DM method. Preliminary studies have indicated that the success rate of the method is 33%, i.e., one out of three 'difficult' shotguns is able to produce at least four reproductive features of impression in three successsive firings. Larger series of test-firings involving ten-to-fifty shots with a specific weapon would be necessary for proper evaluation of the method (cf. ref. 4, 22-25). The disadvantage of the DM method is that the images suffer from distortion due to tilt of the specimen. The firing pin impressions remain to be examined under those conditions which will compensate for the distortion of the image.

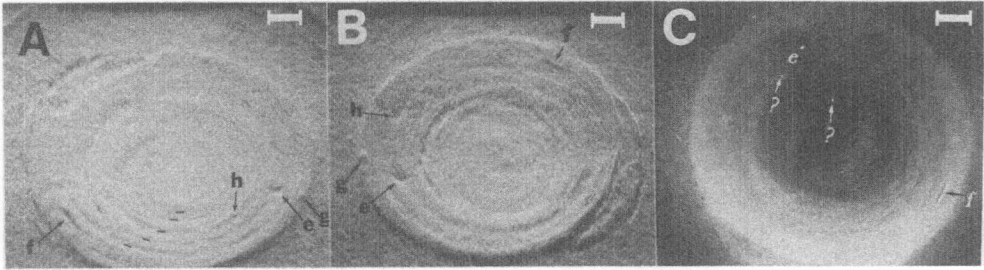

Fig. 7   Images of a firing pin impression from a 20 gauge Ithaca model 37 shotgun by deflection modulation (A and B) and by secondary electron signal (C). Magnification: 45X (See Text for notations). Tilt: 45° (degree) in A & B; Tilt: 0° (degree) in C as per Ref.4. Accelerating voltage: 20kV and working distance 15mm. Specimen uncoated. (Reproduced with permission from ref. 1). Bar: 0.1mm.

Paint Chip Comparison

The layer-like sequence of element distribution in car paint chips is the most significant point of physical comparison in the SEM-EDS, particularly because of the variety of ways in which a car may be refinished. [26-29] The obvious specimens for x-ray dot mapping analysis are those paint chips that remain undifferentiated by microscopic examination of color, tint, type of finish, layer sequence, layer colors, textures and approximate layer thickness. Usually paint chips containing three layers (top coat, primer and undercoat), or more than three layers are mounted on edge on a "glue-lift" disk.[8]

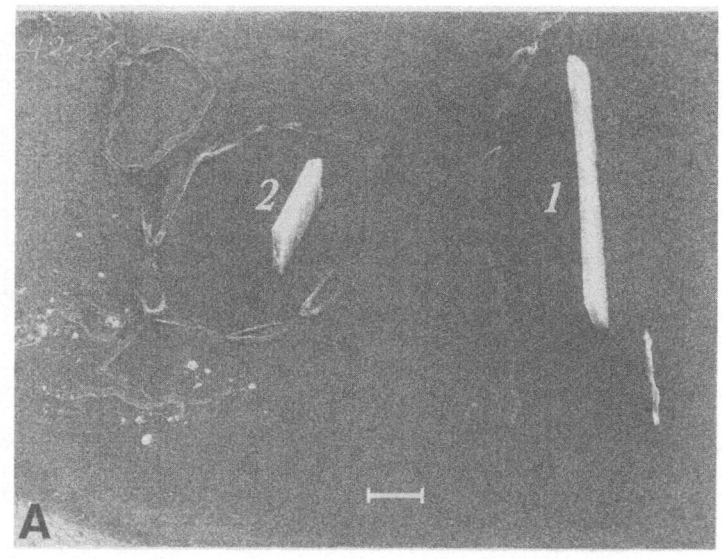

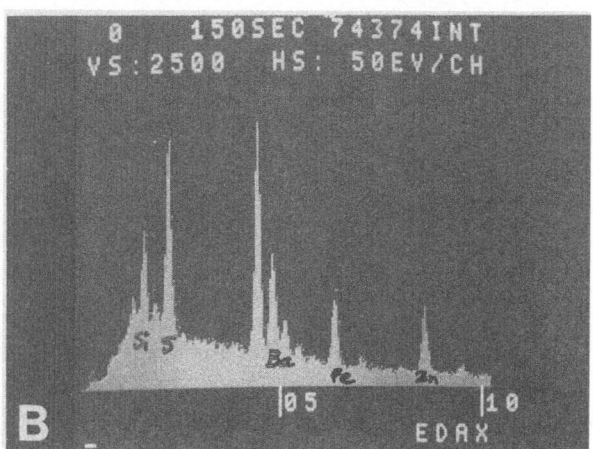

Fig. 8 Elemental comparison of two multilayered paint chips mounted on a
"Glue-lift" disk (ref. 8). A: BSE
image; B: Energy dispersive x-ray
spectra-dotted line for evidence (#1);
Solid line for control (#2).
Magnification in A: 42·5X,
accelerating voltage: 20kV. The two
paint chips had the same general
composition in terms of inorganic
elements (Si, S, Ba, Fe and Zn) and
their relative abundance. Bar: 0.2mm.

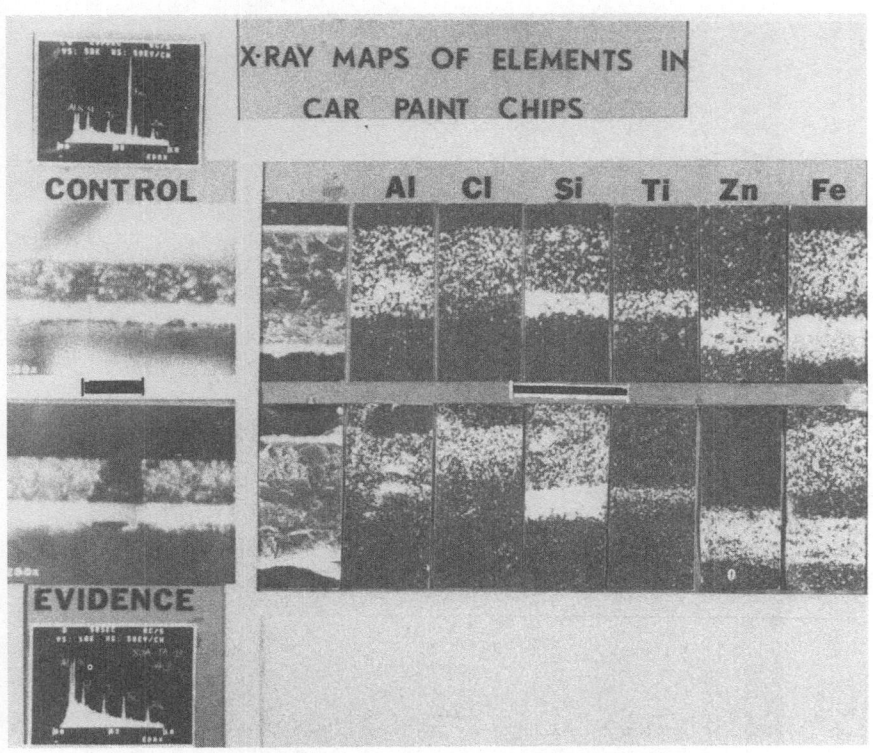

Fig. 9 Matching comparison
between an evidence paint
chip from a victim's clothing
and a control paint chip from
a suspect's car involved in a
hit and run crime. The
sequence of the images are as
follows: light micrographs
→ SEM images → dot maps
of Al, Cl, Si, Ti, Zn and Fe
(body). Tilt: 0°,
accelerating voltage 20kV.
Magnification: 560X (SEM
images); 280 (light
micrographs). Specimen
uncoated. Bar: 0.1mm.
Reproduced with permission
from ref. 1.

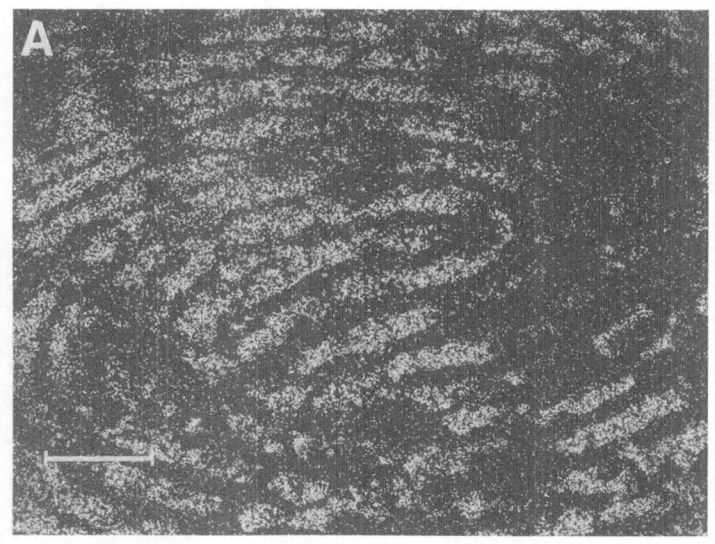

Fig. 10 X-ray dot map of Fe (Kα) (A), BSE image
(B) and EDX spectrum (C) of a fingerprint
of magnetic ink (Flint Ink Co. ℝ). A:
Fe (Kα) 200-460 cts/s, 5 min exposure
(2X); B: BSE (-100 volts), Magnification
17 X; C: Al peak due to substrate;
Signal-to-noise ratio at Fe (Kα) ≈ 11.
Bar: 1mm.

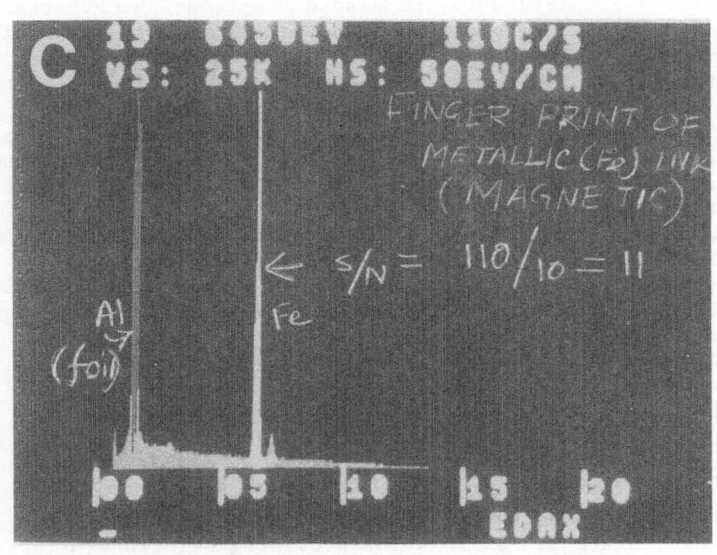

Fig. 8A is the BSE image of a typical mount containing a paint chip in evidence (marked 1 , on right) and a control paint chip (marked 2 , encircled) from a car involved in a vehicular accident. The matching energy dispersive x-ray spectra obtained from the top edges of the paint chips are shown in Fig. 8B. Displays of x-ray dot maps, along with light micrographs and SE images, have been utilized in this case and in other cases for comparative analysis. One such example has been shown in Fig. 9. The stacking sequences of elements in the involved paint chips in this case were 'alike' although the spectral (EDS) distributions of the same elements were slightly different

Fingerprinting with Magnetic ink

The SEM-EDX has been used to examine the technical feasibility of magnetic ink ($Fe_2 O_3$) use in finger printing. The SEM is able to determine the quality of inks and paper in fingerprinting. An example is shown in Fig. 10 A-C. The shown fingerprint was applied on an aluminum paper. The ridge pattern of the fingerprint is visible in the x-ray (Fe[Kα]) dot map (Fig. 10A) and in the BSE image (Fig. 10B) except in some places (right, top). Those places of the ridge pattern, which have fewer x ray (Fe) dots are also darker than the substrate (Al) in the BSE image. This indicated that those darker ridges possibly contained compounds made of lower ($z < 13$) atomic number elements. In other words the latent fingerprint appeared on the aluminum foil but the disbursement of ink (metallic) particles was not optimum. This interpretation was consistent with two other observations; that is, latent fingerprints distinctly appear on aluminum paper[30] and these contain a variety of biological compounds like amino acids, lipids and vitamins that are found in the palmar sweat.[31] Because these compounds disintegrate in the sunlight and on aging, there is the distinct possibility that the metallic (Fe) ink particles are progressively detached from the latent finger print. This problem can be overcome by trial and error experiments. It is anticipated that the uses of a cross-linking agent in the form of a radiation (e.g. Ultraviolet light), or a stabilizing agent (e.g. negative stain, potassium phosphotungstate) that can be added to the ink without interfering with its properties, could alleviate the problem.

Acknowledgements

Thanks are due to Chester P. Sadowski of the Division of Criminal Justice Services, State of New York for providing the fingerprints. The work has been sponsored by New York State Police at the Headquarters Cime Laboratory during the period of 1981-85. Thanks are also due to Robert Miazga for his assistance in photography.

REFERENCES

1 . S. Basu, Forensic uses of deflection (Y) modulation and x-ray dot mapping, in: "Proceedings of the 43rd Annual Meeting of the Electron Microscopy Society of America," G.W. Bailey, ed., San Francisco Press, Inc., San Francisco, P.512 (1985).

2. T. Kelly, W.F. Lindquist, and M.D. Muir, Y-modulation: an improved method of revealing surface detail using the scanning electron microscope, Science 165: 283 (1969).

3. J.C. Russ, X-ray mapping on irregular surfaces, The EDAX Editor 9(2): 10 (1979).

4.   C.A. Grove, G. Judd, and R. Horn, Evaluation of SEM potential in the examination of shotgun and rifle firing pin impressions, J. Forensic Sciences 19 (3): 441 (1974.)

5.   T.E. Everhart and R.F.M. Thornley, Wide-band detector for micromicro-ampere-low-energy electron currents, J. Sci Instrum. 37: 246 (1960).

6.   M. Oron and V. Tamir, Development of SEM methods for solving forensic problem encountered in handwritten and printed documents, in: "Scanning Electron Microscopy/1974, Proceedings," IIT Research Institute, Chicago, Ill., Part I, P.207 (1974).

7.   K.F.J. Heinrich, C. Fiori, and H. Yakowitz, Image-formation technique for scanning electron microscopy and electron probe microanalysis, Science 167: 1129 (1970).

8.   S. Basu, Formation of gunshot residues, J. Forensic Sciences 27 (1): 72 (1982).

9.   S. Kimoto and H. Hashimoto, Stereoscopic observation in scanning microscopy using multiple detectors, in: "The Electron Microprobe," S. Kimoto and H. Hashimoto, eds. John Wiley & Sons, New York, P.480 (1966).

10. R.D. Koons, Sequencing of intersecting lines by combined lifting process and scanning electron microscopy, Forensic Sci. Internat. 27: 261 (1985).

11. P. Morin, M. Pitaval, and E. Vicario, Direct observation of insulators with a SEM, J. Phys. E. 9: 1017 (1976).

12. G.R. Booker, Scanning electron microscopy: The instrument, in: "Modern Diffraction and Imaging Techniques in Material Sciences," S. Amelinckx, R. Gievers, G. Remaut, and J.F. Landuyt, eds., North-Holland Publ. Co., Amsterdam, P.553 (1970).

13. N. Niedrig, Physical background of electron backscattering, Scanning 1 (1): 17 (1978).

14. E.H. Darlington and V.E. Cosslett, Backscattering of 0.5-10 keV electrons from solid targets, J. Phys. D. 5: 1969 (1972).

15. D.A. Moncrieff and P.R. Barker, Secondary electron emission in the scanning electron microscope, Scanning 1(3): 195 (1978).

16. L. Reimer, Scanning electron microscopy-present state and trends, Scanning 1 (1): 3 (1978).

17. P.A. Waeschle, Examination of line crossings by scanning electron microscopy, J. Forensic Sciences 24 (3): 569 (1979).

18. J. Mathyer, The problem of establishing the sequence of superimposed lines, Internat. Crim. Police Rev., Part I: 238 (1980); Part II: 271 (1981).

19. P.J. Nolan, M. England and C. Davies, The examination of documents by scanning electron microscopy and x-ray spectrometry, Scanning Electron Micros. Part II: 599 (1982).

20. J. Mathyer and R. Pfister, The examination of typewriter correctable carbon film ribbons, Forensic Sci. Internat. 25:71 (1984).

21. P.J. Nolan, C. Davies and A.G. Filby, Determination of sequence of writing - a strategy for evaluating new methods, paper presented at the annual meeting of the Intern. Assoc. Forensic Sci., Oxford, 1984; J. Forensic Sci. Soc. 24 (4): 355 (1984).

22. E.J. Korda, H.L. MacDonnell, and J.P. Williams, Forensic applications of the scanning electron microscope, J. Crim. Law Criminol and Police Sci. 61: 453(1970).

23. J.R. Devaney and L.W. Bradford, Applications of scanning electron microscopy to forensic science at the Jet Propulsion Laboratory, 1969-1970, in: "Scanning Electron Microscopy/1971, Proceedings," IIT Research Institute, Chicago, ILL. Part II, P.561 (1971).

24. C.A. Grove, G. Judd and R. Horn, Examination of firing pin impressions by scanning electron microscopy, J. Forensic Sciences 17 (4): 645 (1972).

25. G. Judd, R. Wilson and H. Weiss, A topographical comparison imaging system for SEM applications, in: Scanning ELectron Micros./1973 Proceedings," IIT Research Institute, Chicago, ILL., Part I, P. 167 (1973).

26. R. Wilson and G. Judd, The application of scanning electron microscopy and energy dispersive x-ray analysis to the examination of forensic paint samples, in: "Advances in X-ray Microanalysis, L.S. Birks et. al., eds., Plenum, New York Vol. 16, P.19 (1972).

27. R. Wilson, G. Judd, and S. Ferriss, Characterization of paint fragments by combined topographical and chemical electron optics techniques, J. Forensic Sci 19(2): 363 (1974).

28. B. Collins, Micro-techniques on the examination of problems in surface coatings, Australian Oil & Colour Chemists' Association Proceedings and News 13 (1-2): 6 (1976).

29. P.J. Nolan and R.H. Keeley, Comparison and classification of small paint fragments by x-ray microanalysis, Scanning Electron Micros. Part I; 449 (1979).

30. G.E. Garner, C.R. Fontan and D.W. Hobson, Visualization of fingerprints in the scanning electron microscope, J. Forensic Sci Soc. 15: 281 (1975).

31. R.D. Olsen, Sr., The chemical composition of palmar sweat, Fingerprint and Identification Magazine 53 (10): 3 (1972).

THE USE OF SCANNING ELECTRON MICROSCOPY

IN BITE MARK ANALYSIS

Thomas J. David

Georgia Division of Forensic Sciences
2613 Bolton Road
Atlanta, Ga. 30318

ABSTRACT

The degree of correlation of a particular set of teeth with a certain bite mark is proportional to the number of characteristics common to both. However, individual characteristics are much more significant because they are less likely to occur purely by chance in a given population. Since the SEM can readily demonstrate individual characteristics when they are present, it can be an extremely useful tool for the forensic odontologist.

INTRODUCTION

Bite mark analysis is an attempt to determine the perpetrator of a set of marks found in skin or some other material. This attempt involves the comparison of one or more sets of teeth with the mark in question. If enough data is gathered, the forensic odontologist should be able to prove or dispel the hypothesis that a certain set of teeth made a specific set of marks. Therefore, it is important to obtain as much information as possible, so that the conclusion is not limited by lack of necessary evidence.[1] This is where scanning electron microscopy can play an essential role.

Most methods of bite mark analysis involve recognition of a specific pattern that is compared with a certain dentition.[2] This comparison is usually made by matching the orientation of class characteristics of the teeth in question with those present in the bite mark pattern (figures 1 & 2). As important as these characteristics are, they limit the extent of the conclusion reached. The more common the set of class characteristics, the less likely that an opinion can link only one person with a particular bite mark.[3] However, if individual characteristics can be demonstrated within a set of class characteristics, the likelihood that more than one person could have made a particular bite mark is reduced dramatically.[1] The scanning electron microscope is capable of transforming class into individual characteristics due to its high level of resolution.[4] As a matter of fact, a single class characteristic can be transformed into a set of individual characteristics (figures 3 & 4). The particular analysis involved is somewhat similar to ballistics comparison. By way of contrast, most methods of bite mark analysis rely on matching points of concordance, which is a familiar technique of fingerprint comparison.[4] Therefore, by use of a technique which permits more detailed analysis, it is

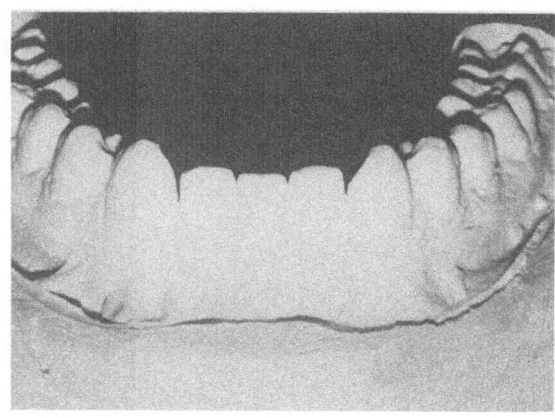

Fig. 1. Lower teeth of suspect in
homicide.

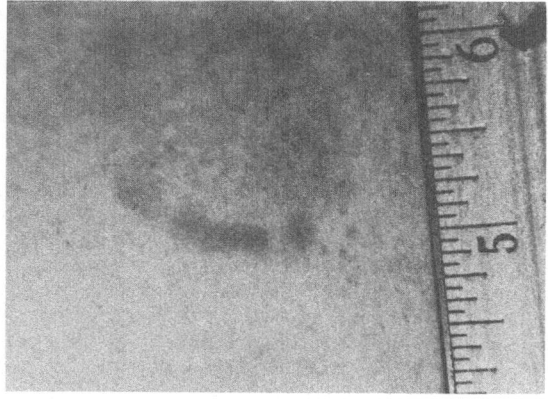

Fig. 2. Bite mark found on homicide
victim.

possible to gather more information upon which to form an opinion. Thus,
an opinion can state with a high degree of certainty[5] that a certain set of
teeth made (or didn't make) a specific set of marks.

MATERIALS AND METHODS

In order to examine bite mark tissue and teeth under the SEM, it is
necessary to produce acrylic or epoxy models small enough to fit onto the
SEM stage (usually less than 3x3x5cm).[5]

Fabrication of Acrylic Models for SEM Examination

Teeth. Stone models are duplicated in acrylic with a dual impression
technique. A stone model of the teeth is covered with a generous amount

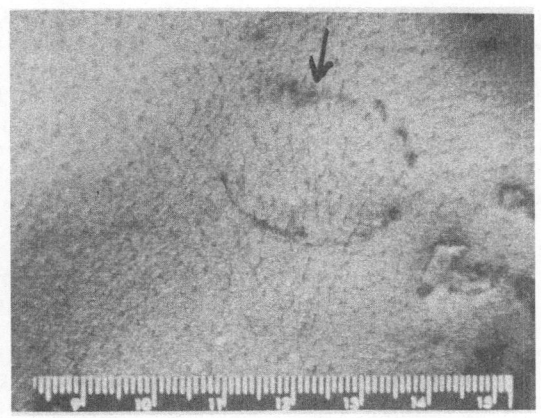

Fig. 3. Bite mark (with class charac-
         teristics) found on homicide
         victim (see arrow).

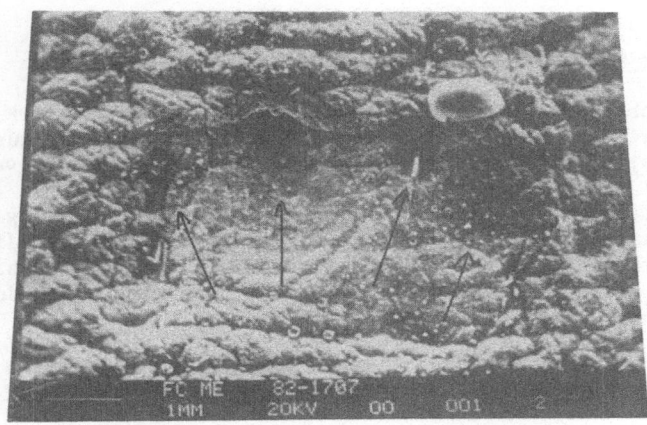

Fig. 4. SEM view of a class characteristic (cf. Fig. 3,
         arrow) showing four individual characteristics
         arrows). Bar: 1mm. Mag.: 13.5x

of a silicone lubricant. An impression tray is then painted on the inside with an adhesive. This impression tray is then filled with a silicon impression material (Optosil) (Unitek Corp.) and seated over the
lubricated stone model (figure 5). A second mix of silicon impression
material (Xantopren Blue)[5] is then injected into the impression tray by
means of an impression syringe (figure 6). The impression tray is then
seated again on the stone model. The tray is removed when the impression
material sets. The internal surface of the impression is cleaned of any
contamination by means of isopropyl alcohol on a sable brush. Now the

internal surface is dried with an air syringe. After drying, the impression is filled with acrylic by gradually mixing monomer (liquid) and polymer (powder). First the internal surface is covered with liquid by means of a sable brush. Powder is then gradually added in small amounts to absorb the liquid present. This procedure is continued until the impression is filled with acrylic (figure 7). When the acrylic has set it is removed from the impression (figure 8). The silicon impression is saved so that duplicate models may be made if necessary.

   Bite mark tissue. The bite mark tissue is first cleaned of surface contamination by means of isopropyl alcohol on a sable brush. Boxing wax is then used to establish a perimeter around the bite mark itself. The boxing wax should be placed approximately 1cm beyond the border of the markings. The boxing wax is attached to the bite mark tissue with sticky wax (figure 9). Silicon impression material (Xantopren Blue) is then mixed and loaded into an impression syringe. This material is injected slowly onto the bite mark tissue (figure 10). All surface irregularities are carefully filled and the impression is removed from the bite mark tissue when set (figure 11). Now the periphery of the impression material is thinned so that the edges are clear and distinct. The boxing wax is put around the edges and attached with sticky wax. Once more, the surface of the impression is cleaned with isopropyl alcohol and dried with an air syringe. The surface of the impression is coated with acrylic monomer (liquid) on a sable brush. Then enough polymer (powder) is slowly added to absorb the liquid present. This process is continued until sufficient thickness of acrylic exists so that the resulting model is firm enough not to fracture easily (6-10mm). The model is removed from the impression when set (figure 12). Impressions are kept to prepare duplicate models if needed.

Preparation of Models for SEM Examination

   The periphery of both acrylic models (teeth and bite mark tissue) is trimmed so that they can fit onto the SEM specimen stage (figure 13). The models are coated with carbon in vacuum to make them conductive to the electron beam.[5] Small pieces of orthodontic wire (1-2mm) are attached to the acrylic model of the bite mark tissue with cyanoacrylate glue (or similar substance). These pieces are aligned around the immediate perimeter of the bite marks at various angles (figure 14). This is done to facilitate orientation of the model once the SEM screen reduces the field of vision.

EXAMINATION AND INTERPRETATION OF SEM EVIDENCE

Examination
   The examination of bite mark evidence requires a comparison of the characteristics found on the edges of the teeth with the markings present in the tissue which has been bitten (figure 15). Although the pictorial representation of this comparison is still only two dimensional, the photographs which result are selected specifically for their ability to depict the third dimension present in the specimens as accurately as possible. Unlike traditional methods of photography, this analysis allows the examiner to carefully scan a specimen in three dimensions and photograph a particular view which yields the most information.

   Teeth. The edges of each tooth on the acrylic model are viewed carefully to determine whether they exhibit class or individual characteristics of the teeth they represent (see Table 1).[6] Individual characteristics do not necessarily represent specific geometric patterns, but rather deviations from those particular patterns (figure 16.).

88

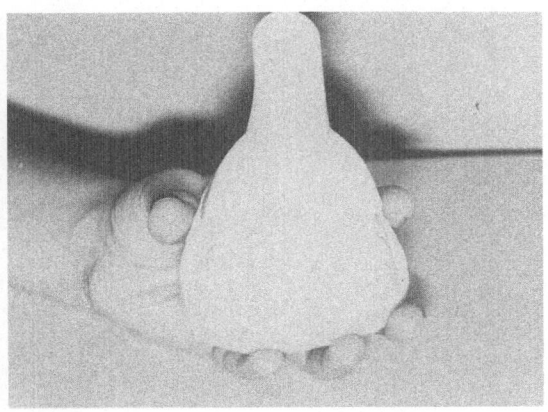

Fig. 5.   Stone model of teeth with im-
          pression tray seated.

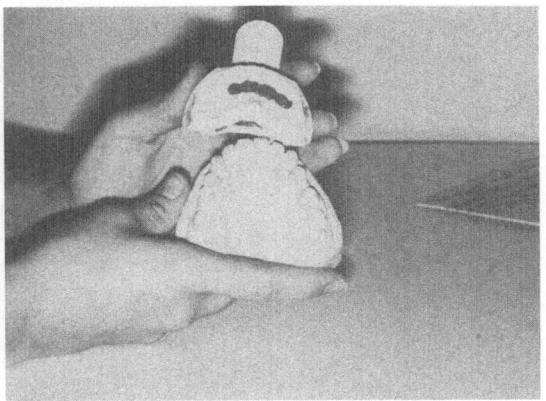

Fig. 6.   Silicone impression with
          Xantopren Blue ready to be re-
          seated on stone model of teeth.

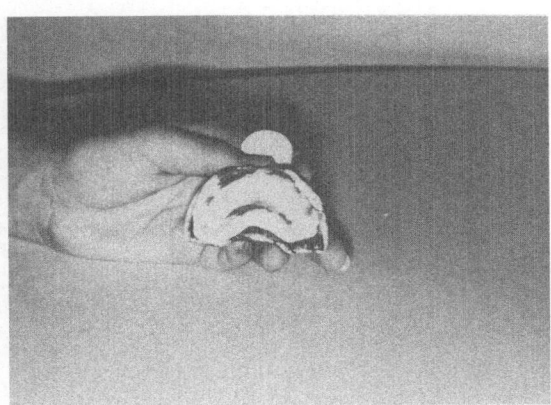

Fig. 7.   Silicone impression of teeth
          filled with acrylic.

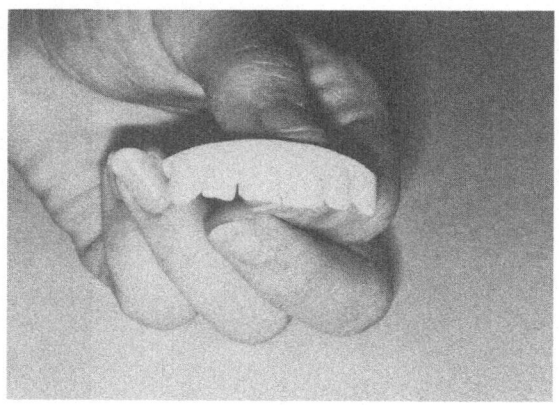

Fig. 8.   Acrylic model of teeth for
SEM examination.

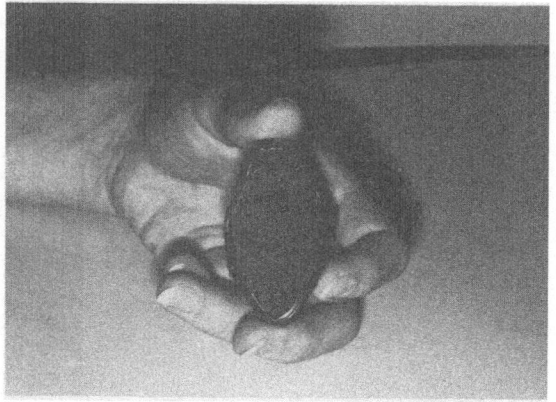

Fig. 9.   Bite mark tissue with boxing
wax perimeter ready for impression.

Fig. 10.   Xantopren Blue being injected
onto bite mark tissue.

Fig. 11.   Bite Mark tissue impression.

Fig. 12.   Acrylic model of bite mark
           tissue for SEM examination.

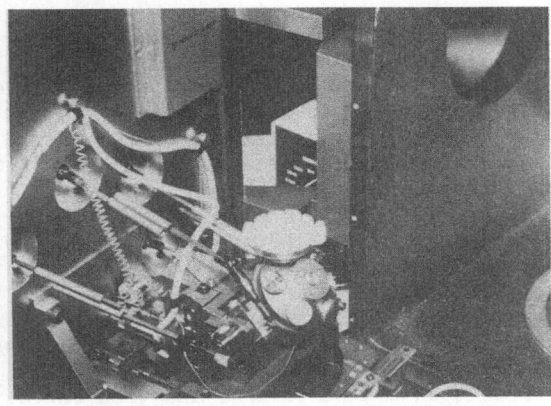

Fig. 13.   Acrylic model of teeth on SEM
           specimen stage.

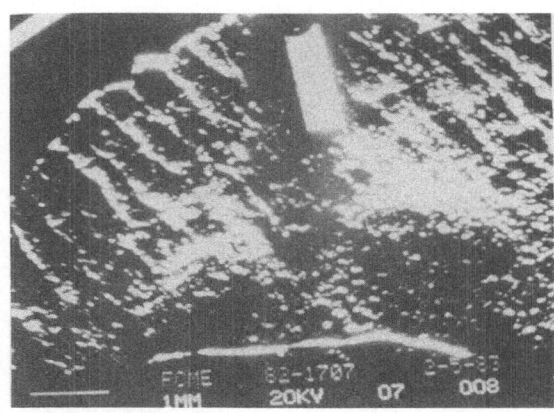

Fig. 14. Orthodontic wire aligned around
periphery of bite mark model to
facilitate orientation.

Table 1. Class Characteristics of Teeth

```
--------------------------------------------------
|                                                |
|  Incisors - Rectangles                         |
|  Canines (Cuspids) - Triangles                 |
|  Premolars (Bicuspids) - Single or Dual        |
|                              Triangles         |
|                                                |
--------------------------------------------------
```

Bite mark tissue. Markings present in the acrylic model of the bite
mark are now viewed with careful attention to the various shapes present
in the bite mark pattern.[4] It should be noted whether the shapes exhibited
in the bite mark are geometric or non-geometric. Regardless of which category the shapes represent, the specific configuration of each should be
recorded for subsequent reference. The sequence in which the various configurations occur should also be recorded.

Interpretation

It should be understood that the use of SEM in bite mark analysis is
not meant to replace more commonly used methods of comparison.[7] On the
contrary, it is intended to enhance the value of evidence gathered by
other means. Therefore, evidence gathered by the use of SEM should cor-
roborate, not conflict with other evidence.[4]

Teeth. At some point, either before or during SEM examination, a
determination must be made as to which teeth are represented in the bite
mark. Once that determination is made, the alignment of these particular
teeth is noted along with the configuration of the edge of each tooth.
Whether the configuration of each tooth represents a class or individual
characteristic is of particular importance at this point.[7] Any individual
characteristics present should be further analyzed so that an exact configuration is known (figure 16).

Bite mark tissue. As was done with the teeth, the alignment of mark-

92

ings present is noted as well as the configuration of each mark as to shape. These shapes are further categorized as to class or individual characteristics with special emphasis given to any individual characteristics.

Analysis of Data. The information compiled from each model is now compared with the other to determine how well they correlate. Several questions are answered to determine the degree of correlation: do the class characteristics of the bite mark correspond to the class characteristics of the teeth: do individual characteristics present match when comparing the configurations of bite mark and tooth: if individual characteristics match in configuration, do they occur in the same sequence in both models?

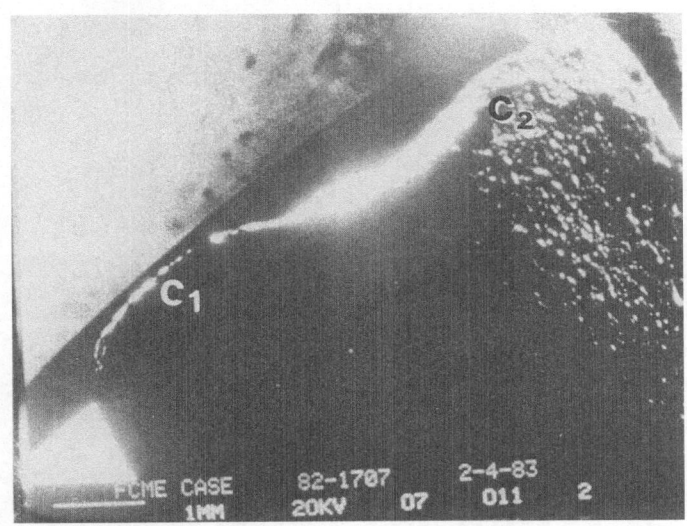

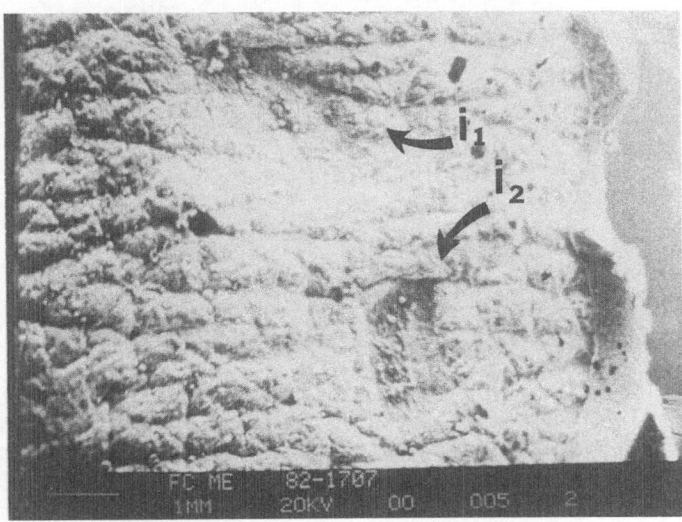

Fig. 15.   Comparison of acrylic model tooth #7 (top) with bite mark tissue model of tooth #7 (bottom). The impressions i_1 and i_2 are due to characteristics C_1 and C_2 respectively.

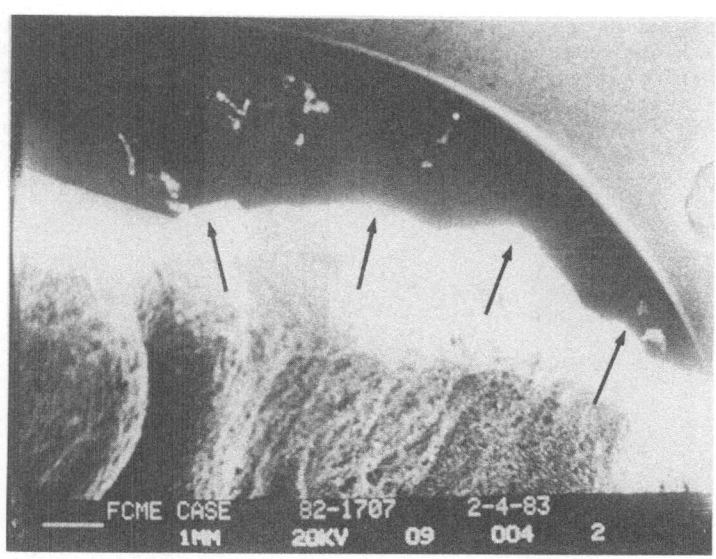

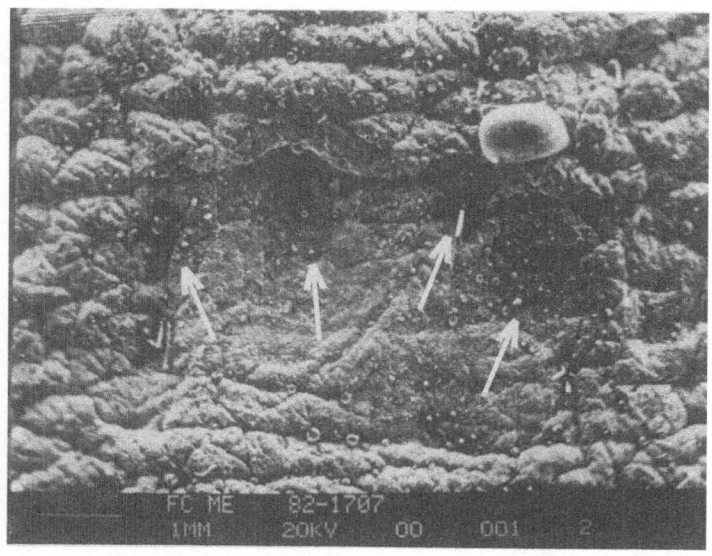

Fig. 16. Acrylic model tooth #9 (top) compared with bite
mark model tooth #9 (bottom) showing individual
characteristics which deviate from basic rectangular shape.

CONCLUSION

Since the SEM has the ability to demonstrate individual characteristics when they are present equivalently in the bite mark and in a particular set of teeth, it can be an useful tool for the forensic odontologist.

ACKNOWLEDGEMENT

Figures 1, 2, 15 and 16 have been reprinted with permission from ASTM, the publisher of the article in Reference 8. These figures correspond to Figures 1, 3, 7 and 10 in the article.[8]

REFERENCES

1. J. W. Beckstead, R. D. Rawson and W. S. Giles, Review of bite mark evidence, Amer. Dental Assoc., 99:70 (1979).

2. J. I. Ebert and H. R. Campbell, Jr., Discussion of "Photographic techniques of concern in metric bite mark analysis,"J. Forensic Sci., 30 (No. 3) :599 (1985).

3. R. F. Sognnaes, R. D. Rawson, B. M. Gratt, and N. B. T. Nguyen, Computer comparison of bitemark patterns in identical twins, Amer. Dental Assoc., 105:449 (1982).

4. G. L. Vale, R. F. Sognnaes, G. N. Felando, T. T. Noguchi, Unusual three-dimensional bite mark evidence in a homicide case, J. Forensic Sci., 21, (No. 3), 642 (1976).

5. R. F. Sognnaes, Forensic identifications aided by scanning electron microscopy of silicone-epoxy microreplicas of calcified and cornified structures, in: Proceedings of the 33rd Annual Meeting of the Electron Microscopy Society of America, Las Vegas, p. 678, 1975.

6. J. A. Cottone and M. S. Standish, "Outline of Forensic Dentistry," Yearbook Medical Publishers, p. 118, 1982.

7. R. F. Sognnaes, Forensic bite mark measurements, Dental Survey, April p. 34, 1979.

8. T. J. David, Adjunctive use of Scanning electron microscopy in bite mark analysis: a three dimensional study, J. Forensic Sci. (in press).

ANALYSIS OF BIOLOGICAL ELEMENTS IN SCALP HAIRS WITH SEM-EDX AND TEM AND ITS APPLICATION TO THE HAIR COMPARISON

Sueshige Seta

National Research Institute of Police Science
Tokyo, Japan

PREFACE

This chapter is largely an elaboration of a paper which was presented at the Forensic, Occupational and Environmental Health Symposium at the 1985 joint national meetings of the Electron Microscopy Society of America and the Microbeam Analysis Society in Louisville, Kentucky. An extended abstract of that presentation has been published by San Francisco Press.

INTRODUCTION

In many forensic science books it has often been pointed out that it is not possible to definitely identify a sample of hair as having originated from one individual's head. As suggested by Kirk[1] and Jones[2], however, the probability of identity grows with every point of resemblance, including chemical and physical characteristics as well as morphological ones. This paper presents a comprehensive discussion of the many varieties of data which are required for succeeding in forensic hair comparison. Examination methods other than morphological ones which are commonly used for crime detection work have been investigated for their potential usefulness in head hair comparison.

Among these examination methods, ABO blood grouping has already been used for Japanese head hair comparison as a routine method. Electron microscopic examination is also used. For over 15 years microscopic studies in the scanning and the transmission electron microscopes (SEM and TEM) have pointed to the utility of morphological analysis of damaged and diseased head hairs. It is clear that morphological peculiarities found in such head hairs can serve to characterize one individual in the forensic hair comparison. In most crime detection work, however, great interest is focused on the comparison of healthy hairs which never show any morphological peculiarities. There is little available data on electron microscopic findings for comparison of healthy hair although structural and chemical analyses of cuticular cells[3] and surface morphology of the hair[4] have been reported using TEM and SEM. Electron microscopy is becoming more and more useful in forensic hair comparison because of the availability of elemental analysis by the SEM/EDX for single head hair comparison and because the access to this analytical instrument is relatively easy in most forensic science laboratories.

It has been suggested[5] that the concentration of an element in head hair may reflect an individual's physiological status, that is, the body constitution. Head hairs from patients with celiac disease,[6] cardiac infarction[7] and multiple sclerosis,[8] in particular, have been investigated in regard to the alteration of the element composition of hair. These reports may well support the validity of hair element analysis in characterizing one individual. In a series of studies, element analysis of some inorganic elements of head hair by SEM/EDX were carried out to form an analytical data base potentially useful in head hair comparison.[9,10] In the present paper the experimental classification of head hairs collected from different persons was undertaken to evaluate the validity of the EDS analytical method in healthy hair comparison. Along with this blind test the distribution of elements in the hair shaft components such as cuticle, cortex, medulla and the melanin granule was also examined in regard to endogeneous elements, the contents of which may intrinsically reflect the body physiological status.

MATERIALS AND METHODS

Head hair samples used for the blind test were taken from adult male and female subjects ranging in age from 25 to 57 years. From each person ten single head hairs were taken from the top on the head by cutting with scissors just on the scalp surface. Two blind test sets were collected by the person not concerned with the test. The examiner was not informed as to how many persons or how many hairs of the same person were contained in each set. Prior to EDX analysis each sample was washed in three changes of an ether-alcohol mixture (1:1) for 15 minutes. For the analysis 0.5 cm long segments at a distance of 3 cm from the proximal end were used. Samples were ashed on a carbon planchet by exposure to a plasma produced by exciting oxygen with a radiofrequency power with a 50 ml/min oxygen flow rate. The ashing time was 60 minutes. All samples contained in one set were treated on the same carbon planchet to eliminate the difference of ashing degree in the plasma chamber. Immediately after ashing, the samples were transferred to double sided tape on a fresh carbon specimen mount under a binocular stereomicroscope and were subjected to carbon coating under vacuum.

For the investigation on the distribution of elements in hair components, both SEM/EDX and STEM analyses were undertaken using the head hair which had been known to show very high X-ray intensity of potassium in the preliminary analysis by SEM/EDX. A cross section was made for SEM/EDX analysis. The hair was directly embedded in low-viscosity Spurr's epoxy resin without any chemical fixation or dehydration. After polymerizing, about one μm thick sections were made with glass knives by a conventional ultramicrotomy. The sections were transferred to a carbon planchet thinly smeared by alon alfa as an adhesive medium. In every sectioning procedure use of water or organic solvent was avoided. Prior to the analysis, the section was exposed to oxygen plasma in the conditions of 50 watts of radiofrequency power, 20 ml/min of oxygen flow rate and 120 seconds of exposure time. Immediately after ashing, the section was subjected to carbon coating in vacuo. For STEM analysis, the ultra-thin section was made from the same Spurr's epoxy resin block and was placed on a copper grid via distilled water. The section was subjected to carbon coating in vacuo. In the case of STEM analysis the plasma ashing treatment was omitted. The SEM was operated at 25KV, 1.5×10^{-10} A specimen current, 13 mm working distance and 20 mm specimen-detector distance. The position of specimen mount was set at 45° incline to the primary electron beam in SEM chamber. The specimen was focused at x1,000 magnification for bulk analysis and at x10,000 magnification for cross section analysis. EDX spectra were recorded in a range of 0 - 5 KeV through all specimens. Analyzing time was 100 seconds for blind test samples, 200 seconds for ashed cross sections and 300 seconds for STEM analysis. SEM/EDX analysis was carried out with a JEOL U3 SEM equipped with a NS-880 X-ray analyzer and STEM analysis was carried out at 80 KV with a JEOL 1200EX TEM equipped with Tracor-5500 X-ray Analyzer.

RESULT AND DISCUSSION

Two blind test sets were designed, one consisted of 20 single head hairs and another of 23 single head hairs. For classifying these samples into each individual group, EDX spectra were collected for biological elements such as sodium, phosphorus, potassium and calcium. As described elsewhere,[10] these elements could be effectively detected by SEM/EDX analysis after hairs were treated with oxygen plasma microincineration technique of Thomas.[11]

Figure 1 shows the results of the first test of separating hairs on the basis of EDX spectra. As a result of classification by EDX spectra, twenty single hair samples tested were classified into five individual groups. Four single hairs classified as individual group 1 were very similar in their EDX spectra to each other, commonly generating sodium, phosphorus, potassium and calcium X-ray peaks showed similar peak heights in all 5 samples. A characteristic feature of this group was that potassium X-ray peak heights were commonly lower than calcium ones. The reciprocal relation between both element peak heights would be a powerful supporting factor in classifying these four single hairs as belonging to the same person. In fact, they proved to have originated from the same person, that is, subject S. However, one single hair sample from this person resulted in a false classification because it generated higher X-ray intensity of each element than the other four single hairs classified as group 1. EDX spectra of two single hairs of group 2 are depicted at 2,000 vertical full scale because of their high generation of X-ray counts. Three single hair samples classified as individual 3 showed common sodium, phosphorus, potassium and calcium X-ray peaks. They were seemingly similar to group 1 hairs but for one feature. Contrary to group 1 hairs, group 3 hairs showed potassium peak heights that were commonly higher than calcium ones. The reversed reciprocal relation of potassium and calcium peak heights between both groups gave a decisive clue to discriminate between both group hairs. In fact, three single hairs of individual group 3 proved to originate from the same person, that is, subject N. One single hair sample from subject N resulted in false classification because of its higher generation of X-ray intensity as in case of one single hair sample from subject S. If X-ray intensity was excluded from the consideration for classification, it would probably have been classified group 3 because of the resemblance of EDX spectrum. This result may suggest that EDX spectrum pattern is more efficient in characterizing one individual than X-ray intensity in routine hair comparison. Three single hair samples classified as coming from individual 4 were well characterized by very high calcium X-ray peaks and no potassium peaks. It was very easy to classify them as belonging to the same person from their peculiar EDX spectra. In fact, they proved to be of the same origin, that is, subject O. As reported previously,[10] high concentrations of calcium in head hair is very common in bleached and permanently waved hairs. In fact, group 4 hairs were proved to be treated with permanent waving. Eight single hair samples were classified in group 5. In spite of general low X-ray intensity they were characterized by relatively high calcium X-ray peaks with clear phosphorus and slight potassium. Basing on such general resemblance these eight single hairs were classified as belonging to the same person. However, they, in fact, consisted of head hairs from two different persons, that was, subject T and subject K. Morphological and blood group examinations should be carried out to see if they would discriminate between the two persons.

Figure 2 shows the result of the second blind test. As a result of the classification by EDX spectra, twenty-three single hair samples tested were classified into five individual groups. Seven single hair samples were classified as from one individual because of common generation of high calcium X-ray peaks with slight but clear sodium X-ray peaks throughout samples. In fact, they consisted of hairs of two different persons, that is, subject O and subject F. They both proved to be treated with permanent

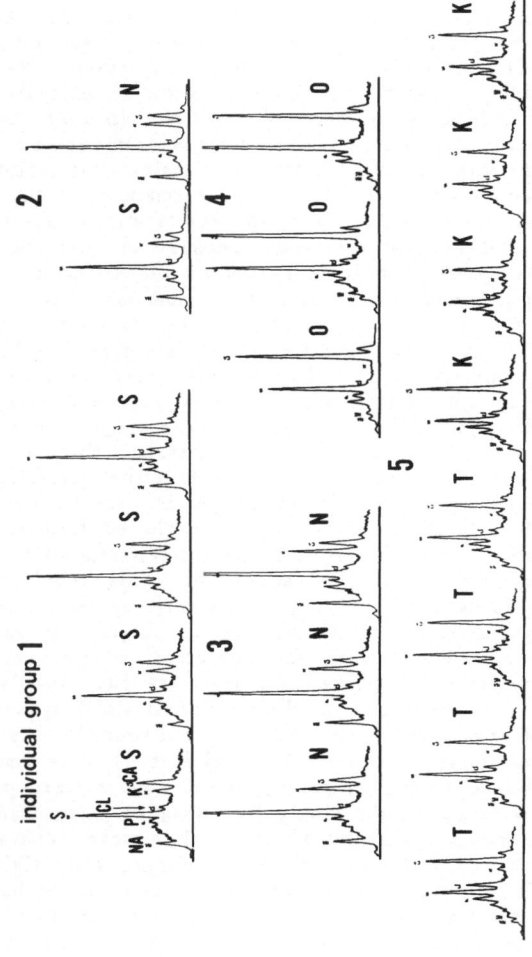

Fig. 1. First test of classification by EDX spectra of head hair samples which were taken from different persons. K, N, O, S and T are codes for the hair donors. 0.5 cm long segment of each hair sample was exposed to oxygen plasma. EDX spectra were recorded at 512 vertical full scale except for two single hairs which were recorded at 2,000 vertical full scale (group 2 hairs). Twenty single head hair samples were mingled as a blind test set and they were classified into five different individual groups from 1 to 5 by EDX spectra in relation to biological inorganic elements such as Na, P, K and Ca in the hair.

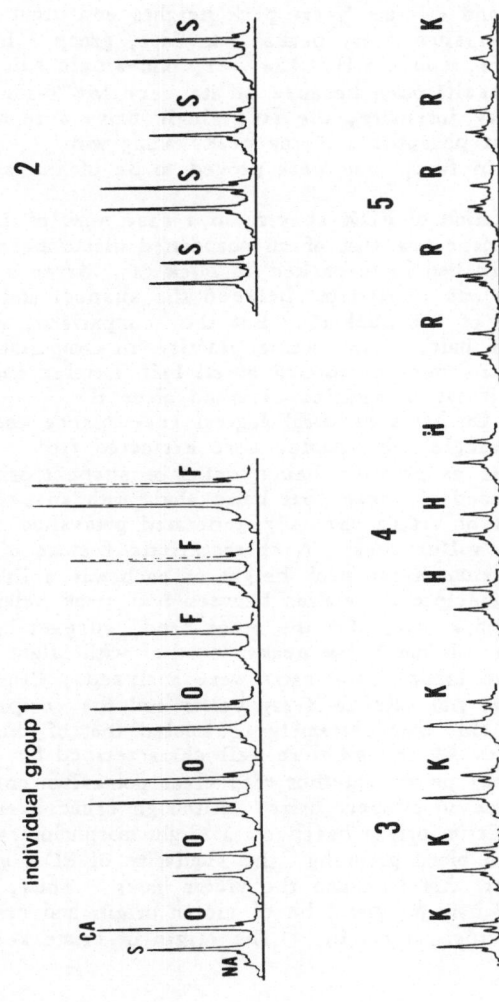

Fig. 2.   Second test of classification by EDX spectra of head hair samples which were taken from different persons. F, H, K, O, R and S are codes of the subjects from which head hair samples were taken. 0.5 cm long segment of each hair sample was exposed to oxygen plasma prior to SEM/EDX analysis. EDX spectra were recorded at 512 vertical full scale. Analyzing time was 100 seconds. Twenty three head hair samples were mingled as a blind test set and they were classified into five different individual groups from 1 to 5 by EDX spectra in relation to biological inorganic elements such as Na, P, K and Ca in the hair.

waving, suggesting that this analytical method is not efficient in discriminating between head hair samples treated with permanent waving. Four single hair samples classified as group 2 also generated very high calcium X-ray peaks as in case of group 1 hairs, but a clear difference was observed between both group hairs in potassium X-ray peaks. Group 2 hairs proved to be not treated with hair bleaching or permanent waving. In fact, they proved to be of the same origin, that is, subject S. The other twelve single hair samples were classified into three individual groups. Although they were seemingly similar to each other in general low X-ray intensity, some features to divide them into different groups were observed. Four single hairs classified as group 3 were well characterized by a similar reciprocal relationship between potassium and calcium X-ray peak heights and group 4 which did not generate clear potassium X-ray peaks. In fact, group 3 hairs were of the same origin, that is, subject H. However, one single hair from subject K resulted in false classification because of its very low X-ray intensity. In spite of low X-ray intensity, the five single hairs were well characterized by slight but clear phosphorus X-ray peaks along with relatively high calcium ones. In fact, they were proved to be of the same origin, that is, subject R.

Figure 3 shows the application of EDX spectra to a case work of forensic hair comparison. The criminal case was that of an abandoned dismembered corpse. The dismembered victim had been packed in buckets. After a suspect was arrested, a hair comparison was carried out between the suspect and hairs which was taken from the cover of the buckets. For the comparison, victim hairs were also added as control hairs. As routine practice in comparison, gross and microscopic examinations were performed of all hair samples and then blood grouping were carried out on samples selected after the morphological comparison work. On basis of morphological resemblance and result of blood grouping, four single hair samples were extracted from fifteen crime scene hair samples as probably being victim or suspect origin. Then, EDX analysis was performed of these four hairs along with suspect and victim hairs. Five single hairs of victim commonly generated potassium and calcium X-ray peaks as well as sulfur ones. A characteristic feature of the victim hairs, was that the potassium X-ray peak height of each was a little lower than calcium one. The reciprocal relation between both peak heights seemed to characterize the victim's hair. On the other hand, suspect hairs were well characterized by high calcium X-ray peaks together with slight but clear potassium ones. Then the crime scene hairs were analyzed. Crime scene hair #1 generated both potassium and calcium X-ray peaks and the reciprocal relation of the elemental peaks this hair extremely resembled that of the victim's hair. Crime scene hairs #3 and #4 were well characterized by considerably higher calcium X-ray peaks together with clear potassium ones, showing a very clear resemblance to suspect hairs. Although crime scene hair #2 was chosen as being from victim origin based on a slight morphological resemblance and accordance with blood grouping, the similarity of EDX spectra was not clearly observed between this hair and the victim hairs. Thus, it was concluded that crime scene hair #1 could be of victim origin and crime scene hair #3 and could be of suspect origin. The origin of crime scene hair #2 could not be clarified.

Because endogenous elements may reflect an individual's physiological status and allow us to characterize an individual, the structural components of hair shaft such as cuticle, cortex, medulla and melanin granule were analyzed. The hair sample used was taken from a 56 year-old male and it had been known to generate a very high potassium X-ray peaks with no calcium in repeated analyses over a long period of time in bulk analysis by SEM/EDX (Fig. 4). Figure 5 shows the secondary electron image of cross section of the hair exposed to oxygen plasma. Cuticular cell layers were very clearly observed allowing an exact count of the numbers of layering and a close look into the cuticular cell structure. It was morphologically shown that the outer surface of each layer, that is, the part of exocuticle remained smooth

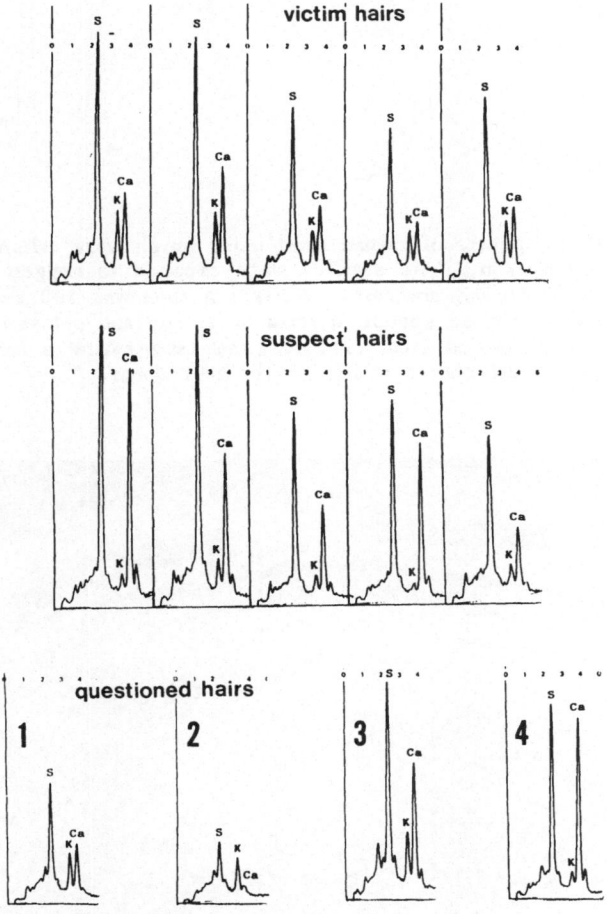

Fig. 3. Application of SEM/EDX analysis of head hair to a criminal case. All samples were exposed to oxygen plasma. EDX spectra were recorded at 512 vertical full scale. Analyzing time was 100 seconds. Victim hairs were well characterized by the appearance of potassium and calcium X-ray peaks in such a relation that potassium peak heights were commonly lower than calcium ones. Suspect hairs were well characterized by considerably higher calcium X-ray peaks with slight but clear potassium ones. Crime scene hair 1 was identified to be coming from the victim and crime scene hairs 3 and 4, coming from the suspect, respectively.

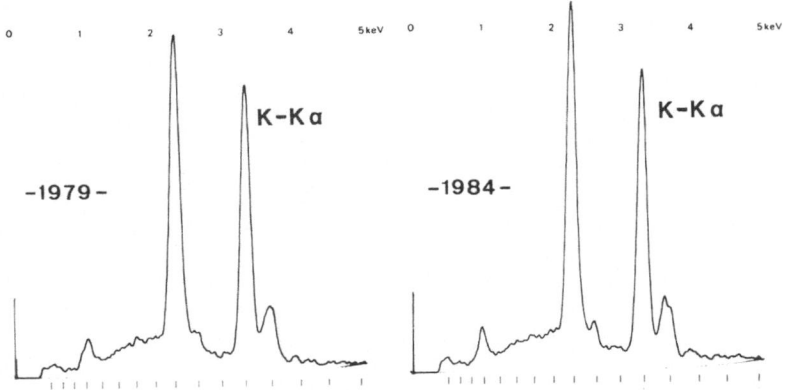

Fig. 4.   EDX spectra of human head hairs which were taken from 56 year-old person. The samples were exposed to oxygen plasma prior to SEM/EDX analysis. Analyzing time was 100 seconds. The hairs yielded similar spectra of sulfur and potassium peaks over a period from 1979 to 1984, suggesting a constant individual characteristic of the hair donor.

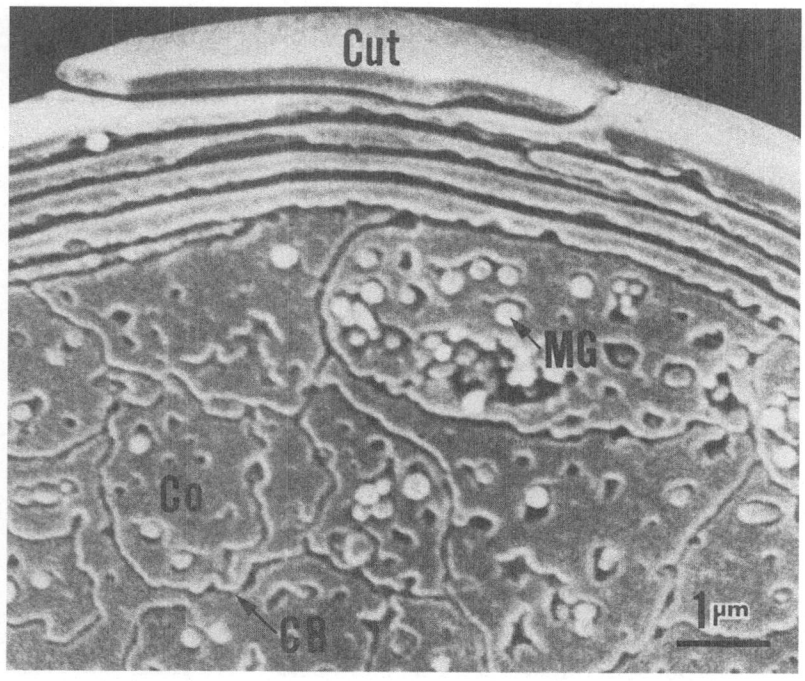

Fig. 5.   Scanning electron image of the hair cross section exposed to oxygen plasma. Each cuticular layer was well discernible and the boundaries of cortical cells were also well defined. Melanin granules were easily observable as round bodies. These morphological images allowed analysis of every structural component of the hair. Cut: cuticle cell, Co: cortex, MG: melanin granule, CB: cortical cell boundary.

but the inner layer, was a rough surface, suggesting that the latter surface is more susceptible to plasma ashing than the former surface. From the cuticular layer and the cortical cell, potassium X-ray peaks were detected but their intensities were not as high as in the case of overall fiber analysis. Much higher potassium X-ray intensity was obtained from medulla and melanin granule. Calcium X-ray peak was hardly detected from the cuticular layer and cortex but it was clearly detected from melanin granule (Fig. 6). These findings may suggest that the melanin granules are the most probable binding or accumulating sites for biological inorganic elements in the hair.

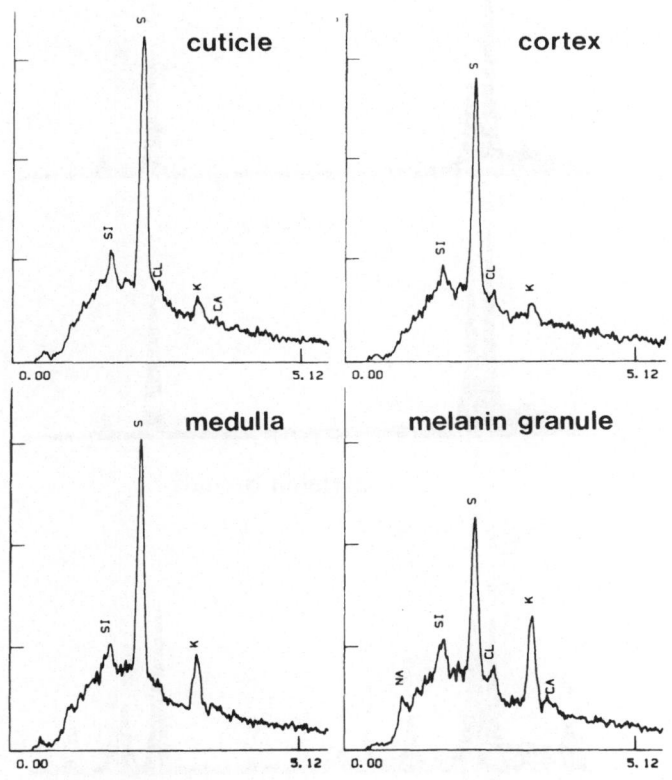

Fig. 6.   EDX spectra of head hair cross section exposed to oxygen plasma: cuticle, cortex, medulla and melanin granule. EDX spectra were recorded at 512 vertical full scale. Analyzing time was 200 seconds. A potassium X-ray peak was detected in melanin granules with the highest intensity. Medulla also showed a relatively high intensity of this peak.

Further evaluation of element distribution in the hair, was carried out on the same hair using STEM analysis. Figure 7 shows the results of this analysis. The cuticular layer, cortical cell and melanin granules did not generate any trace of a potassium peak. Since a potassium X-ray peak was clearly detected from melanin granules in cross sectional analysis of the hair by SEM/EDX, it was concluded that the loss of potassium occurred during sectioning process through water. Marshall reported[12] that potassium was lost to a high degree during the sectioning of other biological specimens. Aside from potassium loss, STEM analysis of hair seemed to detect metal trace elements which could hardly be detected by SEM/EDX. It was, in particular, worth noting that calcium, iron and zinc were clearly detected in melanin

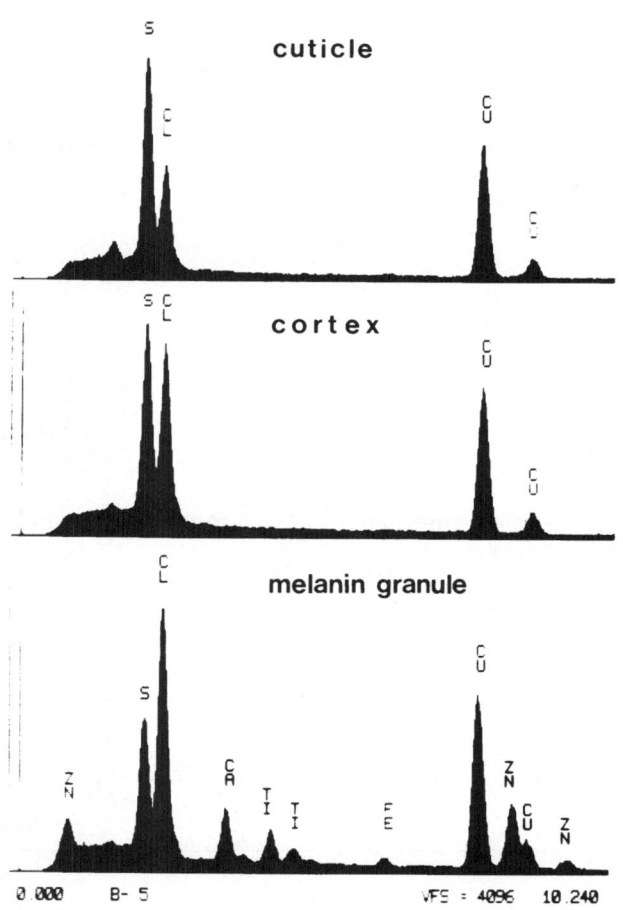

Fig. 7.   STEM analysis of head hair cuticle, cortex and melanin granule. Analyzing time was 300 seconds. From the melanin granule Cl, Ca, Ti, Fe and Zn were clearly detected. From cuticle and cortex no elements other than Cl and S were detected. Cu originated from the copper grid used.

granules. Since Larsen and Tjalve reported[13] that beef eye melanin has a capacity to accumulate metal ions, potassium and sodium in vivo and in vitro, the present STEM analytical data may further suggest that melanin granules offer the most probable binding or accumulating site for metal trace elements as well as biological inorganic ones in head hairs.

Hopps[14] ascribed the endogenous sources of trace elements in head hairs, in a greater part, to the matrix and connective tissue pappila with its blood and lymph vessels and partially to the sebaceous glands and the eccrine sweat glands. If this is true, it is expected that endogenous element contents in head hairs reflect the individual physiological status corresponding to blood trace element contents. Therefore, an elemental analysis within structural components of head hair would give more information for characterizing one individual, which would lead to an enhanced identification probability in forensic hair comparison.

SUMMARY

SEM/EDX analysis was applied to the classification of head hair samples which were taken from different adult males and females. Two blind tests were designed for this purpose. For SEM/EDX analysis each head hair sample was exposed to oxygen plasma. As a result of this blind test it was shown that EDX spectra, comprising of biological inorganic elements such as sodium, phosphorus, potassium and calcium, could be efficiently used for the classification of head hair samples of different individuals. But the technique was not applicable to the individual discrimination between permanent waved hairs. The distribution of elements within structural components of the hair such as cuticle, cortex, medulla and melanin granule was also examined by SEM/EDX and STEM analyses.

ACKNOWLEDGMENT

Figures 3 and 5 have been reproduced from Proc. Abstr. EMSA/MAS-Joint Annual (43rd) Mtg., Louisville, KY Aug. 5 - 9, 1985, pp. - . Reproduction request made to San Francisco Press, San Francisco, Calif.

REFERENCES

1. P. L. Kirk, Human hair studies I. General considerations of hair individualization and forensic importance J. Crim. Law Criminol., 31:486 (1940).
2. D. N. Jones, The study of human hairs as an aid to the investigation of crime, J. Forensic Med., 3:55 (1956).
3. J. A. Swift, and B. Bews, The chemistry of human hair cuticle-I: A new method for the physical isolation of cuticle, J. Soc. Cosmet. Chem., 25:13 (1974).
4. E. Bottoms, E. Wyatt, and S. Comaish, Progressive changes in cuticular pattern along the shafts of human hair as seen by scanning electron microscopy, Brit. J. Dermatol., 86:379 (1972).
5. V. Valkovic, "Trace Element in Human Hair," Garland STPM Press, New York and London (1977).
6. L. Kopito, and H. Shwachman, Alterations in the elemental composition of hair in some disease, in: "The First Human Hair Symposium," A. C. Brown, ed., Medcom Press, New York (1974).
7. J. Basco, P. Kovacs, and S. Horvath, Investigation of some inorganic compounds in human hair, Radiochem. Radioanal. Letters, 33:273 (1978).
8. D. E. Ryan, J. Holzbecher, and D. C. Stuart, Trace elements in scalp-hair of persons with multiple sclerosis and of normal individuals, Clin. Chem., 24:1996 (1978).

9. S. Seta, H. Sato and M. Yoshino, Quantitative investigation of sulfur and chlorine in human head hairs by energy dispersive X-ray microanalysis, <u>Scanning Electron Microsc.</u>, 1979/II:193 (1979).

10. S. Seta, H. Sato, M. Yoshino, and S. Miyasaka, SEM/EDX analysis of inorganic elements in human scalp hairs with special reference to the variation with different locations on the head, <u>Scanning Electron Microsc.</u>, 1982/I:127 (1982).

11. R. S. Thomas, and J. R. Hollahan, Use of chemically-reactive gas plasmas in preparing specimens for scanning electron microscopy and electron probe microanalysis, <u>Scanning Electron Microsc.</u>, 1974:83 (1974).

12. A. T. Marshall, Freeze-substitution as a preparation technique for biological X-ray microanalysis, <u>Scanning Electron Microsc.</u>, 1980/II:395 (1980).

13. B. Larsson, and H. Tjalve, Studies on the melanin-affinity of metal ions, <u>Acta Physiol. Scand.</u>, 104:479 (1978).

14. H. C. Hopps, The biologic basis for using hair and nail for analysis of trace elements, <u>Sci. Total Environ.</u>, 7:71 (1977).

EXAMINATION OF TOOL MARKS IN THE SCANNING ELECTRON MICROSCOPE

R.P. Singh

Department of Science and Technology
Government of India, Technology Bhawan
New Delhi 110 016, India

INTRODUCTION

Examination of tool marks in cases relating to theft, burglary, violence and other criminal offenses is frequently carried out by a forensic scientist. Tool marks are produced on a softer surface by a tool of harder material, while commissioning a crime. The tool is not damaged in the process of cutting or scratching the surface of the material to any appreciable extent. The marks on the surface are reproducible. Thus, a criminal can be linked to a crime through the examination of the tool recovered from the criminal and the tool marks collected from the scene. Although the marks of fairly large size (over a few microns) can be seen under an optical microscope with an optimal magnification of 2000X, the marks of smaller dimensions (below one micrometer) can not be revealed. This is a situation where the application of the scanning electron microscope is called upon.

The scanning electron microscope (SEM) is capable of revealing the details of the size as small as 0.02 μm. The specimen's surface roughness can be 300 times more than that which can be examined with an optical microscope at comparable magnifications. Because of these advantages, the technique has attracted the attention of various investigators during the seventies, who have profitably used it for surface examinations. Van Essen and Morgan[1] have used the split field comparison SEM as a topographical comparison imaging system for the study of surfaces. Later on, Judd et al.[2] used this technique for similar purposes. Meyn and Beachem[3] matched the complementary fracture surfaces under SEM and transmission electron microscope for the study of fracture mechanisms. Matricardi[4] and Matricardi et al[5] have compared the fracture surfaces of broken and cut metal wires to explore the possibility of rematching the separated pieces. Recently, Singh et al[6] have used this technique for the examination of filaments of light bulbs, and extended its application to the cases of railway accidents.

The conventional method of optical microscopy is not suitable for comparing the opposite surfaces of cut ends of a wire because (1) the entire surface cannot be focused, due to inclination to the incident beam, and (2) fine details cannot be revealed, due to poor contrast at low magnifications (200X or below). More recently, Singh and Aggarwal[7] have compared the cut ends for identification of wires and the cutting tool by utilizing an SEM, and have established that the SEM is a potential technique for examination of cut ends.

The present article describes the potential applications of the scanning electron microscope in the examination of tool marks.

EXPERIMENTAL

Indentation marks, scrape marks and cut surfaces were produced by different tools. The surfaces were thoroughly cleaned with acetone. Interesting portions, bearing tool marks, were mounted on aluminum studs with silver paint and examined under an SEM. For comparison of the cut ends of wires, they were first physically matched with an optical microscope, and then compared under the scanning electron microscope.

OBSERVATIONS AND RESULTS

Examination of Indentation Marks

Indentation marks are formed when the material is pressed upon by a tool without a sliding motion, such as hammering or punching against a surface. These marks contain the outlines of the tool surface with its irregularities. Firing pin impressions are an example of such indentation marks, which are frequently encountered in the identification of cartridges in firearms cases. When the trigger is pressed, the striker (hammer) moves forward and hits the centre base of the cartridge, thereby imprinting its surface irregularities. Although these indentation marks (surface irregularities) are visible under an optical microscope, their comparison with trial impressions becomes difficult and sometimes impossible, particularly when the cartridge is punched. The SEM has been a quite useful instrument for such comparisons.[8,9]

Figure 1a shows an electron micrograph of a punched cartridge, bearing indentation marks which have been compared with those on the trial cartridge (Figure 1b) produced by the same weapon. Some of the boundary marks opposite to the punched portion (marked A in Figure 1a) show a close resemblance to those observed at the corresponding portion (marked A in Figure 1b) of the trial cartridge. The exact matching of these indentation marks has been shown in Figures 2a and 2b respectively. Comparison of the firing pin impressions, even on punched cartidges, thus can identify the firing weapon. This conclusion would not have been possible to reach so illustratively with an optical microscope.

Examination of Scrape Marks

Scrape marks in the form of lines (striations) are formed when a surface is scratched by a tool such as chisel, knife, axe, plier, scissors or screw driver. Such tool marks are quite distinct, as they are produced by sliding the edge of the tool on a metal surface. They can be compared with trial scrape marks by an optical microscope. Striations are also produced on cut surfaces. Formation of striations on cut surfaces, however, is not as simple as those produced by scratching the surface. Therefore, it is difficult to compare them under an optical microscope.

Examination of Cut Surfaces

When a single wedged tool, such as a chisel or a blade of plier/scissors, is pressed upon a material, it is finally cut into two parts, each having a cut surface, say X and Y. During this process, the sharp edge of the tool continues cutting with subsequent sliding motion on the cut surfaces. While doing so, the two faces of the tool simultaneously press the cut surfaces X and Y with different pressures, say P_1 and P_2 respectively as systematically shown in Figure 3. Thus, each point of the surface experiences sliding and vertical motions of the tool, thereby producing a complex surface structure consisting of fine striations (parallel lines) and fractured surfaces (irregularities).

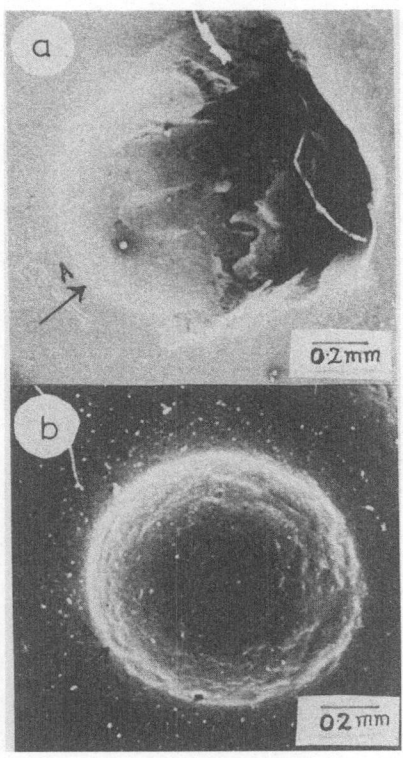

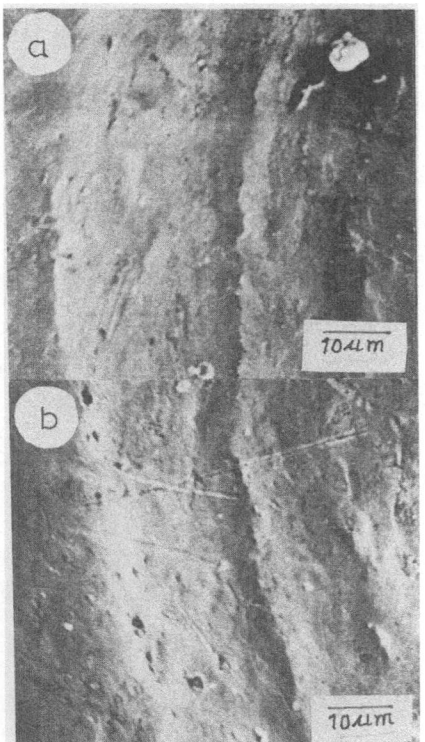

FIG. 1. Indentation marks
     a) Punched cartridge;
     b) Trial cartridge showing
       close resemblance with
       marks in Fig. 1a.

FIG. 2. Exact matching of Indenta-
     tion marks
     a) Punched cartridge;
     b) Trial cartridge.

    Since the two surfaces of such tools are usually not identical with respect to their inclination and smoothness, they produce different irregularities on the respective cut surfaces. Consequently, correspondence does not exist between these cut surfaces. However, the fine striations are produced simultaneously on both the cut surfaces by the common edge of the tool, therefore showing correspondence. Matching of the fine striations on the opposite surfaces of the cut ends under the SEM thus ascertains the two parts having been cut from a single material.

    Surface irregularities and fine striations are reproducible on both surfaces of each consecutive cut made by the same tool because they are produced by the repeated operations of the same faces and common edge of the tool respectively. Correspondence between surface irregularities, as well as between fine striations on the X and Y surfaces, thus can positively identify the cutting tool.

A. Comparison of opposite surfaces of cut ends produced by different pliers.

    Figure 4 represents a pair of cut ends produced from an aluminum wire by a plier A. It is observed that there is a one to one correspondence between the extrusion marks (numbered in the figure) on the cylindrical faces of the cut wires. Irregularities on the cut surfaces do not show correspondence. A close examination of the surfaces shows that there is correspondence between fine striations. Figures 5a and 5b are magnified images of the fine striations at the

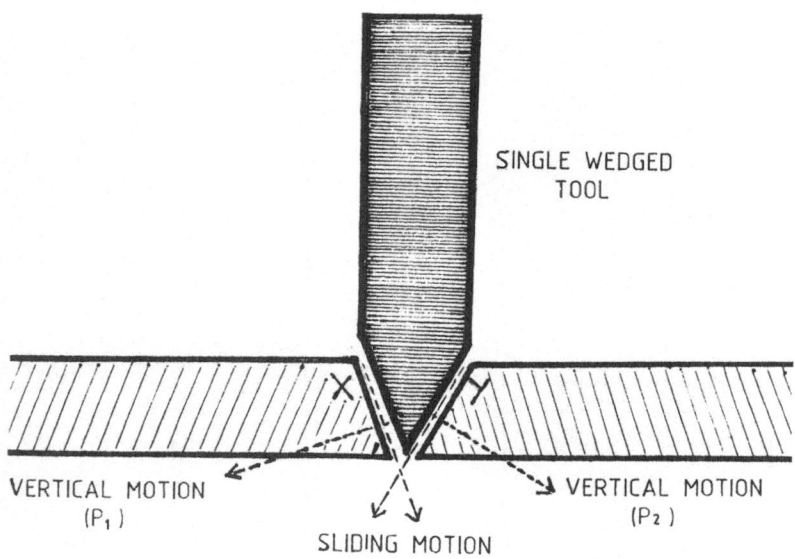

FIG. 3. Systematic diagram showing X & Y surfaces under sliding and vertical motions of the tool.

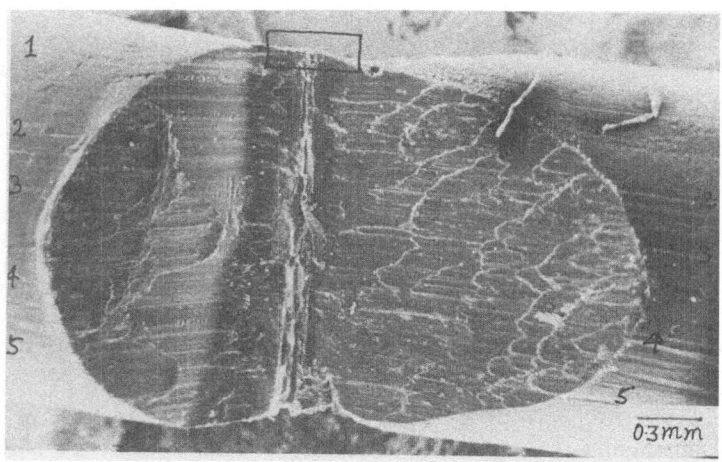

FIG. 4. Pair of cut ends (aluminum wire) produced by plier A showing matching of extrusion marks and fine striations.

corresponding portions of the two surfaces. They show a large number of matching fine striations. These fine striations would not have been revealed by an optical microscope.

Figure 6 represents a pair of cut ends (steel wire) produced by plier B. The wire being harder (than that shown in Figure 4) does not show a long length of fine striations. Close examination, however, reveals that there exists a correspondence between them. Distinct matching of the fractured edges seen in the figure establishes that the two cut ends are counterparts.

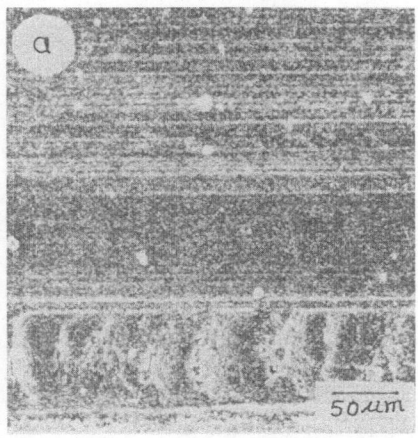

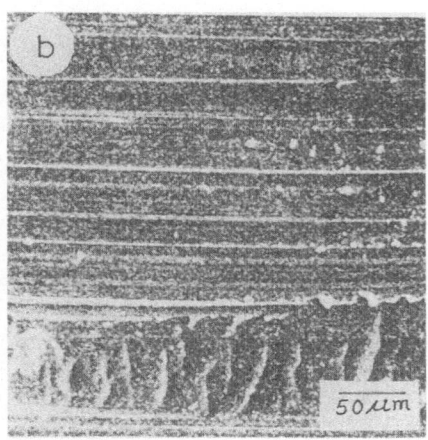

FIG. 5a. Fine striations on left cut
end (Fig. 4)

FIG. 5b. Corresponding fine striations on right cut end
(Fig. 4)

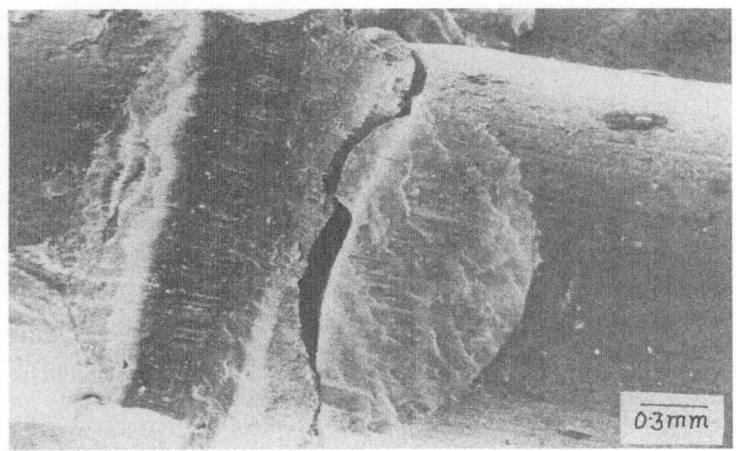

FIG. 6. Pair of cut ends (steel wire) produced by plier B showing matching of edges.

B. Comparison of consecutive surfaces of the cut ends produced by the same
pair of pliers.

A wire (aluminum) was cut at different places using the same portion of
the edge of a plier. The surface of a cut end (Figure 7a) has been compared with
the respective surface of the consecutive cut end (Figure 7b). It may be observed
that surface irregularities and fine striations are reproduced (marked by arrows
in the figures).

It is concluded that as long as the wire was repeatedly cut by the same
portion of the edge of the plier, the surface irregularities, as well as fine striations,
were reproduced. Thus, the comparison of questioned and trial cut ends, based
upon these features, can establish a positive identity of the cutting tool.

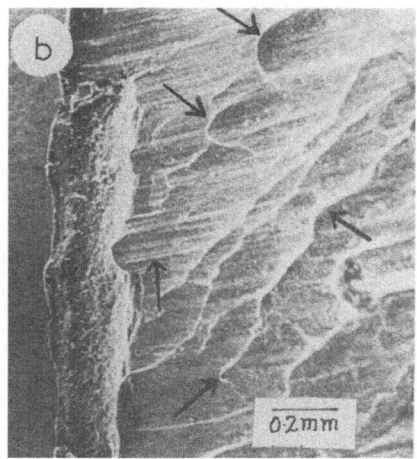

FIG. 7a. Surface of a cut end showing irregularities and fine
striations.

FIG. 7b. Respective surface of
consecutive cut end show-
ing reproducibility of
irregularities and fine
striations.

CONCLUSIONS

Although the optical microscope can be used for the study of indentation
and scrape marks on plain surfaces, the examination of these features under
SEM yields the results of higher reliability, as it reveals even the minute features
on fairly uneven surface. Comparison of tool marks on inclined or broken surfaces
is a unique advantage of SEM. Comparison of firing pin impressions on the punched
cartridges has been possible by SEM and is of great importance in firearm cases.[8,9]

These cut ends of any wire-like material can be conveniently compared
under an SEM. Since this technique records all the comparable parameters,
including extrusion marks, surface irregularities, fine striations and breakage
of edges under the same exposure, it is possible to compare any disputed pieces
with great care and satisfaction. The SEM has been capable of revealing fine
striations on the cut surfaces, thus enabling the establishment of the source
of correspondence between the cut ends (wire) even in the absence of extrusion
marks or fittings of edge breakages.

Reproducibility of surface irregularities and fine striations on the respective
surfaces of the consecutive cut ends produced by a single tool can be distinctly
observed under an SEM. Examination of these surfaces thus assures the results
with greater reliability for identifying the cutting tool in wire theft cases.

ACKNOWLEDGEMENTS

The work was carried out at the Forensic Science Laboratory, Haryana,
Madhuban (Karnal) during the period 1980-84. Thanks are due to Dr. O. P. Chugh,
Director and Shri H. R. Aggarwal, SSO of the Laboratory.

REFERENCES

1. C. G. Van Essen and J. E. Morgan, A split-field comparison SEM, in: "Scanning
Electron Microscopy", O. Johari and I. Corvin (eds.), IIT Research Institute,
Chicago, IL (1973), pp. 159-166.

2. G. Judd, R. Wilson and H. Weiss, A topographical comparison imaging system for SEM applications, in: "Scanning Electron Microscopy", O. Johari and I. Corvin (eds.), IIT Research Institute, Chicago, IL (1973), pp. 167-172.
3. D. A. Meyn and C. D. Beachem, Precision matching of fracture surfaces, in: "Scanning Electron Microscopy", O. Johari and I. Corvin (eds.), IIT Research Institute, Chicago IL (1973), pp. 903-910.
4. V. R. Matricardi, Matching of surfaces, in: "Scanning Electron Microscopy", O. Johari and I. Corvin (eds.), IIT Research Institute, Chicago, IL (1975), pp. 503-510.
5. V. R. Matricardi, M. S. Clark and F. S. Defonja, The comparison of broken surfaces - A scanning electron microscopic study, J. Forensic Sci., 20:507 (1975).
6. R. P. Singh, O. P. Chugh, Harbhagwan, and H. R. Aggarwal, Examination of filaments of light bulbs by scanning electron microscope, Indian J. Criminology & Criminalistics, 3:125 (1983).
7. R. P. Singh and H. R. Aggarwal, Identification of wires and the cutting tool by scanning electron microscope, J. Forensic Sci. International, 26:115 (1984).
8. C. A. Grove, G. Judd and R. Horn, Examination of firing pin impressions by scanning electron microscopy, J. Forensic Sci., 17:645 (1972).
9. C. A. Grove, G. Judd and R. Horn, Evaluation of SEM potential in the examination of shotgun and rifle firing pin impressions, J. Forensic Sci., 19:441 (1974).

FORENSIC EXAMINATION OF BULB FILAMENTS BY THE SEM

Knud Aage Thorsen

Department of Metallurgy
The Technical University of Denmark
Lyngby, Denmark

ABSTRACT

The evaluation of vehicle lighting conditions at the time of an
accident is an important forensic task that is accomplished by examination
of lamp and lamp remains. Optical and scanning electron microscopy are
complementary analytical techniques with growing importance of the latter,
due to its high resolution, large depth of focus and the EDS microanalysis
option. In order to make light bulb analysis less of an analytical art
and more of a reliable investigative technique, a thorough understanding
of bulb design, the materials systems used and the functioning of a bulb
and its behaviour under impact has to be acquired. In this review an attempt has been made to collect the comprehensive but widely scattered literature relevant in these respects. Case stories are presented ob bulb
cases solved with the SEM under reference to the above points of view.

INTRODUCTION

Forensic investigations cover wide ranges of professional interest. In
the last two decades the forensic sciences have developed into an
increasingly complex discipline, partly due to the development of more
sophisticated analytical techniques. At the same time forensic sciences
have grown beyond the original scope of solving criminal cases, into
playing a role in solving occupational and environmental health problems,
as well as in the investigation of product liability cases and accidents
(1).

The present issue is essentially linked up with traffic and aircraft
accident investigations and only rarely with criminal activity, such as in
cases where drunken driving is involved. The following circumstances call
for a lamp examination: In traffic accidents at night it is important to
establish whether head lamps and rear lights were on or off at the moment
of the impact. Daytime accidents can bring up questions as to whether the
turn signal lamps of a car were activated, or as to the state of the lamps
in a signal stand that was brought down by a vehicle. In the air traffic
accident lamps form the panels in the cockpit of a crashed plane are able to
indicate the state of the warning and indicator lights at the moment of impact.

The objectives of the investigations are mainly to find the casual relations which led to the accident. Legal responsibility may be assigned and insurance liability established. With the acquired knowledge at hand the accident may be reconstructed making it possible to eliminate an existing element of risk.

The basis on which the investigations rests is the fact that the stresses generated in a bulb filament by the accelerations and retardations at a collision or a crash produce deformation and fracture phenomena with characteristic differences between hot and cold filaments resulting. Further evidence may arise if the glass envelope is broken at the accident, as the metallic tungsten of the filament will be oxidized if it is hot. Likewise glass splinters are often generated in great numbers when a bulb is broken, and those hitting the hot filament will melt on contact if the temperature of the filament is sufficiently over the softening temperature of the glass.

In principle this looks as a relatively straightforward job. However, many factors may have affected the various phenomena, and as the investigator cannot possibly be prepared for every conceivable combination of factors, he must possess a thorough knowledge of the complex of factors. The information presented in the following has been collected from a widely scattered literature intended for lighting engineers, metallurgists, microscopists, car inspectors and forensic scientists. Lamp systems, the materials systems and physics of bulbs will be presented together with a review of reports on investigations.

LAMP SYSTEMS

With regard to motor vehicle headlights a semi-sealed system with replaceable bulb has been chosen in Europe, whereas sealed beam units are used in North America. In the latter area, however, replaceable bulbs are accepted on imported cars, so the forensic investigators have to be acquainted with both systems. For signal and warning lamps replaceable bulbs are used in any case.

Bicycles present a special problem, their lighting systems being dependent on strongly varying wheel-generator- or battery-fed power systems for the supply of relatively fragile lamps.

In aircrafts especially the warning and indicator lighting in the panels of the cockpits is of interest, and that is a purpose for which special miniature and subminiature bulbs are used.

A detailed account of the lamp types used is given in different national standards. In the USA, the SAE Handbook (2) gives the dimensional specifications for sealed beam headlights (SAE J 571) and for bulbs (SAE J 573g). In Europe reference can be made to German standards (3). Further details can be collected from catalogues from bulb manufacturers such as Osram (4). Descriptions of the miniature and subminiature bulbs for aircrafts can be found in catalogues from GEC (5) and Chicago Miniature (6).

The fundamental features of lighting and lamp systems in general are found to be outside the scope of this review. The IES Lighting Handbook (7) is a good source in such respect.

THE MATERIALS SYSTEMS OF BULBS

The materials systems of bulbs are intricate, however, it is necessary for the forensic investigator to be familiar with them, for instance to be able to identify the extrinsic or intrinsic origin of fragments. Metals and alloys, ceramics and polymers are used, to which could be added gases which are not usually thought of in the materials sense.

Metallic Materials - Filaments

Only tungsten is used for this purpose and due to its outstanding properties it will probably never be replaced by another material.

$\underline{Physical\ Properties}$. Tungsten has the highest melting point, $3410^{o}C$ ($6170^{o}F$), and lowest vapour pressure, $3 \cdot 10^{-14}$ Pa at $2500^{o}C$ ($4532^{o}F$), of all metals. The density is 19.3 g/cm^3, i.e. $2\frac{1}{2}$ times that of iron. The luminous efficacy (the relation between light output and power input) is for normal gas-filled bulbs between 10 and 20 lm/W, the higher values applying to bulbs with large effects. Halogen raises this figure, and for a tungsten-halogen bulb for the head lamp of a car it is about 25 lm/W. As a reference, if the total power input was transformed into monochromatic radiation of a wavelength of 555 nm, where the sensitivity of the human eye is at maximum, the efficacy would be 680 lm/W (8).

$\underline{Chemical\ Composition\ and\ Structure}$. The properties of tungsten wire for incandescent bulbs have to meet primarily the requirement for resistance to creep at elevated temperatures, which on its side is controlled by the recrystallized structure. This is determined by additions of small amounts of second phases such as oxides. Powder metallurgy is the only method by which such alloys can be made. After proper mixing with oxide powder and pressing, the bars are presintered at 1400ºC (2552ºF) before getting the final sintering at 2600-3000ºC (4712-5432ºF). The sintered bar is then hot swaged at 1600ºC (2912ºF) down to about 2.54 mm (0.1 inch), from where it is drawn through different-sizes dies with intermediate anneals until final size is obtained. Coils are produced by winding up the wire on iron or molybdenum mandrels that are later dissolved in acids. For a detailed presentation of the manufacturing processes references can be made to (9-11)

Mounted in the bulb the wire will operate at typically $2600^{o}C$ ($4712^{o}F$) and it will recrystallize in a few minutes. The structure obtained by this process will depend on the oxides added. For the so-called "doped" lamp grade wire, about 0.05 wt% of each of K_2O, SiO_2 and Al_2O_3 is added to the powder mixture. During the final sintering, the dopant particles decompose and Al, Si and O are removed by volatilization. Potassium is insoluble in tungsten and most of it is retained as a fine dispersion of particles. In the wire drawing process the potassium is smeared out in the direction of drawing, and when the wire starts to operate the potassium volatilizes into a linear array of submicron-size bubbles.

The recrystallization temperature is more than $1800^{o}C$ ($3272^{o}F$). On recrystallization the characteristic wrought microstructure of the drawn wire, shown on the micrograph fig. 1, is replaced by new grains without deformation. In the grain growth period that follows recrystallization the aligned bubbles inhibit lateral movement of grain boundaries. The final structure obtained is one of very long, interlocked grains with lengths many times the diameter of the wire as shown on the micrograph fig. 2. None of the grain boundaries are cutting through the complete cross-section of the wire at right angles to the wire axis. This structure has an excellent resistance to grain boundary creep, and with the doped grade no sagging of the filament due to the mere weight of the wire is experienced.

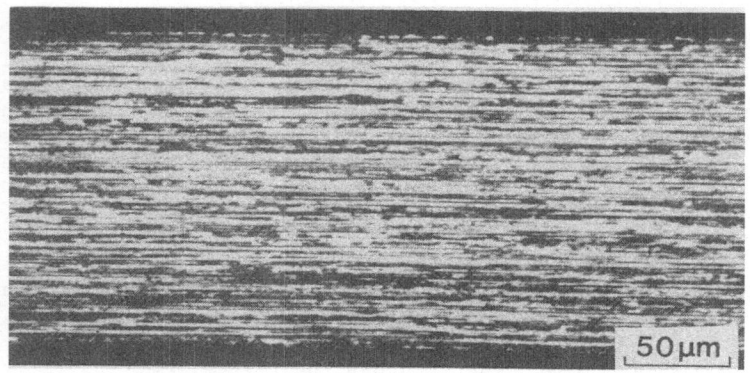

Fig. 1. Photomicrograph of tungsten wire, doped lamp grade,
178 μm (0.007 in) diam; as drawn. Deformed, elongated
grains. Etchant: Murakami (mod). Ref. (12). Reproduced with permission of American Society for Metals.

Refractory oxides such as ThO_2 or ZrO_2 can be added and they will
remain dispersed in the structure during the high temperature sintering.
The recrystallization temperature of such alloys is also more than 1800°C
(3272°F). The dispersed oxides will restrain grain growth by anchoring
the grain boundaries, and consequently a smaller grain size will be obtained. The ThO_2-alloyed grades are less effective for lamp applications,
but due to their high electron emission capability they have found special
applications.

Pure tungsten is unsuitable as a filament material, although it was
used as such formerly. It recrystallizes from 900°C (1652°F) to 1400°C
(2552°F). The recrystallized structure would consist of small, equiaxed
grains as shown on fig. 3, but on further heating the grains would grow,

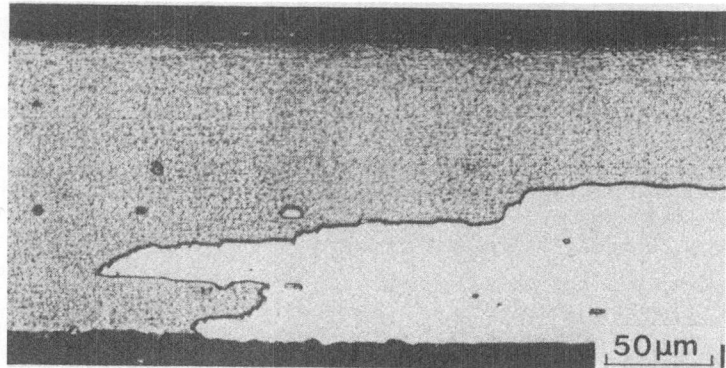

Fig. 2. Photomicrograph of same grade and size of tungsten
wire as for fig. 1, annealed for 5 min. at 2700°C
(4892°F). Recrystallized microstructure; grains are
"finger-locked". Etchant: Murakami (mod). Ref. (12).
Reproduced with permission of American Society for
Metals.

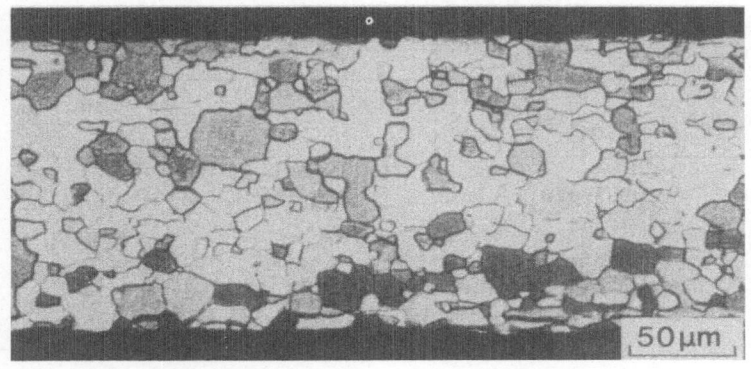

Fig. 3. Photomicrograph of tungsten wire, not doped, 178
        μm (0.007 in) diam; annealed for 5 min at 2700°C
        (4892°F). Fully recrystallized, equiaxed grains.
        Etchant: Murakami (mod). Ref. (12). Reproduced
        with permission of American Society for Metals.

until the diameters of the grains became comparable to the wire diameter.
Due to grain boundary shear this structure would have low resistance to
sagging and to off-setting along the grain boundaries lying in planes perpendicular to the wire axis. The problem of off-setting is illustrated on
the drawing fig. 4.

The properties of doped wire can be further improved by alloying with
up to 5 % rhenium. Over 7 % Re has a negative effect and neutralizes the
positive effects of the dopants. Rhenium additions do not change the
properties of pure tungsten (13). In spite of the attractive effects of
Re for a wide range of filament applications its use is severely restricted
due to its high costs.

Specifications for filament wire laid down by the ASTM are shown in
table I (14).

As regards the chemical composition, typical purity of three
commercial grades of tungsten is tabulated in table II (9).

Mechanical Properties of Tungsten. Tungsten has the highest modulus
of elasticity (E) of all metals, viz. 410 GPa at 20°C (68°F) (15). With
rising temperature E goes down, and at 2400°C (4352°F) it will be about
250 GPa which is close to the magnitude of E for iron at room temperature.
Even at high temperature tungsten is thus a relatively stiff material,

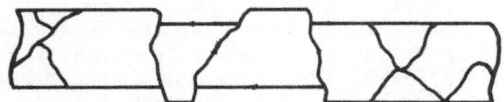

Fig. 4. Drawing of appearance of off-
        setting along grain boundaries
        in pure tungsten.

Table I. ASTM Specifications for Filament Wire (14)

Grade	Type	Chemical Composition
1 A	Commercially pure, non-sag wire	99,95 wt% tungsten
1 B	Commercially pure rod suitable for metal-to-glass sealing	99,95 wt% tungsten
2 A	Thoriated filament wire with 1 % ThO_2	min. 0.75/max. 1.1 wt% ThO_2, bal. W
2 B	Thoriated filament wire with 2 % ThO_2	min. 1.75/max. 2.1 wt% ThO_2, bal. W
-	W-Re alloy wire for electron devices and lamps	2.5-3.5 wt% Re, bal. W

For all grades applies that total contents of other elements must not exceed 500 ppm, the individual element not exceeding 100 ppm.

Table II. Typical purity of three commercial grades of tungsten (9)

Element	Concentration in ppm		
	Electron Beam Zone Refined	Powder Metallurgy Processed Undoped	Doped
Fe	1	10	11
Ni	2	5	5
Si	5	21	47 a)
Al	<2	<5	15 a)
K	<1	12	91 a)
O	10	27	36
C	20	31	24

a) Doping elements.

showing a limited amount of elastic deformation when stressed.

As regards the tensile strength, this property falls with temperature as shown on fig. 5, and it is obvious that the stress required to get a fracture at 29°C (5252°F) is only a small fraction of that required at for instance 1500°C (2732°F). There is not much difference between the various production processes, of which the PM-route is most relevant for filament wire. In this context it could be added that over 1650°C (3002°F) tungsten has the highest tensile strength of all metals. The yield strength, i.e. the stress at which plastic deformation starts, is about 90 % of the tensile strength.

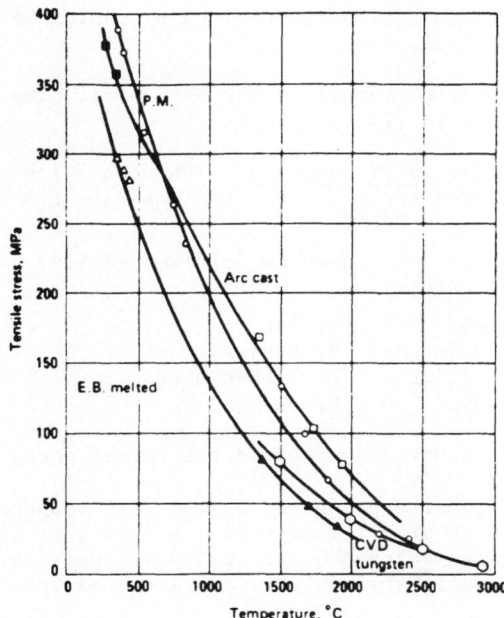

Fig. 5. Temperature dependence of the
tensile strength of annealed pure W,
made by 4 processes, Electron Beam and
Arc melting, Chemical Vapour Deposition
and Powder Metallurgy (15). Reproduced
with permission of American Society for
Metals.

As regards ductility, the elongation and reduction in area measured
by tensile testing of P.M. tungsten is about zero up to 400°C (752°F), that
is tungsten is very brittle in this temperature range. At about 400°C
(752°F) a quite considerable elongation and reduction in area appears, a
maximum in ductility (55 % resp. 90 %) is met at about 1000°C (1832°F),
from where it begins to fall, and it again reaches low values at about
2500°C (4532°F). An improvement in lamp wire ductility at room temperature
can be obtained by alloying with Re (13). After annealing at 2500-2800°C
(4532-5072°F) doped wire without Re is glass brittle, whereas an elongation
of about 25 % is measured on doped wire with 3 % Re, which gives the latter
alloy a significantly better shock- and vibration-resistance.

The change from ductile to brittle behaviour at the ductile to brittle
transition temperature is a consequence of the tungsten being a base centered
cubic (BCC) metal, and all BCC metals have actually this transition. In the
wrought state the transition temperature is 100-250°C (212-482°F), and in
recrystallized tungsten it is about 350°C (662°F) according to (16).

The ductile to brittle transition is one of the factors which makes it
possible to distinguish between impact on a hot and a cold filament as will
be elucidated later on.

With regard to creep properties, slow deformation by creep is the
cause for the sagging of filaments. By producing the wire with the structure

of large, finger-locked grains as obtained by doping (fig. 2) this problem seems to have been eliminated.

For a more detailed report on the mechanical properties of tungsten reference can be made to (17).

Chemical Properties of Tungsten. Tungsten is chemically stable in air at low temperature, there is thus no immediate risk of corrosion damage on the filament of a crushed bulb.

At high temperatures, tungsten oxidizes easily, forming four well-defined oxides (10), (16):

$$
\begin{array}{ll}
WO_3 & - \quad \text{yellow} \\
WO_{2.9} & - \quad \text{dark-blue} \\
WO_{2.72} & - \quad \text{reddish-violet} \\
WO_2 & - \quad \text{brown}
\end{array}
$$

On experiments the author has observed the former three of these oxides.

Tungsten reacts with halogenes at moderate temperatures, the halides formed are unstable at high temperatures, a fact that was utilized in the design of the tungsten-halogen bulb. Halides important in this respect are WBr_2 and the higher bromides WBr_5 and WBr_6, as well as the iodides WI_2 and WI_4.

Metallic Materials - Other Parts than the Filament

A wide variety of metals and alloys is used in bulbs for other applications than filament. In table III the more commonly used metals and alloys and their applications are stated (11).

Table III. Metals and Alloys for Non-Filament Use in Bulbs (11)

Application	Metal or Alloy
Terminal posts	Cu, Ni, W, Mo
Support posts	(Fe, Ni, Ni-alloys, Mo, Ta, W
Internal reflector coatings)	Al
Lamp caps	Al, brass (Cu-Zn)
Internal reflectors) and screens	Mo, Ni
Solders	Sn, Sb

Molybdenum, tungsten and nickel are three of the most common metals used for bulb components that are exposed to high temperatures. Molybdenum has a coefficient of expansion that matches quartz making it the best material to form quartz-to-metal seals.

Complicated solutions can be found in three-piece leads used as conductors through a vacuum-tight glass seal. Three wires are joined end to end by two welded joints, and such components are standardized by the ASTM (18).

On analyzing the leads in a gas-filled R2 and a tungsten-halogen H4 headlight bulb the author found the materials combinations of table IV.

Table IV.   Chemical Composition of leads in headlight bulbs

Component	R2 from Philips 3-piece lead	H4 from Osram 1-piece lead
Inner lead section (terminal post)	Nickel	)
Seal lead section	Copper-plated Fe-Ni wire (Dumet wire)	) Molybdenum
Outer lead section	Copper	)
Screen under lower beam filament	Nickel	Molybdenum
Joining of filament to terminal post	Filament spot welded direct to terminal post	Filament mounted in Mo-tube and this assembly was spot welded to the terminal post

When faced with other bulb types, other combinations might be expected. At an accident only the inner lead sections and the screen under the lower beam filament could be expected to enter into contact with the filament, but the forensic investigator has to be aware of the existence also of other metals and alloys in the system.

Non-Metallic Materials - Glasses and Ceramics

The glass most commonly used as envelope material for general incandescent bulbs is of the soda-lime silicate type (11). The internal glass components such as the terminal and support posts mounting bead and the seal are made from lead-alkali silicate glass, which has a higher electrical resistivity than the soda-lime silicate type. Electrolysis in the seal can thus be prevented. For sealed beam units a borosilicate glass is normally used due to its lower thermal expansion coefficient and thus greater resistance to failures due to changes in temperature is rendered.

For the halogen bulbs which operate at high envelope temperatures and high internal pressures conventional glasses are inadequate and transparent silica was put to use as envelope material. It is essentially pure silicon dioxide having only a fraction of 1 % of other elements present as impurities. It is called "quartz" in the bulb industry, though it is vitreous, not crystalline. The main advantages of it are its transparency, resistance to thermal shock and its ability to operate at high temperature - up to $800^{\circ}C$ ($1492^{\circ}F$). The tungsten-halogen bulb essentially depends on it. The compositions and physical properties of materials for bulb envelopes are presented in table V (11).

Table V. Composition and Physical Properties of
Materials for Bulb Envelopes

Oxide	Composition in weight-%				
	Soda-lime silicate	Lead-alkali silicate	Boro-silicate	Vitreous silica	Vycor
SiO_2	73	56	76	100	96
Na_2O	15	5	4	-	-
K_2O	1	8	2	-	-
CaO	7	-	-	-	-
MgO	3	-	-	-	-
Al_2O_3	1	1	2	-	-
PbO	-	30	-	-	-
B_2O_3	-	-	16	-	3
Thermal expansion coeff. $\times 10^7$ per ^{o}C	92	90	37	5	8
Softening point ^{o}C	710	630	775	~1600	~1530

Non-Metallic Materials - Polymers

Capping cements are usually mixtures of a thermo-setting resin and
an inert inorganic filler applied to the inside of the cap rim, which
attach the cap to the bulb on being heated.

Metallic and Non-Metallic Materials - Getter Materials

To remove residual gases such as oxygen, water vapour and hydrocarbons
that would otherwise affect bulb performance, getters are added in the bulb
production (11). Reactive metals are used for this purpose such as Ba, Ta,
Ti, Zr, and Nb. They are applied in the form of wire or sheet or as surface
deposits on selected bulb components.

Non-metallic getters are also used, such as red phosphorus mixed with
cryolite. In tungsten-halogen bulbs the halogen may be introduced together
with phosphorus, the latter serving as the getter, in the form of halo-
phosphonitriles $(PNBr_2)_3$ and $(PNBr_2)_4$.

The author has not found traces of getter materials in his investigations, but the possibility exists that traces of getters could appear on
the analysis of a bulb.

Non-Metallic Materials - Gases

Bulbs with small effects, say under 15 W for mains voltage bulbs and
3 W for minature and subminature bulbs, operate in a vacuum. At higher
effects the evaporation of tungsten in a vacuum bulb would limit both the
filament temperature and the efficacy considerably. Consequently an inert
gas is used for filling of the bulb in spite of the fact that the gas con-
ducts heat from the filament to the glass envelope. The evaporation rate of
tungsten is typically reduced to 1/70th.

126

The bulbs are filled to just below atmospheric pressure with mixtures of argon and nitrogen with the ratio of Ar to N_2 depending on the voltage rating, filament construction and lead-tip spacing. Typically used is: 99.6 % Ar for 6 V bulbs, 95 % Ar for 120 V coiled-coil bulbs and 90 % Ar for 220 V household lamps (11).

A special gas composition is found in the tungsten-halogen bulb which can be considered as a special variety of the incandescent bulb. Tungsten-halogen bulbs were introduced as recently as 1959. The filament temperature is considerably increased in this type without running the risk of accelerated blackening of the envelope. A chemical equilibrium is estab- lished introducing tungsten halides between the filament and the envelope that totally prevents tungsten vapours from reaching the inner cold wall of the envelope (11). Bromine is the halogen preferred for this purpose, and a simplified equilibrium mechanism is shown in fig. 6.

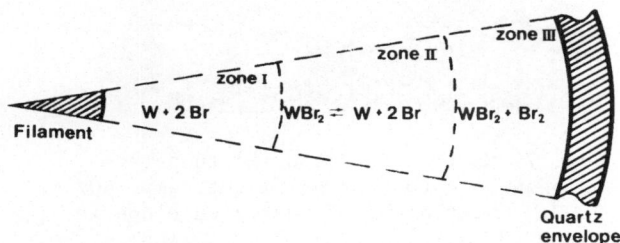

Fig. 6. Simplified equilibrium mechanism in a tungsten-halogen bulb.

In the hot zone (zone I) where the temperature is over $1500^{\circ}C$ $(2732^{\circ}F)$ the halide is completely dissociated, and at the wall (zone III) the recom- bination to W-halides is complete. No tungsten vapour thus ever reaches the wall, and without the risk of blackening a near 100 % lumen maintenance throughout life is found. To prevent condensation of W-halides on the wall, the temperature of the envelope has to be over $250^{\circ}C$ $(482^{\circ}F)$. To keep this high operating temperature the distance from filament to the wall must be reduced and the tungsten-halogen bulbs have characteristic small volumes as it can be seen on the H4 tungsten-halogen bulb 60/55 W shown on fig. 7. The pressure is kept well in excess of atmospheric pressure, viz. at 1.5 to 10 atm. to suppress the transport of tungsten in the gas phase from the hot part of the filament (vaporization of W) to the cold parts (condensation of W).

By introducing halogens the materials system has to be changed. As nickel and iron would react and form unwanted halides, these metals must be kept out of the system. For terminal and support posts amd for lower beam screens of car headlight bulbs, either tungsten or molybdenum has to be used as specified in table III.

Due to the special features of tungsten-halogen bulbs a certain

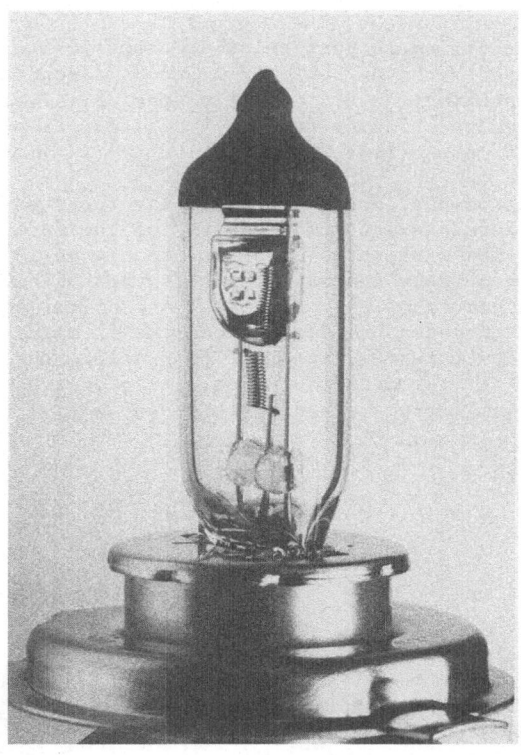

Fig. 7. Macrophoto 2:1 of H4 tungsten-halogen bulb for car headlight lamp, 60/55 W. Outer diam. of quartz envelope is 14 mm (0.55 in).

caution must be exercised when using them. In their catalogue (5) GEC directly puts up the following warnings:

- Do not operate bulb in excess of rated voltage as this will increase lamp pressure. Actually, the author has seen tungsten halogen microscope light bulbs designed for 12 V virtually explode when they were erroneously connected to mains voltage of 220 V.

- Protect envelope against scratches, that could act as notches for cracks, and against sudden quenching by liquids when bulb is operating.

- Do not operate in proximity to substances or materials that are flammable or adversely affected by heat or drying.

THE PHYSICS OF BULBS

The temperatures at which filaments operate depend on the type of bulb. According to (11) a usual gas-filled car headlight bulb operate at about 2500°C (4532°F), whereas a panel indicator bulb which is required to give a dim light at night time and operate virtually for the life of the car has a working temperature of about 2100°C (3812°F).

Bulbs in bicycle lamps operate at temperatures ranging form 2500°C (4532°F) to 1500°C (2732°F) depending on the strength of batteries and the speed of the bicycle.

Tungsten-halogen bulbs work at temperatures between 2600°C (4712°F) and 3200°C (5792°F), which are typically around 2900°C (5252°F).

In twin filament bulbs the filament that is energized will heat the filament that is not energized. Based on resistance measurements the author has registered temperatures of up to 600°C (1112°F) on a thus passively heated filament in a H4 bulb. This is a very interesting point, as it brings the tungsten in both filaments well above the ductile-to-brittle transition temperature and this might confuse investigators.

It is often of interest to know the dynamic behaviour when switching on or off or when pulsing the light. Response to an input is not instantaneous, a short warm up time is required for a filament to reach its working temperature, and a relatively long cool down time will be found before ambient temperature is reached. In a quite simple experimental set-up at our Physics Department in Lyngby, a power circuit and an independent measuring circuit were established through the same filament. On switching off the power circuit, the current in the measuring circuit rises due to the reduction in resistance of the filament as it cools. Recording on an oscilloscope gives the cooling profile. Fig. 8 shows the general appearance of the curves obtained. The author found that the filaments of a H4 bulb required 2-3 seconds to cool off, whereas the filament of a small bicycle bulb cooled down in about 50 msec.

Temperature transients of heating and cooling of filaments have been determined by Bericke et als. (19) who used an infrared radiation pyrometer. The temperature of a H4 lower beam filament fell to 500°C (932°F) in 4 to 5 seconds when the current was switched off, according to their measurements.

The filament temperature of turn signal bulbs turned out to oscillate

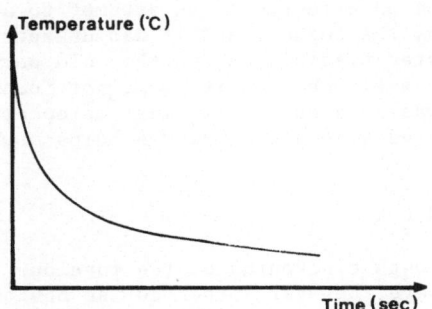

Fig. 8. Drawing of general shape of cooling curves
for filaments on switching off the
power supply.

between $2500^{\circ}C$ ($4532^{\circ}F$) and $1000-1500^{\circ}C$ ($1832-2732^{\circ}F$) when the lamp was activated with 0.4 sec on, 0.34 sec off. This means that the tungsten remains in the ductile state all along the activated period of the turn signal lamp. Thereby an answer has been given to the long outstanding question of whether the turn signal bulb filament became brittle in the periods when the power was off.

The determination of the temperature of a hot filament is no easy matter, direct measurement with a thermocouple being impossible. The spectral effect distribution is measured with a spectrophotometer, the colour temperature is calculated on basis hereof, and as the correlation between colour temperature and real temperature of tungsten is known the latter can be found.

The temperature distribution along the length of a filament has been measured (20). Downward gradients due to end losses, overheating in hot spots and cooling in places with short circuits between turns were shown.

The applied voltage determines life time and light output of a bulb. According to (11) bulb life is inversely proportional to the nth power of the voltage (n = 13 for vacuum bulbs, n = 14 for gas-filled general lighting service bulbs, and n = 11.2 for headlamp bulbs). The light output is directly proportional to the 3.6 th power of the voltage. Thus for a 5% increase in voltage the bulb life is nearly halved and the light output is increased about 19 %.

A normal household bulb is designed for 1000 hours life time. For car headlights different values apply to the upper and lower beam filaments. According to (2), 150 hrs and 320 hrs respectively are stipulated for sealed beam units tested at 14 V, and according to (3) 100 hrs and 200 hrs respectively are stipulated for tungsten-halogen bulbs (H4) tested at 13.2 V. For the signal bulbs in cars the expected average life time is 200 hours measured at a 10-15 % overvoltage.

The dimensions of the filament wires are determined by operating conditions. The wire diameter is increased with increasing current ratings and the length of the wire is increased with increased voltage ratings.

It is generally stated that the power supply is not a critical parameter to lamp operation, it can be AC or DC, pulsed, chopped or rectified. However, as will be seen in the following chapter surface structure changes are functions of the power supply, depending on whether it is an AC or DC supply. This is especially a factor that has to be reckoned with when operating filament wires with diameters less than 18 μm. Even for heavy wire there appears to be an effect. In an attempt to age a H4 bulb an AC power source was used by the author, and it was unexpectedly found that heavy deposits of tungsten precipitated on the cold parts of the filament, screen and terminal posts. Such deposits were not found in a gas-filled R2 headlamp bulb that was aged on AC. It must be concluded that a stabilized DC source must be used when aging halogen bulbs meant for DC operation.

CHANGES IN BULBS DURING USE

Deterioration through blackening of the envelope is a change that can be seen with the naked eye, however, other not so obvious changes take place in the system.

The recrystallized grain structure as shown on fig. 2 is stabilized in rather short time, the surface structure, however, develops very slowly during the whole life of the bulb. Various mechanisms cause the surface structural changes, namely thermal etching, electromigration and the Soret effect.

Thermal etching: Given the time required for a sufficient amount of transport by diffusion, the surface will in the end consist of low energy crystal planes {110} and {100} (21). This leads to the development of planar surfaces and terracing. Balancing of the different surface energies around grain boundaries leds to grooving where grain boundaries emerge on the free surface. Thermal etching is a pure temperature function and consequently there is no difference between AC and DC. Thermal etching was observed on the filament from a burnt-out 220 V AC 60 W household bulb, fig. 9.

Electromigration: In a bulb energized by DC the filament is polarized and there seems to be a drift of tungsten ions in the field along the filament. This statement was made as early as 1938 (22). Electromigration leads to a sawtooth appearance of the surface, also called DC notching. The steep side of the notch points towards the negative polarity. This phenomenon is of course only met in DC operated systems. Reference to figs. 10 & 14.

The Soret Effect: Also mentioned in (22) a surface structure identical with the DC notching can be caused by a temperature gradient in the filament, as the gradient leads to surface diffusion in the gradient. The effect of this process only comes forward where there is a temperature gradient, as it is the case at terminal and support posts. The Soret effect could be effective in AC as well as DC systems.

The electromigration and Soret phenomena together have proved to be especially critical for subminiature bulbs designed for indefinitely long life time for applications where replacement could be exceedingly difficult. By running them below about 2000°C (3632°F) the deleterious effect of tungsten vaporization and thermal etching on life was thought to be nearly eliminated, however, it had been overlooked that the DC-notching processes were still very active at the lower temperatures (23-26). The depth of the sawtooth surface structure is independent of the diameter of the wire,

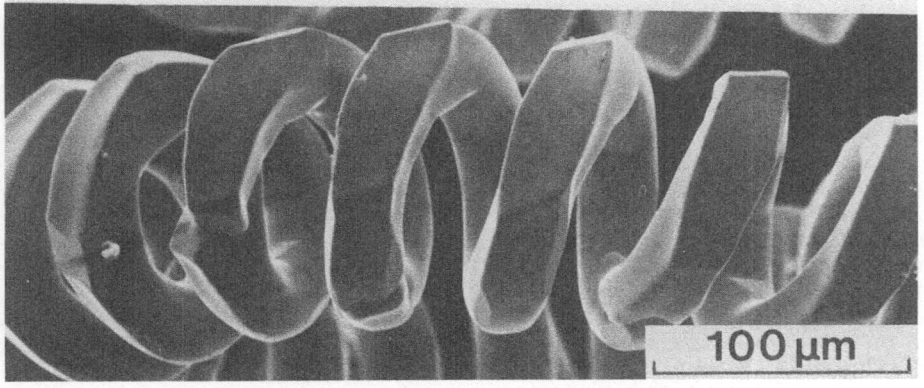

Fig. 9. Thermal etching on surface of filament from 220 W AC,
60 W household bulb. Planar surfaces and grooves at
grain boundaries have developed.

and for the very thin filaments used in subminiature bulbs the notches meant a severe reduction in cross sectional area leading to "hot spots" and failure. The effect on life becomes insignificant on wire diameters >18 μm, since the effect of the surface changes becomes negligible as the wire diameter increases.

EFFECT OF IMPACTS ON BULBS

One of the main variables to be considered is the type and nature of the impact loading.

The intensity of the shocks in a crashing aircraft was estimated by the RCAF (27) to be about 70 g with a during of the shock of about 95 msec. In 1985 Transport Canada (28) reports 500 g and 3 msec for a typical impact.

Laboratory tests aiming at simulation of crashes by applying shock impulses on bulbs were performed by the RCAF (27) with an impact shock machine having a range of 0-70 g, by Heaslip et als (29) with a machine having a range of 50-1000 g, and finally Transport Canada (28) refers to a machine that reached 3500 g. The glass envelopes did not break in any of the cases. The tests demonstrated that the components of the bulbs could react in two ways when impact loaded (29). Firstly, the filaments and support posts of the bulbs investigated can attain resonance at various frequencies. A sharp pulse (a high g loading over a short period of time) may hit the natural frequency of the support posts and filament and lead to severe excitation of these components, leading to snaking, deformation and possibly fracture. Secondly, in the absence of resonance (g loading is applied over a relatively long period of time) the inertia of the filament can cause deformation and fracture. The paper by Heaslip et als (29) reports on an extensive experimental work on these phenomena on two types of aircraft bulbs, and some of the details are excellent illustrations of the effects of resonance and inertia. Loading at 650 g for 0.8 msec on a hot new bulb caused significant hot deformation due to excitation of supports posts and filament. After 600 g for 0.75 msec a cold new filament exhibited cold deformation and a fracture, again due to excitation. Impacting at 850 g for 1.5 msec on a hot, aged filament did not cause resonance; the inertia alone, however, caused the beginning of uncoiling and stretching but no fracture.

With regard to car accidents no work has been done to measure shock impulse loading on head lamps, e.g. when a test car collides with a wall, nor has there been any investigations of the effect of impact loading on car bulbs, as far as the author knows of.

There are some differences between the subminiature bulbs referred to in (29) and car headlamps, so that a certain caution must be exercised when reasoning from one field to the other. The subminiature bulb filaments operate at 1500-1700°C(2732-3092°F) whereas, the car bulb filaments operate at 2500°C (4532°F), and at 2900°C (5252°F) if they are of the halogen type. In the car bulbs there are no support posts, the filament wire has a much larger diameter and only few turns as compared to the filament wire of the subminiature bulb. The natural frequency of the filament is probably very high, which means that only the inertia of the filament will be active in the deformation and fracture processes. At road accidents bulbs often get hit directly so the envelope is broken. Significant phenomena such as oxidation and melting of glass splinters taken place. The minature and subminature bulbs mounted in panels are better protected: their glass envelopes do not usually break. Within car headlamps there are differences between sealed beam units and semi-sealed units with either normal or tungsten-halogen bulbs. It is the experience of the Danish car inspectors that the cylinder-shaped quartz envelope of tungsten-halogen bulbs (fig. 7) is a mechanically very strong design that is not easily broken. Usually only the tip comes off or the envelope breaks at the base.

EXAMINATION OF BULBS AFTER AN ACCIDENT

Quite a number of reports on how to perform such examinations exists. The absolute condition for obtaining a correct result is to make a proper investigation at the site of the accident of lamp abnormalities resulting from the accident, direction of impact, short circuits, light switch position, etc. Lamps and lamp fragments must be carefully removed, labelled and packaged. Special reference is made to the report by Baker & Lundquist (30) which gives exhaustive instructions on how to proceed. Attention should be paid to such phenomena as: insulating corrosion products on lamp contacts, use of non-standard sockets, traces of soldering operations on lamp base which might have caused softening, deformation or melting of the solder joining the lead wires to the inside of the base, thus destoying the joint, and which might also have loosened the capping cement, etc.

In the forensic laboratory a stereomicroscope and a macrophotostand may be adequate in producing reliable evidence of heavy layers of oxides in a broken bulb, heavy stretching of filament turns, etc. Higher magnifications were previously obtained with an optical microscope, if required, however, by magnifications over 50 x the depth of focus was generally too small.

Results obtained with these techniques have been reported by Kremmling & Schöntag (31), Mathyer (32), Dolan (33), and Coldwell & Melski (34), as well as by the previously mentioned reports (30), (27).

When scanning electron microscopes became generally available they were put in use for this type of forensic work. The first reference found on the use of the SEM was by Thornton (35). In 1975 Goebel (36) reported on the investigation of filaments by SEM that this method made it easy to see scratches and contact marks, glass splinters and marks of loose connections. In 1980 Thorsen (37) paid special attention to the examination in the SEM of fracture surfaces. Further reports on the use of the SEM were given by Kroll & Schlagenhauf (38) and Thorsen (39).

The SEM must be considered to be the ideal microscope for investigation of bulbs due to its high resolution, depth of focus and the availability EDS analysis. Various flow diagrams showing how to correlate observations to be able to draw the right conclusions are given in (30), (33) and (38). The broad outline of table VI presents the main principles of the various diagrams.

Hagström & Söder (40) have reported on the use of the specific surface analysis technique of Auger spectrometry to solve the particular question of whether a filament was suddenly exposed to air when working (i.e. hot), or whether it was exposed to air already when the current was switched on. It was found that oxygen concentration was the lowest on the surface of the wire then it was lit after the air was let in and that the tungsten oxide layer was thinnest on the same wire. In the author's opinion, this method might work well in laboratory experiments but it is doubtful if the method will give reliable results on accident bulbs die to lack of reference curves for oxygen profiles. To these points must be added that the method is very expensive and time-consuming.

Table VI. Simplified table of observations and related conclusions

Observation	State of filament on impact
Envelope not broken	
Filament not broken	
- straightening of turns	- hot
Filament broken	
- globular fracture, no straightening of turns	- normal burn-out before fracture a)
- globular fracture, with straightening of turns	- hot, power on
- brittle fracture	- cold b)
- fibrous fracture	- hot, power off c)
- rounded, smooth fracture	- hot, power on d)
Envelope broken	
As above, adding	
- oxidation of wire	- hot e)
- no oxidation of wire	- cold
- tungsten oxides on envelope, screen, terminal posts, etc.	- hot
- spherical glass particles on filament	- hot f)

a) ref. fig.10   c) ref. fig.17   e) ref. fig.21
b) ref. fig.11   d) ref. fig.23   f) ref. fig.19
   12,15,22

CASE STORIES AND EXPERIMENTAL RESULTS

Preparation of materials for the SEM

Handling should be kept at a minimum in order not to damage the very fragile filaments. If vibration of free ends of thin filaments becomes a problem, it may be necessary, however, to cut off a minor piece and mount it in a conductive cement.

If filaments and other components have to be removed from an intact envelope, the envelope can be sectioned with a diamond wafering blade on a low speed saw. Tungsten-halogen bulbs can be sectioned without blowing up, they are apparently depressurized rather slowly when the cutting wheel opens up to the atmosphere.

Before mounting in the SEM, sputter coating with gold is favorable, and actually essential if glass particles and oxide layers are present.

Case Stories

Globular Fracture of Filament burned out normally. An R2-bulb had
served for about 30,000 km before it broke. It appeared that the lower
beam filament was broken in two places so that a piece was lying loose in
the still intact, but heavily blackened envelope. The coil of the filament
had not been straightened out. One of the fractures is shown on the SEM
micrograph fig. 10. Evidently the filament had been working at full voltage
when local heating began in a hot spot. Melting occurred and the wire
broke, leaving a typical globular appearance of the fracture surface, a
feature that is characteristic of a normal burn-out. Perfect spheres may
also be observed. On the wire surface DC notching can be seen. A few
longitidunal grains boundariews are visible and the microstructure is
thought to correspond to the structure in fig. 2.

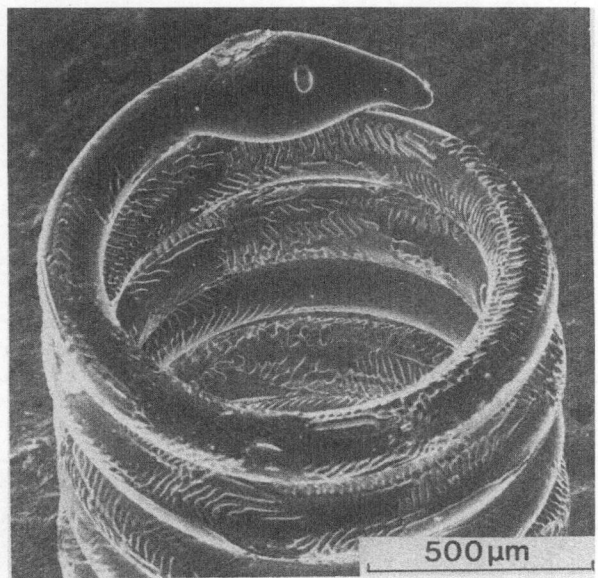

Fig. 10. Globular fracture of filament after a
normal burn-out. DC notching of wire surface.
(37). Reproduced with permission of Canadian
Society of Forensic Science.

In the same filament quite different types of fracture were observed
as discussed in the next case.

Brittle Fracture in Cold Filament. By pressing the blade of a knife
between two turns of the loose piece of filament from the previous case, the
filament broke into two pieces as if it was made of glass. The SEM micrograph
in fig. 11 shows the surface of the induced fracture, which can be clearly
identified as a brittle fracture. This is in agreement with the fact that
the induced fracture was made at room temperature which is well under the
ductile-to-brittle transition temperature. At A the fracture is intercrystalline (following boundaries between the grains) but at B it has cut
through the interior of a grain following certain cleavage planes
(transcrystalline fracture).

The original fracture, of which part is seen at high magnification
on fig. 12, is clearly of the brittle, intercrystalline type.

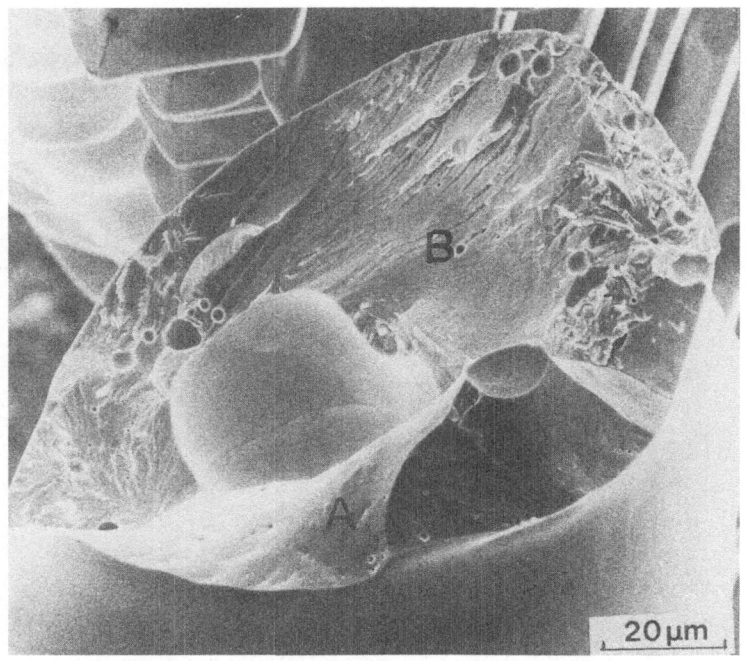

Fig. 11. Brittle fracture of cold, aged filament. At A: intercrystalline. At B: transcrystalline. (37). Reproduced with permission of Canadian Society of Forensic Science.

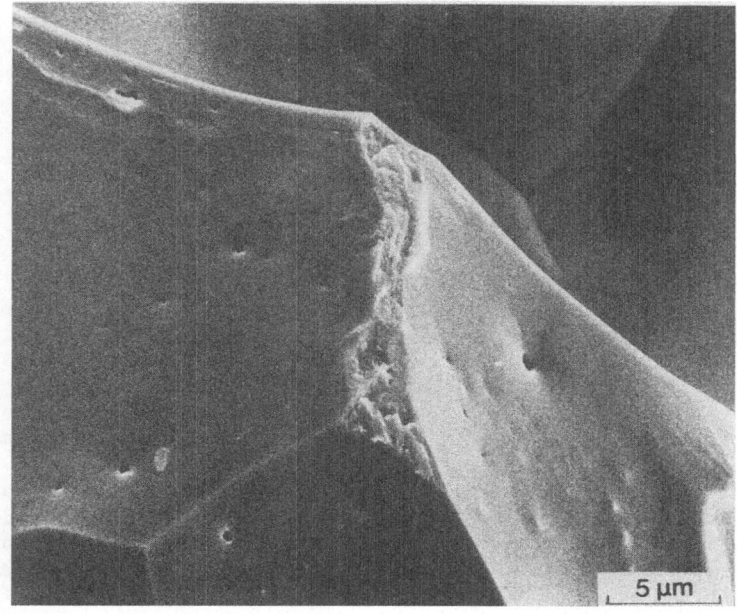

Fig. 12. Brittle fracture of cold, aged filament. Inter-crystalline. (37). Reproduced with permission of Canadian Society of Forensic Science.

Both intercrystalline and transcrystalline brittle fracture surfaces are thus observed in fractures at room temperature in an aged wire. The sequence of events in the bulb being the object of the first two cases is as follows: after the wire had burned out and cooled to ambient temperature the wire pieces started to vibrate and the stresses thus generated caused the second original fracture.

Brittle Fractures in Rear Lamp Filaments. A parked car was run into from behind by another car. The glass envelopes of the bulbs from both rear lamps of the parked car were intact but the filaments were broken.

At low magnification the very uniform spacing between turns was noted, fig. 13. At higher magnification, fig. 14, the severe thermal etching and DC notching were noted, the bulb had clearly been in service for a long time. The fracture was identified as being of the brittle, transcrystalline type characteristic of a cold fracture, fig. 15.

The filaments of both bulbs were severely notched, and the extreme reduction in cross sectional area at the fracture face is evident when looking at figs. 14 and 15. If the bulbs mounted in the rear lamps had been new, the filaments might have been able to withstand the impact without breaking.

It was concluded that the parking light of the parked car was not lit when the collision took place. There was no doubt that the filaments were cold when they broke, and most probably that happened when the bulbs were exposed to the shock of the collision.

Two Cases of Hot Fracture in Rear Lamps of Bicycles. The cyclists were killed in these accidents by cars coming from behind. In both cases the glass envelopes were broken at the accident, the filaments were broken as well but remnants of them were still sitting on the terminal posts.

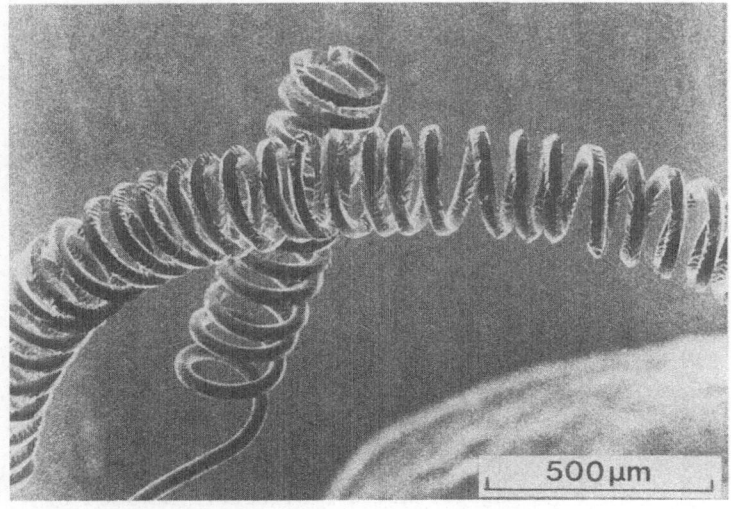

Fig. 13. Broken filament in bulb from rear lamp of parked car. No straightening, still very uniform spacing between turns. (37). Reproduced with permission of Canadian Society of Forensic Science.

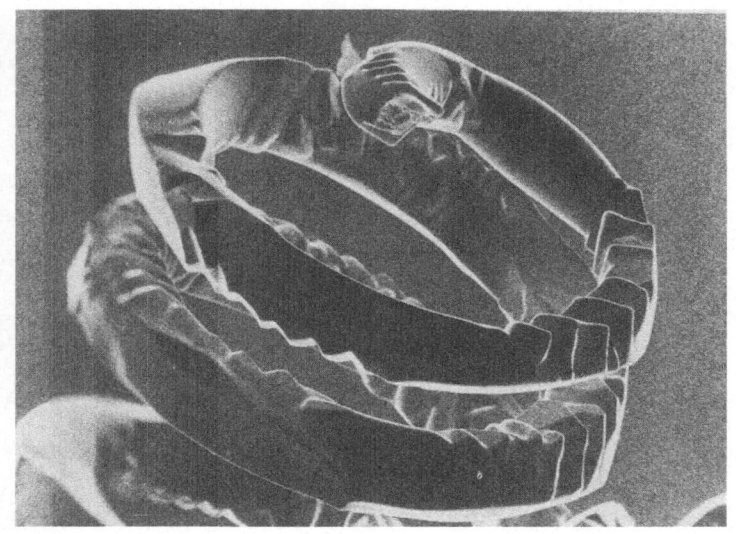

Fig. 14. Higher magnification of wire shown on fig. 13. Severe thermal etching and DC notching. (37). Reproduced with permission of Canadian Society of Forensic Science.

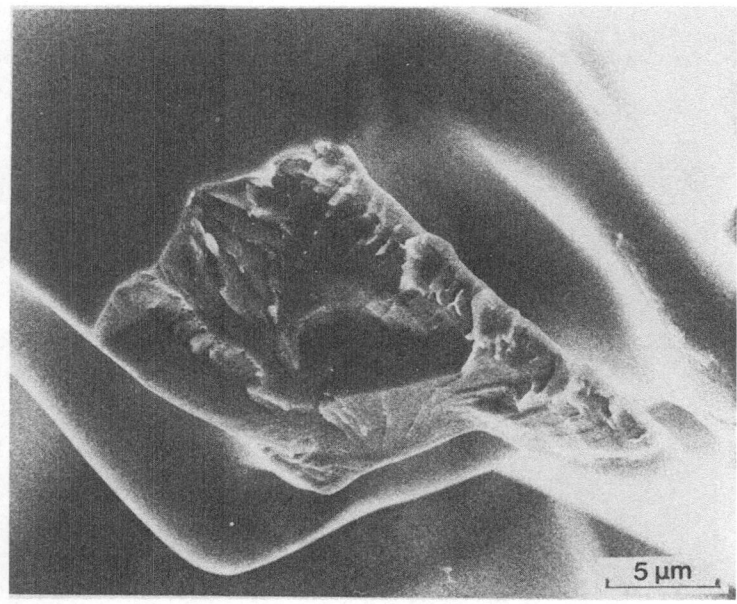

Fig. 15. High magnification of fracture surface of wire shown on fig. 14. Identification: brittle, transcrystalline, cold fracture. Extreme reduction in cross sectional area at fracture face. (37). Reproduced with permission of Canadian Society of Forensic Science.

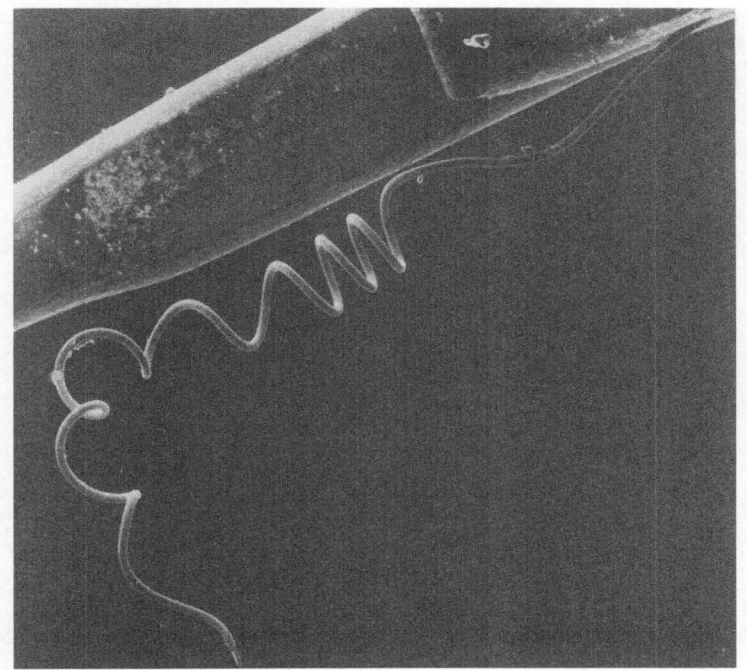

Fig. 16. Bulb from bicycle rear lamp. Terminal post with piece of filament. Coil straightened out. (37). Reproduced with permission of Canadian Society of Forensic Science.

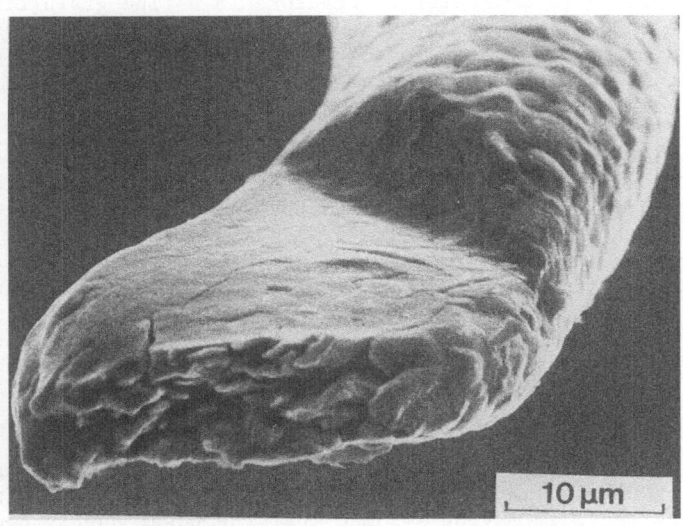

Fig. 17. High magnification of fracture surface at end of wire on fig. 15. Flattened out at end. Fibrous fracture surface. (37). Reproduced with permission of Canadian Society of Forensic Science.

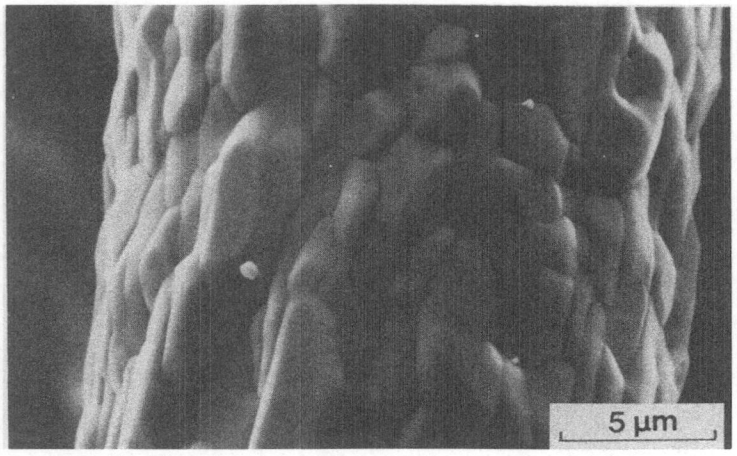

Fig. 18. Filament from bicycle bulb run for 132 hrs at
6 V. Thermal etching shows fine grain size, similar to
structure on micrograph fig. 3.

In the first case the coiled filament had been straightened out as
shown on fig. 16. At the fracture face the wire had been flattened out,
fig. 17, probably because a foreign object had hit the wire at the
accident and hammered it flat. The fracture face itself is fibrous,
pointing to the fact that the fracture was ductile and that the wire was
hot when the accident took place. No melting had occurred at the fracture
face, either because the power supply was cut off before the wire broke,
or because of the cooling effect of the foreign object that hit the filament.

Thus the existence of a fibrous fracture in a fine grain filament
(microstructure as on fig. 3) has been demonstrated. Whether a coarse
grain filament (microstructure as on fig. 2), with the power cut off, but
still at high temperature, would exhibit fibrous fracture as well, still
remains to be seen.

In a similar bulb run for 132 hrs at 6 V in a laboratory test the
filament had the surface structure shown on fig. 18. Thanks to thermal
etching the grain size can be seen. On comparison of fig. 17 with fig.
18 it is noted that both filaments are fine grained (as on fig. 3),
apparently due to a low operating temperature of this type of filaments.
The blurred outline of the surface of the filament from the accident is
thought to be caused by oxidation. The oxide layer is very thin, and a
plausible explanation might be that the power was cut off when the envelope
was crushed. In this way the filament would only get a small fraction of
a second at high temperature and in contact with the atmosphere, due to
the fast cooling of the thin wire.

In the second case it was very doubtful whether there was any
straightening of the coil at all. However, as it will be noted from
the following evidence it may be assumed that the power supply had been
cut off, so that the wire was cooling down when it was shocked. According
to fig. 5 the tensile strength of tungsten rises steeply as the temperature
drops, and the stresses generated in the filament by the impact might have
been too small to exceed the yield strength of the wire at the actual
temperature.

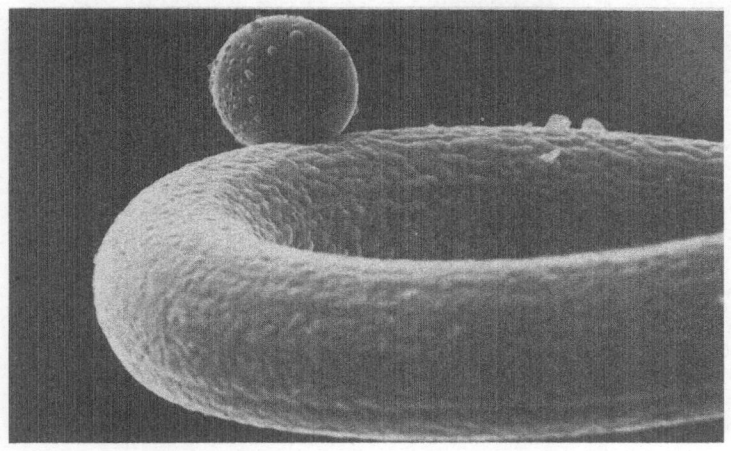

Fig. 19. Bulb from bicycle lamp. Filament with spheroidal
glass particle (identified by Si) that comes from a glass
splinter which melted on contact with the hot filament.
(39). Reproduced with permission of San Francisco Press,
Inc.

Spheroidal glass particles were found on the filament, one of them is
shown on fig. 19. On hitting the hot filament glass splinters will melt
and form perfect spheres, because the molten glass do not wet the surface
of the wire.

A copper particle had also hit the filament and melted on contact,
but this particle wetted the surface of the filament as one would expect
a molten metal (cf. fig. 20). A very probable source of this copper was
found to be one of the terminal posts that had been damaged. A chip had
been cut off and a part of this chip could quite well have hit the filament.
The surface of the terminal post consisted of pure copper, it was either a
pure copper wire or a Dumet wire.

The outlines of the fracture surface shown on fig. 20 are blurred by

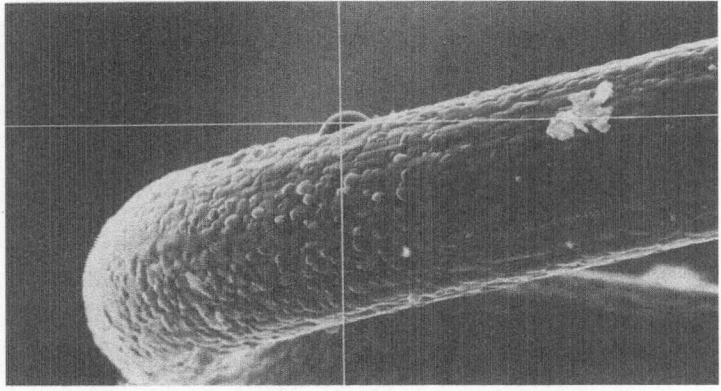

Fig. 20. Same bulb as in fig. 19. A particle with the
shape of a spheroidal segment on the filament. Identified as copper, coming from one of the terminal posts.

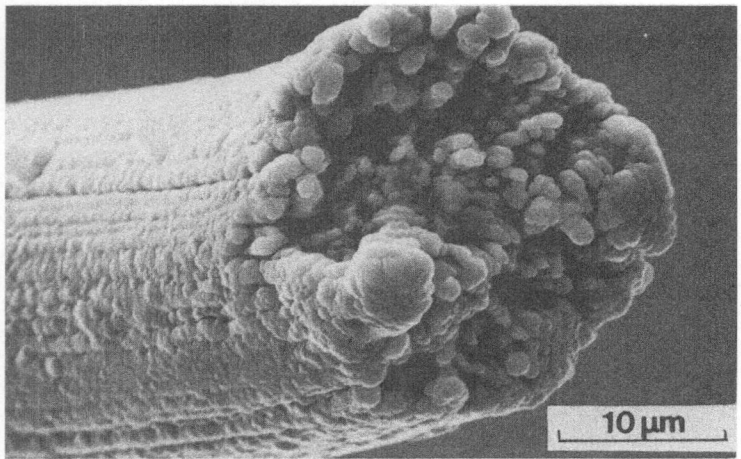

Fig. 21. Same bulb as on fig. 19. Fracture surface
blurred by oxidation, however, most likely fibrous
fracture as on fig. 17.

oxidation, however it is found to be very like the fibrous fracture surface shown on fig. 17.

   In both cases it was concluded that the rear lights of the bicycles
were lit when the accidents took place and that is based on the following
evidence: deformation of turns (only one of the filaments), fibrous
fracture, oxidation, melting of glass splinters and copper chips (only
one of the filaments).

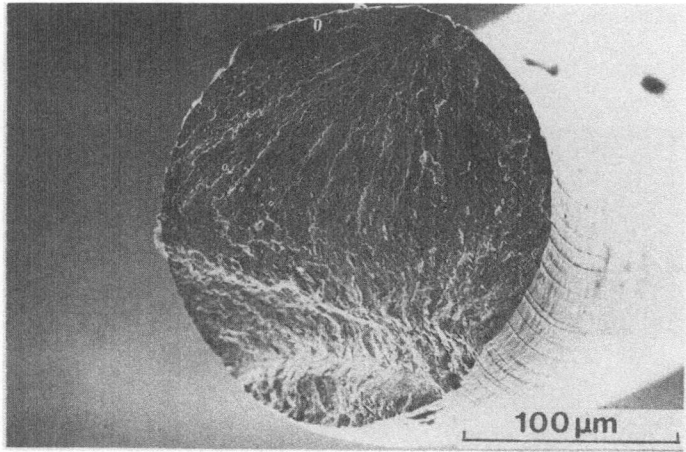

Fig. 22. Fracture at room temperature in new tungsten-
halogen bulb filament. Brittle and transgranular,
displaying a rugged surface as compared to the fracture
surfaces in aged wire.

Fracture at Room Temperature in New Tungsten-Halogen Bulb Filament.
A new H4 bulb was sectioned and the filaments removed. On stretching the
coils there was an insignificant plastic deformation before fracture took
place. The fracture was brittle and transgranular, displaying a rugged
fracture surface as compared to the fracture surfaces in aged wire; fig.
22. Furthermore, all fractures in the new wires were transgranular, in
opposition to fractures in wires that had been in service. In the latter
case both inter- and transgranular fractures were found. Working for a
long time at high temperatures apparently weakens the grain boundaries in
the filament wire.

Hot Fracture in Energized Filament. In an experimentally simple
spring set-up, aged H4 bulbs were energized and impacted without switching
off the connection to the power supply before the impact, and without
crushing the envelope.

Using a sufficiently strong spring the filament broke on impact with
the smooth, rounded fracture surface shown on fig. 23 resulting. Apparently
an arc was struck at the moment the fracture occurred leading to melting of
the tungsten at the fracture surface. According to the literature fractures
producing spheroidal surfaces and coil-to-coil short-circuits could be met
under these conditions as well. The type of defect met might be a function
of the direction and the duration of the impact and the magnitude of the
g's generated.

In an attempt to produce a hot fracture in H 4 bulbs with the power
supply switched off, the same set-up was used. The connection to the
power source was cut off fractions of a second before impact, but it was
impossible to produce the effect aimed at. The cooling rate of the filament on switching off is extremely steep (fig. 8) and consequently the
tensile strength increases steeply after switch-off (fig. 5), and that

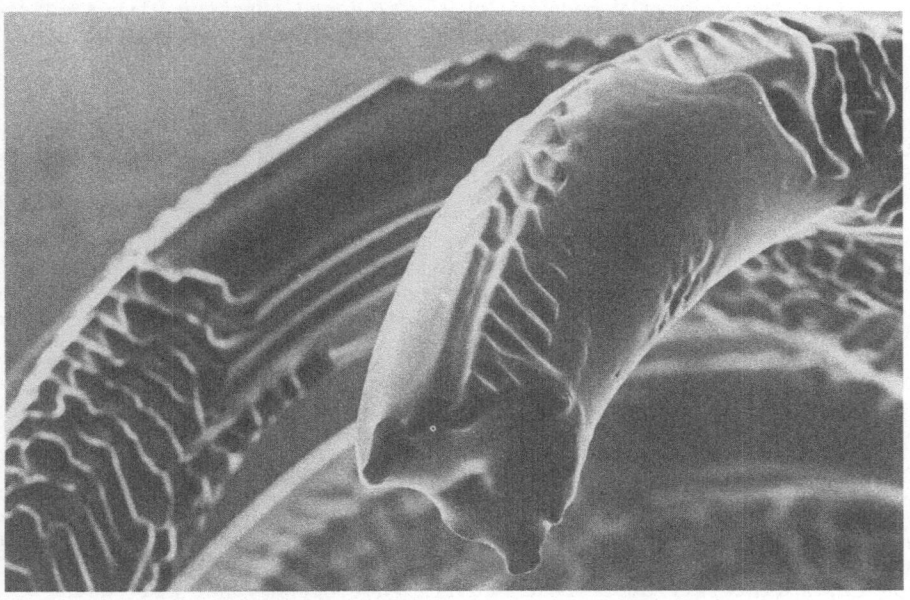

Fig. 23. Hot fracture in aged H4 bulb filament, impact
loaded with power on. Smooth, rounded fracture surface
apparently caused by arc formation at the moment of
fracture.

apparently to such a value that the wire cannot be broken with the spring system available.

The experiments are now being continued on an AVCO shock loading machine that produces well-defined pulses with a bell-shaped g versus time curve, which is very close to the experimental conditions used by (28, 29). Hot R2 and H4 bulbs aged for 120 hrs started to show deformation of coils at about 300 g, and had developed distinct deformation at 400 g, which is the maximum g tested so far. Apart from a single non-conclusive fracture, no fractures have been produced at the present state of testing.

SUMMARY AND CONCLUSIONS

1. Based on characteristic differences in behaviour of hot and cold filaments it is in most cases possible to ascertain whether a bulb or a sealed beam unit damaged at an accident was on or off when the accident took place.

2. A basic knowledge of incandescent light systems and of the effects of impact on bulbs are necessary conditions of obtaining optimum results of the investigations and to arrive at reliable conclusions.

3. Such knowledge has been collected from sources that are widely scattered in literature but when structured actually present a comprehensive amount of knowledge on all aspects of bulbs.

4. The many bulb types existing are described in standards and manufacturers' catalogues. The bulbs are properly marked to permit identification. The various types behave differently on impact or when shock loaded due to differences in design and mounting. Tungsten-halogen bulbs are mechanically very strong and are not easily broken. In contrast sealed beam units can be expected to be very fragile.

5. Filaments are generally made of 99.95 wt% doped tungsten with a coarse recrystallized structure of elongated finger-locked grains. A structure of fine grains has been observed in thin wire, it might be caused by a low operating temperature.

6. The tensile strength of tungsten decreases as temperature goes up. Recrystallized tungsten is brittle under about 400°C (752°F) and over that temperature it is ductile. This is one of the factors that makes it possible to distinguish between impact on a hot and a cold filament.

7. During operation, thermal etching, electromigration and the Soret effect cause roughening of the surface of filament wire. The cross sectional area can locally be severely reduced thus weakening thin wire considerably. A given stress that can cause fracture in an aged wire might be insufficient to break a new wire.

8. Tungsten oxidizes readily at elevated temperatures which is another factor making it possible to distinguish between hot and cold filaments provided the envelope is broken at the accident.

9. For other metallic parts than the filament, Cu, Ni, Ta, Fe, W, and Mo are used; in tungsten halogen bulbs, however, only W and Mo are used.

10. Different glasses are used for envelopes. For halogen bulbs quartz is used. Due to the softening temperature of about 1500°C (2732°F) of quartz, splinters from a broken quartz envelope may not melt when hitting a hot filament, as glass splinters would do.

11. Temperature transients arise when switching on or off and when pulsing. Heating is nearly instant, and the cooling depends on the dimensions of

the wire. Thick wire requires several seconds to reach the ductile-to-brittle transition temperature. The filament of a turn signal lamp of a car will remain over 1000°C (1832°F) when pulsing, and the filament will thus remain ductile as long as the lamp is activated.

12. The magnitude of the stresses generated in car bulbs at accidents is not known. The behaviour of filaments in car bulbs when subjected to well-defined shock curves have not been investigated. Such investigations have been made on subminiature bulbs for aircraft but direct comparison is not possible.

13. For the examination of bulbs after accidents the optical microscope and the Scanning Electron Microscope are complementary methods, the latter instrument especially being required for complicated cases and for thin filaments, filament fragments, and for fracture surface identification.

14. In cases and experimental work the author has identified following phenomena: spheroidal fracture on normal burn-out of filament; inter- and transcrystalline brittle fractures in cold filaments; smooth, rounded fracture due to arc formation in a filament fractured with power on; fibrous fracture in a filament with power off but still hot; straightening of coils in hot filaments; melting of glass splinters and copper particles that hit a hot filament, and finally oxidation of tungsten. Other phenomena can be identified according to the literature in some cases.

15. There are still unsolved problems which should be investigated further: Magnitude of stresses required to cause deformation of filaments and fracture in different bulbs. The possibility of distinguishing between various types of hot fracture (globular, smooth, fibrous) and to connect these observations to the state of filament at impact and the type of impact need to be studied.

ACKNOWLEDGMENTS

The author would like to express his thanks to Dr. Samarendra Basu of the New York State Police Crime Laboratories, Albany, N.Y. for the invitation and the encouragement to write this paper. The author is indebted to Canadian Aviation Safety Board, Ottawa, and to Bill Dean, Hamilton County Coroner's Lab, Cincinatti, for invaluable support with the providing of literature.

REFERENCES

1. C.H. Wecht, and J.W. Hicks, Forensic Science, ASTM Standardization News, p.24, March (1985).
2. Chapter 21, Lighting, in:"1985 SAE Handbook - Vol. 2", SAE, Warrendale (1985).
3. Kraftfahrzeug-Glühlampen, in: "DIN 72601", Deutsches Institut für Normung e.V., Berlin (1969-1983).
4. Licht für Fahrzeuge, Catalogue, Osram GmbH, Berlin (1982).
5. Miniature and Subminiature Lamp Catalog, General Electric (1981).
6. Miniature and Subminiature Lamps, Catalogue, Chicago Miniature.
7. IES Lighting Handbook, Illuminating Engineering Society, New York, latest edition.
8. Ib Ovesen, "Elektriske Lyskilder 3 - Lys og Belysning", Aschehoug Dansk Forlag, København (1967).
9. Metals Handbook, 9th ed., 3:325 (1980); 7:152, 629, 765 (1984). American Society for Metals, Metals Park.

10. S.W.H. Yih and C.T. Wang "Tungsten, Sources, Metallurgy, Properties and Applications", Plenum, N.Y. (1979).

11. S.T. Henderson and A.M. Marsden, "Lamps and Lighting", 2nd ed., Edward Arnold, London (1972).

12. Metals Handbook, 8th ed., 7:197. American Society for Metals, Metals Park (1972).

13. J.W. Pugh, L.H. Anna and D.T. Hurd, Properties of Tungsten-Rhenium Lamp Wire, Transactions of the ASM, 55:451 (1962).

14. F 73: Tungsten-Rhenium Alloy Wire for Electron Devices and Lamps, F 288: Tungsten Wire for Electron Devices and Lamps, "Annual Book of ASTM Standards", vol. 10.04, ASTM, Philadelphia (1985).

15. Metals Handbook, 9th ed., 2:816, American Society for Metals, Metals Park (1979).

16. Encyclopedia of Science & Technology, 5th ed., 14:161. McGraw-Hill, N.Y. (1982).

17. F.F. Schmidt and H.R. Ogden, "The Engineering Properties of Tungsten and Tungsten Alloys", DMIC Report 191, Batelle Memorial Institute, Columbus (1963).

18. F 10: Miniature Electron Tube Leads, F 29: Dumet Wire for Glass-to-Metal Seal Applications, Annual Book of ASTM Standards, vol. 10.04, ASTM, Philadelphia (1985).

19. D. Benicke, R. Goebel and U. Puchner, Das Temperatur- Zeit Verhalten von Glühwendeln aus Fahrzeuglampen, Verkehrsunfall und Fahrzeugtechnik, 23,12:317 (1985) and 24,1:9 (1986).

20. O. Schob, J.R. de Bie and W.G.A. Klomp, The Use of Television Equipment in Research on Incandescent Lamps, Lighting Research and Technology, 5,1:29 (1973).

21. D.A. Porter and K.E. Easterling, chapter 3 in "Phase Transformations in Metals and Alloys", Van Nostrand Reinhold Co. N.Y. (1981).

22. R.P. Johnson, Construction of Filament Surfaces, Physical Review, 54:459 (1938).

23. G.M. Neumann, W. Hirschwald and I.N. Stranski, Surface Structure of Tungsten Wires heated by Direct Current in a Vacuum in "Crystal Growth", H. Steffen Peiser, ed., Pergamon Press, London (1967).

24. S.C. Ackermann and M.A. Mortensen, Notching of Incandescent Lamp Filaments and its Significance. Presented at SAE Aircraft Lighting Sub-Committee A-20A Meeting, Los Angeles (1970).

25. S.C. Ackermann and M.A. Mortensen, Notching Limits Life of Very Thin Filaments, Automotive Engineering, 79,4:36 (1971).

26. N.P. Demas, Direct Current Life of Subminiature Vacuum Incandescent Lamps, Illuminating Engineering, april:267 (1971).

27. M.R. Lundgreen, Lightbulb Analysis as an Aid in Accident Investigation, Report Gen. Test 13/66, RCAF Materiel Laboratory, Ottawa (1966).

28. D. Hanchet, A Guide to Light Bulb Analysis in Support of Aircraft Accident Investigation, Transport Canada Report No. TP 6255E. Canadian Aviation Safety Board, Ottawa (1985).

29. T.W. Heaslip, M. Vermij and M.R. Poole, Advances in the Analysis of Aircraft Crash Impacted Light Bulbs. The International Society of Air Safety Investigators Forum, 16,4:56 (1983).

30. S. Baker and T. Lundquist, Lamp Examination for on or off in Traffic Accidents 1977 revised edition, Traffic Institute, Northwestern University (1977).

31. G. Kremmling and A. Schöntag, Zur Systematik der Untersuchung von Fahrzeuglampen nach Verkehrsunfällen, Archiv für Kriminologie, 128:1 (1961).

32. J. Mathyer, Evidence obtained from the Examination of Incandescent Electric Lamps, Particularly Lamps of Vehicles involved in Road Accidents, International Criminal Police Review, 29:2 and 29:34 (1975).

33. D.N. Dolan, Vehicle Lights and Their Use as Evidence, J.Forens.Sc. Soc., 11,2:69 (1971).

34. B.B. Coldwell and T.B. Melski, Motor Vehicle Lights as Evidence, Royal Canadian Mounted Police Gazette, May:2 (1961).

35. J.I. Thornton, G. Mitosinka and T.L. Hayes, Comparison of Tungsten Filaments by Means of the Scanning Electron Microscope, J.Forens. Sc.Soc., 11,3 (1971).

36. R. Goebel, Examination of Incandescent Bulbs of Motor Vehicles after Road Accidents, in "Scanning Electron Microscopy/1975", II:547 (1975).

37. K.Aa. Thorsen, Examination of Bulb Filaments by the Scanning Electron Microscope, Royal Canadian Mounted Police Gazette, 42,11:9 (1980) and Can.Soc.Forens.Sci.J., 14,2:55 (1981).

38. W. Kroll and M. Schlagenhauf, Rekonstruktion des Brennzustandes von Fahrzeuglampen bei Verkehrsunfällen, DEKRA-Fachschriftenreihe, 18:119 (1982).

39. K.Aa. Thorsen, Forensic Examination of Bulb Filaments by the SEM, p. 90, in Proceedings of the 43rd Annual Meeting of The Electron Microscopy Society of America, G.W. Bailey, ed. San Francisco Press, 1985.

40. A.L. Hagström and S. Söder, Light Filaments of Incandescent Lamps studied by Auger Electron Spectroscopy. Report 5, Dept. of Physics and Measurement Technology and Forensic Science Center, Linköping University, Linköping (1979).

ASBESTOS MEASUREMENTS IN WORKPLACES AND AMBIENT ATMOSPHERES

Eric J. Chatfield

Chatfield Technical Consulting Limited
2071 Dickson Road
Mississauga, Ontario, Canada L5B 1Y8

INTRODUCTION

The observation of respiratory disease among asbestos workers led to the promulgation of the Asbestos Industry Regulations in 1931 by the United Kingdom. This appears to be the first regulation specifically directed towards control of asbestos exposure in the workplace (Royal Commission on Asbestos, 1984). Because of the very long period between the first exposure of a worker to asbestos and the observation of adverse health effects, it has been difficult to establish airborne asbestos levels which, during a worker's lifetime, would not give rise to clinically-detectable effects. Control limits which were initially established have been periodically revised downwards, in response to refinements in the epidemiological data as they became available. In the United States, the control limit initially set at the equivalent of 30 fibers/milliliter (fibers/mL) has been reduced to 2 fibers/mL, and proposals have been made to reduce this still further. One proposal, if adopted, would reduce the control limit to 0.1 fiber/mL. Throughout the rest of the world, the control limits have been set at levels ranging generally from 1 fiber/mL up to 5 fibers/mL.

During the period since control limits were first recommended, the method of measurement for airborne asbestos fibers in workplaces has also been refined. From initial methods based on the use of a midget impinger and a total particle count, the method now in use is based on collecting the airborne dust on a membrane filter and then counting the fibers by optical microscopy.

In the 1970's, the occupational diseases which had been observed in asbestos workers gave rise to concern that the general population was perhaps at some risk also, because asbestos-containing products had been used so widely in modern society. It was considered that most of the population was exposed to airborne asbestos to some extent, and on the basis of some extrapolations of the epidemiological data from asbestos workers, an alarming forecast for asbestos-related disease in the general population could be made. In response to this type of forecast, measurements of asbestos fiber concentrations in ambient air were made, in order to assess to what extent the general population were exposed. In the 1970's, however, the measurement techniques for ambient air measurement were somewhat primitive, and along with the reported measurements there

---

*Present Address: 2071 Dickson Rd., Mississauga, Ontario, Canada L5B 148.

were also improvements in the analytical methods used. The method best suited for this measurement was the subject of considerable debate, and eventually collection of airborne dust on a membrane filter and examination of the collected sample by transmission electron microscopy (TEM) emerged as the method of choice. Even now, the methods by which an air sample filter can be transferred to an electron microscope specimen are still under discussion. During the 1960's and 1970's, changes in building construction techniques resulted in a requirement for structural steel in high-rise buildings to be insulated against heat from any major fire which might occur. One of the methods for achieving the required fire-rating was to spray an asbestos-containing material on to the structure. Steam and hot water pipes were also insulated using various materials, ranging from corrugated chrysotile asbestos millboard to mixtures of asbestos with cellulose or Portland cement. In addition, asbestos-containing products were used as decorative or acoustical coatings for ceilings and walls. The presence of large quantities of asbestos in general purpose buildings, particularly in those cases where the asbestos-containing materials were exposed in the air handling system or located so that they could be contacted by the occupants, led to concern in many countries that occupants of such buildings may be exposed to airborne fiber emissions from these materials. Some measurements of airborne asbestos levels made in buildings which contain asbestos insulation have shown elevated fiber concentrations, and other measurements have not. Nevertheless, public pressure has caused asbestos removal programs to be established, particularly in school and government buildings, and it is likely that this asbestos removal program will continue, probably with the inclusion of many privately-owned buildings.

Naturally-occurring asbestos is found not only in asbestos mines. Many mineral deposits, such as talc, vermiculite, iron ore and nickel ore, may contain sometimes large proportions of acicular particles of asbestos-forming mineral varieties associated in the ore body with the material being mined. When the ore is crushed or ground, the asbestos-forming minerals may yield fibers, and many of the other mineral varieties not currently under legislative control may yield acicular particles. These acicular particles often exceed the minimum 3:1 length to width aspect ratio commonly used to define a fiber, and because of the complexity in discrimination of these particles from actual asbestos fibers, they are treated by some government agencies as asbestos fibers. Unfortunately, the amphibole minerals are among the most common minerals on the surface of the earth (Ross, 1978), and a large proportion of the particles produced by crushing these minerals will have aspect ratios exceeding 3:1, although the amphibole variety may be mineralogically different from those amphiboles which are known to form asbestos. Where a mineral product is contaminated by any of these types of minerals, and when any dust is generated, there is the potential for exposure of workers and, in the case of vermiculite or talc, end users of the material may also be exposed to airborne fibers.

In all of these situations there is a requirement for measurement of asbestos fiber concentrations in air. In some cases, such as the asbestos industry, very simple methods have been used which incorporate no identification of the fibers found. In other situations, it is necessary to use the most sophisticated analytical tools, such as a modern transmission electron microscope with energy dispersive X-ray analysis capability, in order to identify each fiber. This is particularly the case for "non-asbestos" mineral products, where a whole range of acicular mineral particles may be present, most of which are not considered mineralogically to be asbestos, neither are these fibers legislated as asbestos.

150

MINERALOGICAL ASPECTS, PROPERTIES AND DEFINITIONS OF ASBESTOS

Some of the analytical techniques used in the identification of asbestos are very sophisticated, and in order to appreciate the analytical requirements, it is necessary to understand how asbestos is defined, and the chemical properties of the various types of asbestos.

Asbestos is not a single mineral. The term is used for several fibrous silicate minerals which have found commercial application. Figure 1 shows the common varieties of asbestos, although there are a number of other minerals which, had they been available in commercial amounts, might have been included. The majority of the asbestos used commercially is chrysotile, the fibrous variety of serpentine, the nominal composition of which is $Mg_3(Si_2O_5)(OH)_4$. Other elements may also be present in the structure: the silicon may be partially substituted by iron, nickel, manganese or cobalt, and in material from some sources particles of magnetite are present embedded in the fibers. Sodium and chlorine may also be present. Chrysotile has an unusual crystallographic structure in which a dimensional misfit between parallel sheets of silicon tetrahedra and magnesium octahedra gives rise to a curved cylindrical structure in the form of a scroll. Other varieties of serpentine accommodate the dimensional misfit in different ways, leading to two other types of structure for antigorite and lizardite. A detailed discussion of the structures of the serpentine minerals has been published by Wicks (1979).

Chrysotile is readily attacked by acids (Speil and Leineweber, 1969), even by the very weak organic acids such as acetic acid, and all or part of the magnesium may be leached from the structure. Chrysotile is also degraded by heat. In the range of $600°$ to $780°C$, dehydroxylation occurs, which is an irreversible process. At higher temperatures of $800°$ to $850°C$ the anhydride decomposes to yield forsterite and silica (Hodgson, 1979). Finely divided chrysotile samples have been shown to dehydroxylate at much lower temperatures (Berry, 1971).

The other types of asbestos are the fibrous varieties of some amphibole minerals. Amphiboles are iron-magnesium silicates of general composition $A_{2-3}B_5(Si,Al)_8O_{22}(OH)_2$, where magnesium, ferrous iron, calcium, sodium or potassium can occupy site A, and magnesium, aluminum and iron can occupy site B. In some minerals these elements can be partially substituted by manganese, chromium, lithium, lead, titanium or zinc. The amphiboles, both asbestos and non-asbestos varieties, all have similar

Chrysotile	$Mg_3(Si_2O_5)(OH)_4$
Amosite	$(Fe,Mg)_7(Si_8O_{22})(OH)_2$
Crocidolite	$Na_2FeII_3FeIII_2(Si_8O_{22})(OH)_2$
Anthophyllite	$(Mg,Fe)_7(Si_8O_{22})(OH)_2$
Tremolite	$Ca_2Mg_5(Si_8O_{22})(OH)_2$
Actinolite	$Ca_2(Mg,Fe)_5(Si_8O_{22})(OH)_2$

Fig. 1. Varieties of Asbestos

crystal structures, characterized by a cross-linked double chain of silicon tetrahedra with a silicon:oxygen ratio of 4:11 (Whittaker, 1979). The crystal structure of most amphiboles is monoclinic, but a few are orthorhombic. The common varieties of amphibole asbestos are crocidolite (blue asbestos), and amosite (brown asbestos). The term "amosite" was derived from a trade name, and would be described mineralogically as cummingtonite-grunerite asbestos. Anthophyllite is now rarely used commercially. The wide range of composition possible within the amphiboles has given rise to a considerable nomenclature problem. Many of the reported varieties with specific names were simply minor compositional variations of an existing well-known variety. This problem has been considered by a committee of the International Mineralogical Association (1978), and a precise classification system has been developed. As a result of the new classification system, about 200 previously used mineral names were recommended for extinction.

The amphiboles are generally quite resistant to attack by acids. The reaction on heating is complex, yielding decomposition products including pyroxenes, cristobalite and iron oxide. Some amphiboles, such as cummingtonite-grunerite and crocidolite, react to form oxy-amphiboles when heated in an oxidizing atmosphere (Hodgson, 1979).

It should be recognized that after either chemical or thermal decomposition of asbestos, the products may retain the fibrous morphology of the original minerals. The fibers which remain consist of finely-divided mixtures of the reaction products, and can therefore no longer be classified as asbestos, but their mean composition will be similar to that of the original asbestos mineral.

RELEVANT MEASUREMENT PARAMETERS

Assuming that a sample of asbestos fibers representative of those inhaled by an individual has been obtained, and is available for evaluation by any of the analytical techniques, what parameters should be measured? It is obviously important that the measurement parameters correspond to those aspects of the material which are thought to be responsible for the biological effects. There are many different properties which could be measured, the simplest of which are the total mass or the number of fibers per unit volume of air. The dimensions of each fiber could also be measured. A further refinement would be to measure the composition of each fiber, which may be important if a mixture of different fibers is encountered or if different varieties of fiber were known to display different levels of toxicity. Surface properties of the fibers, or their ability to persist in the lung tissue for long periods, may also be relevant to the measurement requirement.

Currently, it is thought that the biological effects are related to the number concentration of fibers, on the basis of the hypothesis that these effects originate from the interaction of a fiber and a single cell. It is also known that fibers longer than about 5 μm, and of diameters less than 3 μm, are the most effective for inducing biological effects in laboratory animals (Stanton and Layard, 1978). The medical community has so far been unable to discount the possibility that health effects may result from inhalation of fibers shorter than 5 μm. Pott (1978) considers that there may be a continuous scale of toxicity with fiber length, and that although the shorter fibers may display a lower carcinogenic effect than the long fibers, their effect is not actually zero. If this is the case, a high concentration of short fibers may be just as effective as a low concentration of long fibers, so far as the carcinogenic potential is concerned. Given the fact that in airborne dispersions, the short fibers

are usually very much in the majority, Pott's hypothesis would indicate
that the carcinogenic effect of a particular fiber dispersion may not be so
strongly weighted towards the long fibers as Stanton and Layard's work
would suggest.

For workplaces where asbestos is in use, legislation in most countries
is based on measurement of the numerical concentration of airborne fibers,
using phase contrast optical microscopy (PCM). This is a technique which
will be discussed in detail, but essentially it allows measurement of only
a small proportion of the fiber size distribution. Figure 2 shows a fiber
length distribution, and the shaded area represents the proportion of
fibers which are detected when the PCM technique is used. Only those
fibers longer than 5 μm are included in the fiber count, and the optical
resolution limits the minimum detectable diameter to about 0.2 μm.

Measurements made in the ambient atmosphere, or in the atmospheres of
buildings, are more difficult because the assumption cannot be made that
all fibers detected are asbestos fibers, as is done when the PCM technique
is used. Experience has shown that only a minority of the fibers found
will actually be asbestos fibers, and in the majority of cases those
asbestos fibers which are found will be too small in diameter to be
detected by the PCM technique. The fiber definition adopted for use in
occupational fiber counts clearly specifies limits on the fiber dimensions
to be included in the measurement, and the resolution of the microscope
further limits the size of fibers reported. It has been accepted that a
fiber count based on this restricted range of fiber dimensions represents
only an index of exposure, which can then be compared with some legislated
standard. No generally accepted standards of this type exist for the

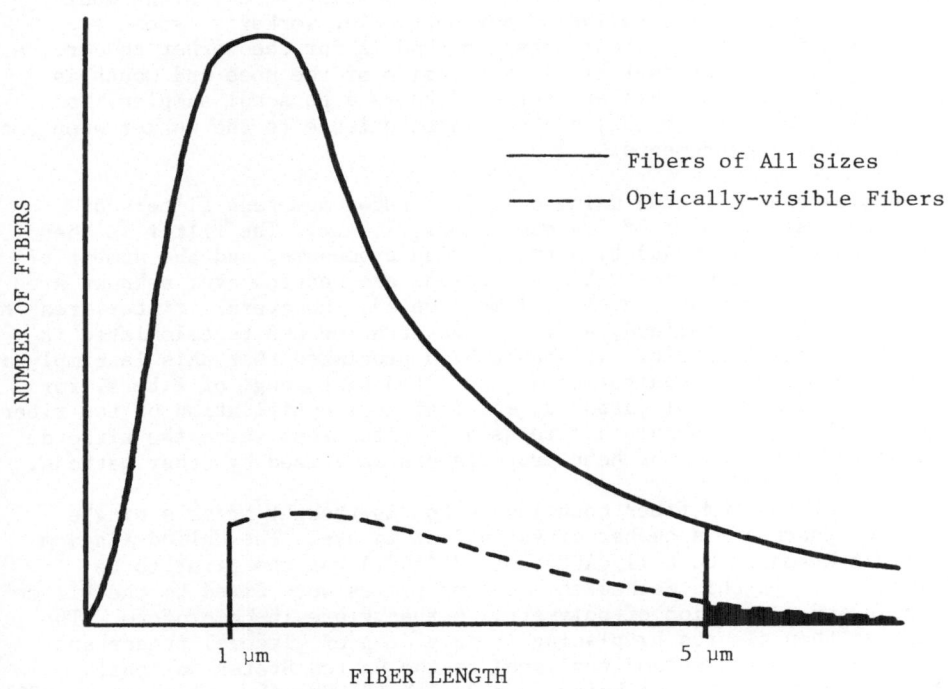

Fig. 2. Fiber Length Distribution of Typical Airborne
Asbestos Dispersion

ambient atmosphere or for building atmospheres. The medical community has not been able to agree on the minimum size of fiber which the analyst should attempt to measure and identify. Accordingly, it has not been possible to place any limits on the dimensions of fibers that should be incorporated into an environmental fiber count. Until further guidance is forthcoming, it is desirable that analytical methods for determination of mineral fibers in ambient air should be capable of measurement and characterization of all fibers within the dimensional ranges of the Pott hypothesis. It has been suggested that the best approach to fiber exposure measurement in interpretation of environmental data would be to use an integrated carcinogenicity index, produced by weighting each fiber with a carcinogenicity factor from Pott's model (Chatfield, 1984). This approach would not require any arbitrary fiber definition, and would take proper account of the short fibers in accordance with the biological data as currently understood. No guidance is available as to the minimum numerical fiber concentrations which are of concern, but this, together with the minimum fiber dimensions, is to some extent defined by the instrumental limitations. Essentially, in the absence of other guidelines, the procedure used for ambient air samples is to identify and measure the dimensions of all asbestos fibers within the detectable range of the instrument.

MEASUREMENT OF AIRBORNE FIBER CONCENTRATIONS IN WORKPLACES

When airborne particulate samples are collected in workplace atmospheres, the intention is to provide a control over the airborne asbestos fiber concentrations to which workers are exposed. Many workers perform a range of different tasks during the work day, involving frequent movement from one location in the plant to another, and it is therefore unlikely that a fixed air sampling station placed at a point in the plant would collect a sample representative of any particular worker's exposure. Because of this problem, the standard method is for the worker to carry a portable sampler, such that air in the region of the nose and mouth is sampled. The portable sampler, referred to as a personal sampler, is battery-operated, and is only a minor inconvenience to the worker when used for occasional measurements.

Samples are collected using cellulose ester membrane filters of nominal pore size 0.8 μm or, in some cases, 1.2 μm. The filter is then cleared (made transparent) by a preparation procedure, and the number of fibers which have the prescribed dimensions are counted over a known area of the filter. Knowing the air volume sampled, the overall filter area and the area of filter examined, a fiber concentration can be calculated in terms of fibers/mL of air. It should be appreciated that this is simply an inexpensive method of monitoring a restricted size range of fibers, for routine exposure control purposes, and that no identification of the fibers is made. The method should not be used in situations where the airborne particulate matter has not been properly characterized by other methods.

Although the PCM fiber counting method is thought of as a single technique, there are a number of variations in use. The United Kingdom Asbestosis Research Council (ARC) method (1971) was the first to be published, and in this method the dust particles were fixed to the filter surface using a solution of polymethyl methacrylate in chloroform. The filter was then cleared by placing it on a drop of glycerol triacetate (Triacetin). In the method published by the United States National Institute for Occupational Safety and Health (NIOSH) (Leidel et al., 1979), usually referred to as Method P&CAM 239, the mounting and clearing medium used is a mixture of dimethyl phthalate and diethyl oxalate, with the addition of cellulose ester material to increase its viscosity. Methods used in other jurisdictions were generally similar, until the emergence of the Asbestos International Association (AIA) method RTM1 (1979). This method specifies the filter mounting and clearing technique originally developed by the Australian Department of Health. The filter structure, originally with a sponge texture, is first collapsed to a thin film of plastic by exposing it to a jet of acetone vapor. A drop of Triacetin is then applied to the filter, followed by a cover slip. The AIA method was designed to minimize all of the factors which had been identified as sources of variability, and all preparation parameters were standardized as far as possible. Fiber counting rules were also specified in much greater detail in the AIA method than had previously been the case. A Draft International Standard method, based very largely on the AIA method, has been issued by the International Organization for Standardization (ISO) (1981) for comment, and eventually this should emerge as an internationally agreed method. The AIA method has also been adopted by the Commission of European Communities as the European Reference Method. One alternative technique for clearing and mounting of the membrane filter has been described by LeGuen and Galvin (1981), in which the immersion medium is a mixture of dimethyl formamide, acetic acid and water. More recently, NIOSH Method 7400 (NIOSH, 1984) has been published, which adopts many of the features of the AIA method and the ISO Draft International Standard.

Fiber counts made by the PCM method are subject to a great deal of variability. This variability prompted a number of investigations to establish the cause, and although some instrumental factors were found which could seriously affect the results, most of the variability now appears to be a consequence of operator subjectivity. In early work, Beckett and Attfield (1974) showed that incorrect alignment of the phase rings in the optical microscope could result in serious reductions in the fiber counts obtained. In the same study, they observed that, when counting chrysotile fibers, use of a graticule defining a field area of 5 - 20% of the total field of view resulted in 300% higher fiber counts than those obtained using full-field counting. When additional time was spent in rigorous examination of each field of view, this discrepancy was greatly reduced, indicating that the cause was subjectivity on the part of the operator, rather than of instrumental origin. The Walton/Beckett graticule is the only one designed specifically for fiber counting, and in both the AIA and the ISO methods it has been specified. The detection limit for chrysotile fibers on membrane filter samples has also been the subject of some investigation. LeGuen and co-workers (1980) established that chrysotile fibers having diameters less than about $0.2 - 0.25$ μm were not visible in the phase contrast microscope. The effect of the clearing methods on the minimum detectable fiber diameter has been studied by Teichert (1980) and Ogden (1981), and it appears that only minor differences in fiber counts occur when different clearing methods are used. Both the acetone-Triacetin and the dimethyl formamide-acetic acid clearing techniques produce slides which are permanent and therefore available for later evaluation. The NIOSH Method P&CAM 239 produces slides which must be counted within 48 hours and are useless after that time. The clearing and mounting medium used in P&CAM 239 is also stable for only a limited storage period, after which acicular crystals develop which can be mistaken for fibers.

The PCM fiber counting method is an analytical technique which differs in principle from those used to measure many other hazardous agents, in that changes of fiber definition or fiber counting criteria affect not only the precision but also the reported value. For PCM fiber counting there are currently no absolute standard samples which can be used to certify the performance of an operator. Thus any change in the definition of a fiber, the treatment of fiber aggregates or fibers which are in contact with particles, or any changes in the sample preparation which change the visibility of fibers, will have the effect of changing the actual result which is reported. This absence of any absolute standards has other serious consequences when attempts are made to compare the fiber counting performance of different operators.

Measurement of fiber concentrations by the PCM method is subject to both systematic and random errors. Systematic errors associated with sample collection include those of flowrate determination, sampling period and less controllable random aspects such as non-representative sampling and sample contamination. Errors introduced during analysis of the filter include those incurred in determination of the active filter area and the analysis area, non-uniformity of the collected material on the surface of the filter, and those introduced by variability in the filter mounting. These errors, however, are usually much smaller than those introduced by subjective effects such as variation in visual acuity, and the state of fatigue of the operator. When the performance of experienced counters is compared, even for counting of ideally-loaded samples, the results generally display coefficients of variation of 0.4 or greater (Chatfield, 1982). The absence of reference standards and the high variability of PCM fiber counts place serious limitations on the minimum fiber concentration which can be detected, and also the ability to demonstrate compliance with established control limits, particularly when the control limit is lowered to a value close to the analytical detection limit.

The minimum fiber concentration which can be detected has been the subject of considerable discussion. Theoretically, the detection limit can be reduced indefinitely, simply by increasing the volume of air filtered, increasing the area of filter examined, and perhaps also reducing the active area of the collection filter. All of these methods to improve the detection limit have been embodied in the NIOSH Method 7400, which claims a lower limit of reliable quantitation of 0.02 fiber/mL. Undoubtedly, in an atmosphere such as that in a semiconductor processing clean room, this could theoretically be achieved, but the method is intended for use in some very dusty occupational environments where fibers are a minority particle species and there are other practical limitations.

A fundamental limitation on the detection limit, of course, is imposed by the fact that non-zero fiber counts are often reported from blank filters. When a membrane is cleared by any of the variety of techniques available, some artifacts are created, even when an unused filter is prepared. Many of these artifacts are interpreted by fiber counters as fibers, giving rise to a non-zero fiber count. Background counts of up to 22 fibers in 100 Walton/Beckett graticule areas have been reported by Crawford (1986), a count which corresponds to a fiber concentration of 0.1 fiber/mL if it is assumed that 240 liters of air had been sampled using a 37 mm diameter filter cassette. This volume of air could be increased in some situations, theoretically reducing the detection limit, but in many industrial situations this would not be possible. Firstly, at higher air volumes the filter may become overloaded by other particulate material; and secondly, the presence of large concentrations of particulate material is likely to give rise to an increased fiber count for reasons related to operator subjectivity.

There is also the question of natural fiber background. In ambient air, by far the most common fiber species found is gypsum, not necessarily collected from the air, but often generated actually on the filter surface by the reaction of atmospheric sulfur dioxide with particles of calcite or dolomite. At some times of the year, depending on locality, there are also other fibers present in the ambient air which originate from plant or insect life. Workplaces in which asbestos is in use are usually not well isolated from the ambient air, and so some contribution to the fiber count

156

can be expected from the natural background, quite apart from the presence of other fibers such as textile fibers from the clothing of the workers. In measurements made in southern Ontario, PCM fiber counts as high as 0.5 fiber/mL have been found in suburban locations where parallel samples analyzed by transmission electron microscopy showed that no asbestos fibers were present (Chatfield, 1983c). These practical limitations of PCM fiber counting were recognized by the ISO working group ISO/TC146/SC2/WG5, and the Draft Method produced by this working group contained the specific statement that the detection limit lies between 0.1 and 0.5 fiber/mL, depending on the particular situation, and that this detection limit cannot be reduced simply by increasing the volume of air filtered per unit area of the filter.

Although there is no evidence that the variability is different for different sampling conditions which yield the same numerical density of fibers on the filter when fibers are the majority particle species present, this is not the case when fibers are a minority species in a variable proportion with other particulate. Jones and Johnstone (1985) have reported that a low density of fibers, less than 100 fibers/mm^2, is subject to bias, substantially due to interpretation of artifacts as fibers. Moreover, there is a tendency to produce high fiber counts at low fiber densities, although the tendency is not consistent. These authors conclude that the evaluation of low density samples is unreliable. The combination of low fiber density and a high particulate loading will further degrade the reproducibility of such fiber counts.

In view of the difficulties encountered when attempts are made to measure asbestos fiber concentrations lower than about 0.5 fiber/mL, when such measurements are required it is obviously necessary to supplement the PCM determinations using an analytical method which is capable of discriminating particles of different species, is not subject to preparation artifacts, is not working at the limit of resolution and is less influenced by the subjective judgement of the operator. The dispersion staining technique (McCrone and Stewart, 1974) is a useful technique for identification of particles using the optical microscope, but this cannot be used unless the particle is isolated and available for immersion in a series of different refractive index media. Moreover, the resolution of the technique is not adequate for use in identifying fibers less than about 1 μm in diameter. For measurements in workplace atmospheres the scanning electron microscope (SEM) with energy dispersive X-ray analysis (EDXA) equipment is suitable for this purpose. In workplace atmospheres, the primary interest is in those particles with aspect ratios of 3:1 or greater which fall into the optically-visible size range, namely those longer than 5 μm and greater than 0.2 μm in diameter. At a magnification of 5000, a fiber of these minimum dimensions appears on the SEM display as an image 1 mm in diameter and 25 mm long, and the resolution of modern instruments is adequate to produce a satisfactory image with sufficient contrast. The SEM can be used to examine each fiber which falls into the specified size range, and the fiber can be classified on the basis of its composition using the EDXA spectrum. In an asbestos workplace, the materials in use are known, generally large fibers are present, and there is a very limited compositional range of fibers which may be found. In this situation, classification on the basis of composition alone is usually quite adequate.

Ideally, if SEM examination is the primary analytical method, air sampling would be conducted using a Nuclepore polycarbonate capillary pore membrane filter. The preparation of an SEM specimen from this type of filter is direct, consisting simply of vacuum coating the collected particles and filter surface with a thin film of gold or carbon. However, the analytical requirement is usually to refine data already obtained from

157

a PCM fiber count, and it is therefore not possible to re-sample using this type of filter. The cellulose ester filter, used for PCM determinations, has a great deal of surface structure, the fibers are at various angles on it, and in some cases are partly embedded in the filter structure. The principal difficulty facing the microscopist who wishes to examine a sample collected on a cellulose ester filter is how to prepare a specimen suitable for SEM examination.

A method has been described (Ortiz and Isom, 1974) in which the structure of the cellulose ester membrane filter is collapsed into a thin film of plastic with little surface detail. This technique, which will be discussed later in connection with transmission electron microscopy (TEM), permits preparation of a sample suitable for SEM examination if the larger fibers are the only ones of interest, which for this application is the case. The preparation technique is simple to perform. Using adhesive tape at the edges of the filter, the filter is held in position on a glass slide, and it is then exposed in a container to acetone vapor. The filter is observed to go transparent as the sponge structure collapses. The surface of the plastic film, with the particulate attached to it, can then be vacuum coated using either gold or carbon. Unfortunately, some of even the larger fibers become partially engulfed in the plastic film, and imaging of these in the SEM is difficult. The conventional cellulose ester filter, prepared in this manner, has another disadvantage in that the polymer is not stable under bombardment by the electron beam, and the examination in the SEM must be performed very carefully with minimum beam currents. The vinyl co-polymer Gelman DM800 and DM450 filters do not suffer from this instability, and they can be prepared for SEM examination in a similar way. LeGuen et al. (1980) have described a method for preparation of these filters, which is a more controllable procedure than the acetone vapor method. In this technique, a drop of a mixture of cyclohexanone and 1-4 dioxane is placed on a microscope slide, and the filter is placed on the liquid, after which the slide and filter are heated in an oven for a short time. The filter structure collapses very gently, and there appears to be much less disturbance of the particulate deposit than there is when the acetone vapor procedure is used. The problem of partial engulfment of fibers is solved by treatment of the plastic surface in a low temperature oxygen plasma for a few minutes. This treatment etches away some of the plastic film, leaving the fibers supported on mesas of plastic and available for analysis. Care must be taken that the oxygen plasma treatment is not carried out for too long a period, however, because this will result in the loss of particulate. The oxygen plasma treatment may also destroy organic fibers which are present in the sample, thus the proportion of asbestos fibers relative to total fibers cannot be determined. The SEM specimen produced by this technique is much more stable under electron beam bombardment. The etching procedure can also be applied to collapsed cellulose ester filters to improve the visibility of fibers.

The SEM techniques described above provide the facility to investigate further those filters, from asbestos workplace monitoring, which yield PCM fiber counts close to or exceeding control limits. The SEM techniques can be used to exclude fibers which are not of the designated minerals in use, and also to discriminate between the asbestos varieties if more than one kind is in use. The use of mixed fiber varieties is quite common, particularly in the manufacture of asbestos-cement products. The imposition by some jurisdictions of different control limits for different varieties of asbestos gives rise to a monitoring problem when PCM is the only technique in use. In the SEM, chrysotile, crocidolite and amosite can be readily distinguished from each other, and from fibers of many other materials which may conceivably be present.

Another situation in which SEM examination of the airborne dust would be advisable is in the monitoring of workplaces where asbestos is known to be present as a contaminant in other materials being processed. This is particularly appropriate when the non–asbestos material yields acicular particles as it is being processed. When only particles within the optically–visible size range are being monitored, and when there is sufficient compositional difference between the asbestos and the other material, then SEM examination of the samples to discriminate between the two particle species is appropriate and useful. However, in some cases, there is insufficient compositional difference between the two species, and the choice is either to assume all acicular particles are asbestos, or to examine the sample using an analytical electron microscope (AEM), which may be capable of discriminating the asbestos fibers. The application of the AEM is described later, in the discussion on asbestos measurements in ambient atmospheres.

MEASUREMENT OF ASBESTOS FIBER CONCENTRATIONS IN AMBIENT AIR

In the general environment, asbestos fibers are usually only a minority of the total fiber concentration. Fibers, in turn, represent only a small proportion of the total numerical concentration of particles. The types of fibers which may be present in an ambient air sample may originate from a wide variety of sources, and some fibers, although they may not be asbestos, may have compositions which are very close to that of one of the known asbestos varieties. The diameters of asbestos fibers found in the general environment are usually very small, below the minimum diameter which can be detected by optical microscopy. All of these features of the measurement in ambient air are different from those which relate to measurements in workplace atmospheres, in which most of the fibers are known to be asbestos and many have diameters such that they can be detected by optical microscopy. Therefore, assessments of exposure to airborne asbestos in the general environment and in building atmospheres cannot be obtained using the PCM methods developed for monitoring of asbestos in workplaces.

The precise techniques which should be used for measurement of the fibers in ambient atmospheres are still under discussion. Much of the controversy is because the medical and biological community have not been able to specify unequivocally what should be measured. Measurement of exposure to asbestos, and to other fibers, is different from that of practically every other potentially hazardous pollutant, in that a simple measurement of one concentration parameter is insufficient. In the case of asbestos fibers, it is known that the biological response is a function of fiber species, fiber diameter and fiber length (Stanton and Layard, 1978; Pott, 1978). The measurement problem is also compounded by the fact that fiber species other than the commercially–used varieties of asbestos have been shown to be carcinogenic. The fibrous zeolite erionite is now thought to induce mesothelioma (Baris, 1980), and animal studies have shown that the fibrous clay mineral attapulgite can produce mesothelioma after intra-peritoneal injection (Pott et al., 1976). A complete analysis should therefore include a statement of fiber identity, length and diameter for each fiber which has dimensions within the ranges thought to be of biological significance. The analytical method must also be designed to permit measurement of a specific minimum fiber number concentration, and little guidance is available as to the minimum concentration considered to be of concern by the medical community. A working group of the International Organization for Standardization (ISO, 1982) recently agreed that for measurements in ambient air a detection limit of 1 fiber/liter should be the target for urban atmospheres, and 0.1 fiber/liter in remote rural locations. In view of the relative total airborne particulate loadings, these represent realistic detection limits.

Figure 2, discussed earlier, shows the relationship between the total fiber distribution and the optical fiber count. Before an environmental fiber count is attempted, there should be a clear idea as to what size ranges are going to be incorporated into the fiber count. If the microscopist is instructed to count all fibers, depending on the type of sample the required information may not be forthcoming. As is shown in Figure 2, the fiber size distribution is usually strongly weighted towards short fibers, and since the usual fiber counting rule applied is to terminate the count after 100 fibers have been counted, there will be very few of the longer fibers in the data. If the primary interest, as it is in some jurisdictions, is in the fibers longer than 5 μm, the fiber count will not be satisfactory because of statistical limitations when only a few fibers in the category of interest have been counted. The microscopist also needs to know whether the interest is in all fibers, or just asbestos fibers. If the intention is to determine the asbestos fiber concentration, a count of 100 fibers which only includes a few asbestos fibers does not represent the most efficient use of the fiber counting labor. It is therefore important for the intention of the analysis to be clearly specified. When a direct specimen preparation is to be used, the analytical requirements should be known before the air sampling is conducted. Although the use of an indirect specimen preparation technique permits some flexibility for sample collection, it is important to ensure that a sufficient volume of air is filtered, and that the analytical requirements are specified before the specimens are prepared. For any analysis, adjustments in the procedures can be made, so that the required information is obtained, while at the same time allowing the most suitable specimen to be prepared for examination in the TEM.

Characteristics of SEM and TEM Instrumental Methods

For measurements of asbestos fiber concentrations in the ambient air, or in the atmospheres of buildings, the applicability of the SEM and TEM has been the subject of extensive discussion. Most analysts would agree that the method of choice, assuming that cost and time were of no concern, is the TEM. However, because of the requirement for inexpensive analyses in building insulation programs, and the relative availability of the two types of analysis, a number of studies have been made using the SEM.

For the simplest SEM technique, the air sampling is conducted using a Nuclepore polycarbonate capillary pore membrane filter. The filter is then gold coated by vacuum evaporation, and examined directly in the SEM. Recently, there have been some variations on this method. The collection filter may be a conventional cellulose ester or a vinyl co-polymer filter, and the preparation technique must then include the collapsing and etching steps discussed earlier in connection with workplace measurements. A methodology has been published (Asbestos International Association, 1984), in which the Nuclepore filter which is used for sample collection is gold coated prior to sample collection. The sample filter is then etched in a low temperature oxygen plasma, which results in oxidation of all organic material on the surface, leaving only inorganic particles and fibers. The purpose of the gold coating in this method is to protect the filter surface from the action of the oxygen plasma, and to permit SEM examination without further vacuum coating after sample collection. This also permits a more accurate determination of the fiber diameters, since the fibers are not coated with the gold film.

Although the modern SEM has a nominal resolution of about 5 nm, which should be adequate for detection of the smallest fibrils of chrysotile, (usually about 30 - 40 nm in diameter), the practical resolution is somewhat degraded, depending on the type of sample under examination. In order to obtain the optimum resolution, it is necessary to operate the instrument using low beam currents, and this requirement leads to images which contain a substantial amount of electrical noise. The electrical noise places limits on the ability to image small fibers in real time at realistic field scan rates. Test micrographs, illustrating instrumental resolution, are obtained by using long exposure times which permit the noise component to be averaged, and the effect of the noise is then not noticable in the final micrograph. Figure 3 shows an SEM micrograph of chrysotile fibers on a polycarbonate filter, obtained in this manner. However, in order to perform a routine fiber count, it is necessary to work with a real time image so that fibers which are located can be classified by obtaining an EDXA spectrum. Under these conditions, a suitable image can only be obtained by use of higher beam currents, which degrades the resolution. The higher beam currents are also required if EDXA spectra are to be obtained in reasonable periods of time. It has been demonstrated that under typical operating conditions, chrysotile asbestos fibers below about 0.2 μm in diameter are not readily visible in the real time SEM image. It has also been demonstrated that EDXA spectra, of sufficient quality for fiber classification, require unacceptably long acquisition times (Steel and Small, 1985). These authors also concluded that the rastered viewing mechanism is extremely tiring for the operator, unless TV scan rates are used, and this further degrades the visibility limit. These fiber visibility limitations of the SEM technique were recognized in the published AIA methodology RTM2 (Asbestos International Association, 1984), and therefore the method specifies inclusion of only those fibers exceeding 0.2 μm in diameter, and in the length range 2.5 μm to 100 μm. This is claimed, with some justification, to represent a large proportion of the envelope of fiber sizes included in the Pott (1978) hypothesis. A similar methodology has been published in draft as the German Reference Method (VDI, 1983).

There are, however, fiber identification problems associated with any SEM-based technique. A major problem is presented by the limitation of spatial resolution when EDXA spectra are acquired (Chatfield, 1979). The electron beam placed on a small fiber scatters within the substrate and the

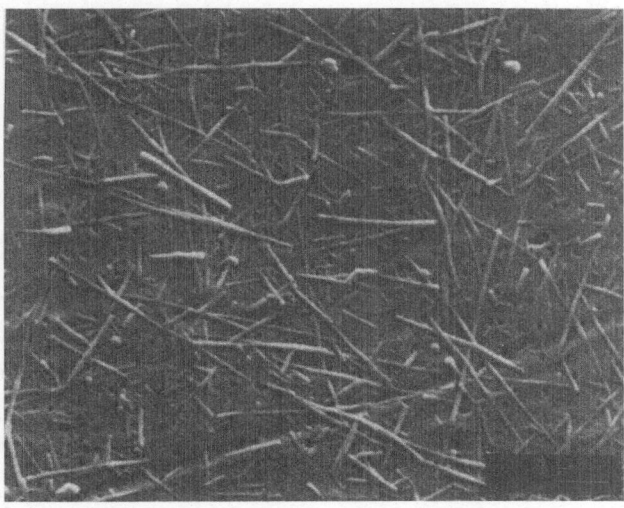

Fig. 3. Scanning Electron Micrograph of Chrysotile Asbestos
         Fibers on Gold Coated Nuclepore Filter

net effect is that other particulate close to the fiber being analyzed may contribute to the EDXA spectrum obtained. When samples contain large amounts of other particulate, as is usually the case with ambient air samples, this can lead to considerable ambiguities of fiber classification. The use of gold coating, either applied before sampling as described in these two published methodologies, or applied after sample collection, further degrades the ability to identify fibers reliably (Chatfield, 1983b).

The modern TEM has a resolution of about 0.2 nm, which is readily attainable under routine operation. The image obtained, because of the high operating voltage, permits observation of any internal structure in the fiber, and this is of great assistance in the identification of chrysotile. When the instrument is interfaced with an EDXA system, it offers the most powerful method of particle characterization which is available, and this instrument configuration is usually referred to as an analytical electron microscope (AEM). Unlike the SEM, the TEM does not suffer to the same extent from the identification limitations created by the presence of other particulate close to the fiber being analyzed, because the TEM specimen is very thin and the electron beam passes through without significant scattering. This is not to say, however, that there is no contribution to an EDXA spectrum from sources other than the particle being analyzed. There is generally a contribution from the support grid, in particular, and if the particle being analyzed is not sufficiently separated from other material on the specimen some contribution will be observed from this other material. A great deal of care must be taken to ensure that the specimen orientation in the TEM and the type of support grid in use are appropriately chosen. If the presence or absence of sodium in the analysis is critical to the interpretation, as it is for the discrimination of crocidolite from amosite, the use of copper grids will interfere with reliable observation of the sodium X-ray peak. The copper Lα peak is partially overlapped with that of sodium Kα, and since detection of X-rays in this energy range is already very inefficient, this overlap does not allow sodium to be measured accurately. When gold support grids are used, there are no peaks from the support grids which interfere with the detection of sodium, but the gold M peaks interfere with the detection of sulfur.

Sample Collection and Preparation for TEM Examination

Two types of membrane filter are available for collection of air samples to be analyzed for asbestos fibers. These two filter types are the Nuclepore polycarbonate capillary pore membrane filter and the cellulose ester or vinyl co-polymer membrane filter. All the techniques used for preparation of TEM specimens from collection filters require that the particulate on the filter be transferred quantitatively to a thin carbon film, supported on a fine metal mesh. The preparation of the TEM specimens can be direct, in that the particles on the filter surface are transferred to the TEM specimen such that they are retained in their relative positions to each other. Alternatively, preparation of the TEM specimens can be by an indirect technique in which the particles on the air sampling filter are removed from that filter and deposited on another filter from which TEM specimens are prepared by the direct technique. The particles can be removed from the sampling filter surface by ashing a fraction of the filter; the residual ash is then re-dispersed in water and this dispersion is filtered to provide an analytical filter which can be prepared by the direct method. As an alternative to ashing, the particulate can be washed off the sampling filter by agitation in water which is then filtered to provide an analytical filter. The indirect techniques have the advantage that organic material can be oxidized and water-soluble material is removed, both of which improve the analytical detection limit. In

162

addition, suitable filter loading does not need to be determined in the field; the particulate can be diluted or concentrated on the TEM specimen. The TEM specimens can be prepared so that they are suitable either for optimum determination of the overall fiber size distribution and concentration, or for determination of the concentration of fibers exceeding a specified length. The ability to adjust the particulate loading of the final TEM specimen allows much more flexibility in the conditions selected for the air sampling. If the direct preparation methods are used, the particulate loading cannot be adjusted after the sample has been collected. The indirect preparation methods have the disadvantage that there is some modification of the fiber size distribution as measured on the microscope. It is not yet clear as to whether this modification is actually due to physical breakage of fibers when they are re-dispersed in water, or whether the different fiber counts are a consequence of the different appearance of the particulate on the TEM specimen.

The preparation method in most common use in North America is referred to as the direct carbon-coated Nuclepore procedure. The particulate is collected using a 0.4 μm or a 0.2 μm pore diameter polycarbonate filter. After sample collection, an area of the filter is carbon coated in a vacuum evaporator. The carbon film replicates the surface of the filter, and also traps all of the particles on its surface. Portions of the carbon-coated filter are placed on 200 mesh metal grids, and then the original filter is dissolved away in a Jaffe washer (Jaffe, 1948) with chloroform as the solvent, to produce the TEM specimen. The direct preparation of a sample collected using a cellulose ester filter has already been described in the discussion on the use of the SEM to evaluate workplace atmosphere samples. For TEM examination, the surface of the collapsed, or collapsed and plasma etched, sample collection filter is carbon coated to replicate the filter surface and trap all of the particles on the surface. Portions of the coated filter are placed on metal grids and the filter material is dissolved away. The method for direct preparation of TEM specimens from cellulose ester (Millipore) filters reported by Ortiz and Isom (1974) was tested by Chatfield et al. (1983) to determine the efficiency with which asbestos fibers were transferred to the TEM specimen. This was done by preparing parallel Nuclepore and Millipore filters from fiber dispersions of known concentration. A transfer efficiency could be calculated from the fiber counts, expressed as a function of fiber length. The transfer losses were found to be a function of both filter pore size and fiber length. These data are shown in Figure 4, in which it can be seen that for even 0.45 μm pore size filters, significant losses of the longer fibers occur, and nearly all of the short fibers can fail to transfer to the TEM specimen. Direct preparation of samples collected on Millipore filters has been reported by Burdett and Rood (1982), who used a filter collapsing technique based on dimethyl formamide and glacial acetic acid, and also incorporated a low temperature oxygen plasma etching step. The collapsed and etched filter is then processed in exactly the same way as for the Nuclepore procedure, except that acetone must be used as the solvent. This new procedure appears to offer a good alternative to the direct preparation of a Nuclepore filter. There is, however, some difficulty in specifying the operating conditions for the plasma etching step, and complete loss of fibers has been observed when this treatment was allowed to proceed too far. Nevertheless, the procedure seems to be comparable with the carbon-coated Nuclepore procedure. The performance of the direct preparation of a cellulose ester filter when an etching step is included was compared with that of the Nuclepore method, using two different aerosols. A single fibril aerosol was generated by spraying a methanol dispersion of short chrysotile fibers from an atomizer, and sampling the aerosol using parallel Nuclepore and Millipore filters. A highly aggregated aerosol was generated using the vibrating bed aerosol

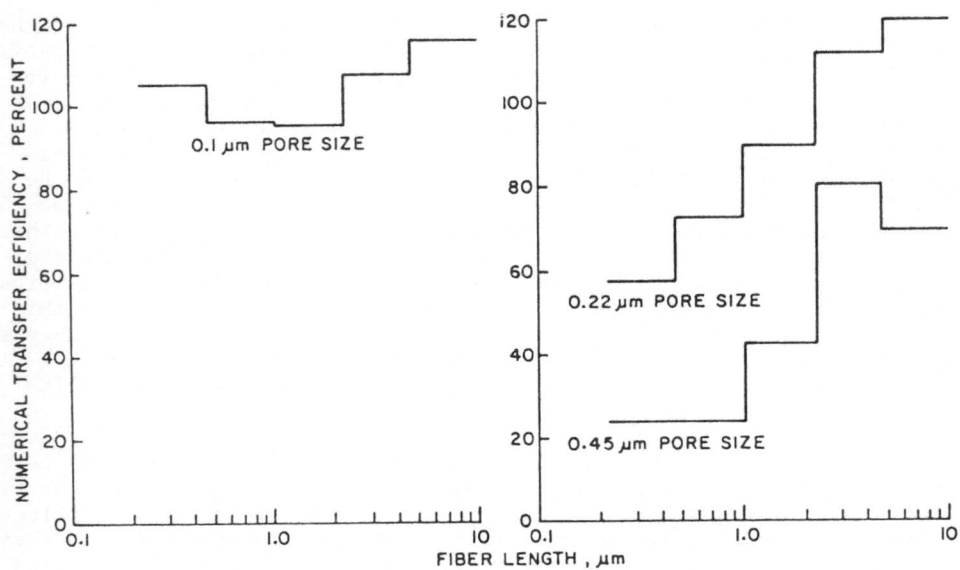

Fig. 4. Ortiz and Isom Collapsed Membrane Filter Preparation
Technique: Transfer Efficiencies for Chrysotile.

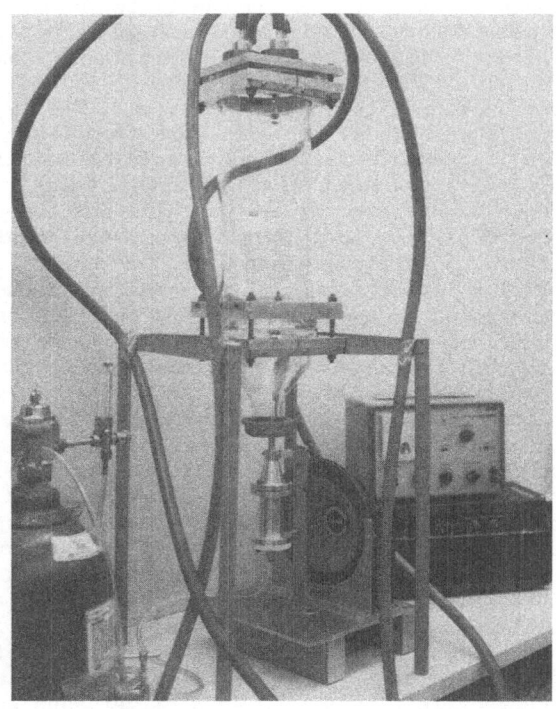

Fig. 5. Vibrating Bed Aerosol Generator

Table 1. Comparison of Direct Nuclepore and Millipore TEM Specimen
Preparation Techniques for Airborne Chrysotile

(Numbers of Fibers Counted)

Fiber Length Range μm	Single Fibril Aerosol Generator		Vibrating Bed Aerosol Generator	
	0.2 μm Pore Size Nuclepore	0.8 μm Pore Size Millipore	0.2 μm Pore Size Nuclepore	0.8 μm Pore Size Millipore
<0.5	53	66	18	25
0.5 – 1.0	111	74	57	73
1.0 – 2.5	48	29	75	68
2.5 – 5.0	4	4	54	65
>5.0	1	0	22	34
Total Fibers	217	173	226	265

generator shown in Figure 5 which is similar to that developed by Spurny et al. (1976), and this aerosol was sampled in a similar manner. The performance of preparation methods of this type is thought to vary with the character of the aerosol, and so these two extremes were used in the evaluation. The results are shown in Table 1. It can be seen that there is no reason to believe that the techniques produce different results, and those differences which exist are within the bounds of the fiber counting statistics.

If an indirect preparation method is intended, the sample can be collected on either a Nuclepore or a conventional cellulose ester filter. A known fraction of the filter is then ashed in a low temperature plasma asher, a procedure which takes some hours to complete. The residual ash is then re-dispersed, using an ultrasonic probe, and a known proportion of the liquid dispersion is filtered to yield an analytical filter for preparation by one of the direct techniques described above. TEM specimens can also be prepared by an indirect method in which the sample is collected on a Nuclepore filter, the deposit is washed from the surface of the sampling filter and is re-deposited by filtration to yield an analytical filter with a more suitable loading for a direct preparation.

The methods based on the direct preparation of a Nuclepore filter have been compromised to some extent by the existence of some surface contamination of unused filters by asbestos fibers. Moreover, the current materials available have often been difficult to dissolve completely in chloroform. It is most important that the filter be completely removed from the TEM specimen, because even a thin film of residual plastic can make the sample unsatisfactory for analysis. Until filters are available with no surface contamination, the contamination question can only be controlled by examination of filters from specific lot numbers to locate filters which are of acceptably low contamination level. The difficulty of dissolution is solved by incorporating an additional dissolution step, using a condensation washer with chloroform as the solvent, into the procedure. When using the direct preparation technique of Burdett and Rood (1982), there does not appear to be difficulty in achieving satisfactory blank results on unused filters. The methods based on indirect preparation also are compromised by asbestos fiber contamination of the air sampling filters. When either type of air sampling filter is ashed, asbestos both on the surface of the unused filter and within the filter material itself will be redispersed along with the collected airborne particulate and is indistinguishable from fibers originating from air sampling. When the particulate deposit is washed from the collection filter, asbestos fiber contamination on the surface of the filter will also be included in the transfer, but contamination within the collection filter material will not be transferred to the analytical filter.

When the direct analytical methods are in use, the target for the analytical sensitivity, that fiber concentration corresponding to one fiber detected in the analysis, must be specified in order that appropriate sampling conditions can be used. Assuming that an air volume of about 5 cubic meters is sampled using a 37 mm diameter disposable filter cassette, and fibers on twenty 200 mesh grid openings are counted, an analytical sensitivity of about 0.001 fiber/mL will be obtained. If the fibers are assumed to be distributed on the sample according to a Poisson distribution, this analytical sensitivity leads to a detection limit of about 0.004 fiber/mL, this being the upper 95% confidence limit for zero fibers detected. The detection limit can only be lowered by filtering a greater volume of air during sample collection, or by examining a greater number of grid openings on the TEM specimens. The indirect methods, as indicated, allow concentration of the sample, or dilution, and provided that the total particulate concentration is not the controlling factor, the preparation can be adjusted to yield any desired analytical sensitivity.

Indirect preparation methods have often been criticized on the grounds that they modify the size distribution by causing breakdown of the fibers, and therefore lead to much increased fiber counts. At the same time, the indirect technique was seen by some analysts as the only one by which any realistic measurement could be made in ambient air, because of the large amount of organic material found in airborne particulate. Because it was thought that the fiber breakdown phenomenon rendered the fiber counts meaningless, it was common to express the results of such analyses in terms of mass concentration. Until recently, no comparisons of the results obtained by direct and indirect preparation techniques had been reported, except in terms of total fiber concentrations. Some studies were recently completed (Chatfield, 1986) in which the comparison between direct and indirect preparation techniques was made for a range of fiber lengths, in order to determine whether the large increase in fiber numbers obtained by the indirect technique was related to fiber length. The studies were made using an aerosol of short single chrysotile fibrils and a highly aggregated aerosol of chrysotile produced by the vibrating bed aerosol generator. For each aerosol, parallel Nuclepore and Millipore filters were collected and analyzed. The Millipore filters were ashed in a low temperature plasma asher, the residual ash was dispersed in double-distilled water by use of an ultrasonic probe, and the dispersion was filtered through a 0.1 μm pore diameter Nuclepore filter. This filter was then prepared by the direct method. The Nuclepore sample filters were prepared simply by hand shaking a fraction of the filter in double-distilled water, after which the dispersion was prepared by the direct Nuclepore procedure. No ultrasonic treatment was used in the preparation from the Nuclepore sampling filters. Specimens were also prepared by the direct carbon-coating technique from each Nuclepore sample filter, and by the direct collapsing and etching technique from each Millipore sample filter. Tables 2 and 3 show the results. In Table 2, as might be expected, it can be seen that there is no

166

Table 2. Comparison of Direct and Indirect TEM Preparation
Techniques for Chrysotile Single Fibril Aerosol

(Numbers of Fibers Counted)

Length Range μm	Washed Nuclepore Filter		Ashed Millipore Filter	
	Direct Preparation	Indirect Preparation	Direct Preparation	Indirect Preparation
<0.5	53	53	66	85
0.5 – 1.0	111	85	74	94
1.0 – 2.5	48	31	29	27
2.5 – 5.0	4	1	4	5
>5.0	1	0	0	0
Total Fibers	217	170	173	211

Table 3. Comparison of Direct and Indirect TEM Preparation
Techniques for Aggregated Chrysotile Aerosol

(Numbers of Fibers Counted)

Length Range, μm	Washed Nuclepore Filter		Ashed Millipore Filter	
	Direct Preparation	Indirect Preparation	Direct Preparation	Indirect Preparation
<0.5	18	319	25	281
0.5 – 1.0	57	485	73	412
1.0 – 2.5	75	307	68	270
2.5 – 5.0	54	89	65	89
>5.0	22	33	34	24
Total Fibers	226	1233	265	1076
Fibers >2.5	76	122	99	113

statistically significant difference between the two preparation techniques. For single fibrils of chrysotile which cannot easily be broken down further, the two procedures yield essentially identical results for all of the fiber length ranges. In Table 3, it can be seen that the results from the aggregated chrysotile aerosol are very different, and there is about a 4 or 5 fold increase in the total number of fibers reported by the indirect method over the values obtained by the direct method. The increase also is apparently independent of the use of ultrasonic treatment, since this increase is also found for the indirect preparation of Nuclepore filters which were simply hand-shaken in water. The increase in the number of fibers seems to be confined to those fibers shorter than about 2.5 μm, and if only fibers longer than this are considered, the fiber numbers reported by the two techniques are statistically similar. This is a very useful observation, because for those applications in which only fibers longer than 5 μm are considered, it may be possible to use indirect specimen preparation methods without compromising the results. More investigation of the effect is required, but it does seem that early criticism of the indirect methods may not be completely justified. The observation that the increase in fiber numbers, produced when the indirect preparation method is used, is confined to the very short fibers, leads to the question whether this increase is due to actual physical disintegration of fibers or is simply a disaggregation phenomenon which makes the shorter fibers more readily seen and/or identified in the TEM.

In discussions related to the features of the direct and indirect preparation methods, it is often stated that the direct preparation yields a TEM specimen which is representative of the aerosol as it exists in the air. This is not necessarily the case; even the sample collection is a sample modifying step, in the sense that an aerosol which existed in 3 dimensions has been confined to 2 dimensions. This has the effect of allowing more obscuration to occur. Where the total particulate loading of the filter is low, this effect is clearly minimal, but at high particulate loadings this can be a very serious effect, and fiber counts may be much lower than is really the case. Fiber counts made in very dusty environments, such as a subway, are seriously affected by this obscuration effect. Table 4 shows the results of measurements made in the Toronto Subway System. Air samples were collected on Nuclepore filters, and initially analyzed by the direct method. Later, portions of the same filters were washed in double-distilled water, and the detached particulate was ashed to remove organic particulate. After it had been re-dispersed in water, the residual ash was then prepared by the direct Nuclepore method for TEM examination. In Table 4 it can be seen that some samples showed a large increase in total fiber count when they were re-analyzed using the indirect preparation, whereas others showed much less of an increase. The increase to be expected would depend not only on the amount of total particulate but also on the amount of asbestos in the sample; if there is little asbestos in the sample, the direct and indirect analyses should both produce results close to the detection limit. Where there is a larger amount of asbestos in the sample, obscured in the direct preparation, an increase in the fiber concentration should be observed when the indirect preparation is used. Table 4 shows that this appears to be the case, and this conclusion is supported by the fact that the mass concentrations also generally show an increase, with one exception. The fiber length data for two of these samples are shown in Table 5, in which it can be seen that two apparently similar samples according to the direct preparation are clearly different when the indirect preparation was used. As was the case in the laboratory experiments discussed earlier, the majority of the increase in fiber counts was in the fiber length range below about 2.5 μm. The observation of more long fibers in the case of the indirect preparation is probably a consequence of the obscuration effect.

168

Table 4. Toronto Subway: Comparison of Results Obtained by Direct and Indirect Analytical Methods

CHRYSOTILE FIBER CONCENTRATION

Sample	Direct Preparation				Indirect Preparation			
	Mean Fiber Concentration Fibers/mL	Estimated Mass Concentration ng/m^3	Number of Fibers Counted All Lengths	Longer Than 5 µm	Mean Fiber Concentration Fibers/mL	Estimated Mass Concentration ng/m^3	Number of Fibers Counted All Lengths	Longer Than 5 µm
LA39	0.14	14	20	0	2.6	47	99	0
LA47	0.064	0.14	9	0	0.23	15	30	0
SH43	0.072	0.33	8	0	1.3	35	156	2
SH51	0.11	51	15	2	3.0	190	109	2
LA23	0.016	0.17	3	0	0.11	2.5	23	1
LA41	0.013	0.015	2	0	0.17	11	24	0
LA49	0.018	0.043	3	0	0.039	0.24	6	0
QP45	0.025	19	4	0	0.12	1.2	17	0

Table 5. Toronto Subway: Fiber Size Distributions of Samples
Prepared by Direct and Indirect Methods

(Number of Fibers Counted)

Fiber Length Range, μm	Sample SH43		Sample LA49	
	Direct Preparation	Indirect Preparation	Direct Preparation	Indirect Preparation
0.50 – 0.73	4	70	2	4
0.73 – 1.08	1	56	1	1
1.08 – 1.58	1	14	0	1
1.58 – 2.32	0	9	0	0
2.32 – 3.41	2	3	0	0
3.41 – 5.00	0	2	0	0
5.00 – 7.34	0	2	0	0
7.34 –10.77	0	0	0	0

The obscuration effect has not been investigated completely; more work
is required to characterize the effect in both real world, and laboratory
samples. It is not an easy experiment to design, because the obscuration
effect is not caused simply by the addition of two or more aerosols.
Simulation of the effect by successive sampling of asbestos fibers and
another aerosol on the same filter would probably yield misleading
results. Moreover, any calculations based on this experimental design
would also not be useful. The mixture of particle species which may be
encountered in a subway system almost certainly interact with each other in
the air, and aggregates of the mixtures, including the asbestos fibers,
have already formed before they are collected on the filter during air
sampling.

Identification of Asbestos Fibers in the TEM

Identification of asbestos fibers is achieved in the TEM by a
combination of morphology, selected area electron diffraction (SAED) and
energy dispersive X-ray analysis (EDXA). The extent to which fibers must
be identified depends on how much is known about the particular location in
which samples have been collected, and what use is to made of the fiber
counting data. Where the material in use is known, simple morphological or
qualitative analysis may be adequate; for example, in a building atmosphere
in which final clearance is sought after removal of chrysotile insulation,
morphological classification would often be satisfactory and could be
achieved by relatively inexpensive analyses. In contrast, if an
identification is likely to be the subject of a serious challenge in court,
the most complete and unequivocal identification should be carried out.
This kind of analysis requires recording of the SAED pattern and
quantitative analysis of the EDXA spectrum, and in the case of fibers which
might be classified as amphibole asbestos, zone-axis SAED analysis is
required. The latter procedure is very time-consuming, and may require
several hours of work for each fiber.

In order to derive quantitative analyses of particles from the EDXA
spectra, it is necessary to calibrate the instrument. The use of
"fingerprint" methods, in which EDXA spectra are obtained from the
commercial varieties of asbestos, and the spectra of all fibers are simply
compared with these, is a somewhat primitive approach which is inadequate
because there are compositional variations between fibers of the same

variety, and from one deposit to another. It cannot therefore be assumed that a particular mineral will always yield precisely the same EDXA spectrum, and the range of compositional variation is difficult to interpret in terms of the EDXA spectrum unless all possible varieties have been examined. Moreover, the SAED spectra obtained are specific to the particular instrument, and cannot be reproduced if the X-ray detector, in particular, is changed. The characteristics of each detector are different, and although the spectra from two different detectors are superficially similar, there are usually differences in the relative heights of the peaks from different elements. The inclusion of additional reference standards other than those of the common asbestos varieties increases the flexibility of the approach, but a large library of reference spectra is then accumulated which is not transferrable to another instrument. The author has published (Chatfield and Dillon, 1983; Chatfield, 1983a) a quantitative EDXA technique based on the work of Cliff and Lorimer (1975), with a modification for the effect of fiber diameter. Because mass absorption and fluorescence matrix effects are small for most of the range of particle diameters involved, the ratio of the X-ray peak area to the mass concentration is constant for each element. There is, however, some variation for the larger particle diameters (Small et al., 1979). This approach permits the system to be calibrated against only a few reference minerals, so that the EDXA spectra of unknown fibers can be converted to quantitative analyses. The main difficulty of calibration is in the selection of suitable standards. Minerals are not often homogeneous materials which have the nominal stoichiometry, and properly-characterized samples of suitable silicate compounds are not available in the form of fibers with the diameter ranges required. The approach used was to divide a selected small single crystal of the mineral; one half was properly embedded and polished for an accurate microprobe analysis, and the other half was crushed and an aqueous dispersion of the powder used to prepare TEM specimens. The mineral standards used were riebeckite, chrysotile, halloysite, phlogopite, wollastonite and bustamite. This range of minerals provided calibration data for the elements sodium, magnesium, aluminum, potassium, calcium, manganese and iron, expressed as element to silicon ratios. This calibration procedure was performed for groups of particles in a range of diameter classifications. The technique is capable of some refinement, particularly for analyses involving sodium, but has been found to yield satisfactory results. These analyses can then be compared with a library of composition data for all relevant mineral varieties. The method is transferrable to other instruments, simply by carrying out the calibration procedure against the reference standards. The library of minerals contains published data only, and is not specific to the characteristics of any particular instrument.

In some cases, the classification of a fiber may depend on a precise analysis of sodium, which is one of the more difficult elements to analyze accurately by EDXA. Some refinement of the calibration procedure is still required in this area.

The identification of chrysotile was at one time considered to be relatively simple, since the morphology was quite characteristic and the SAED pattern also had some distinguishing features. This is still the case when the absence of certain other minerals can be specified. However, attapulgite may in some cases cause confusion if the SAED pattern is not examined carefully, or if the EDXA spectrum is not interpreted quantitatively. Some varieties of vermiculite have been shown to display a scrolling effect, which produces structures which can be mistaken for chrysotile (Chatfield and Lewis, 1980). Figures 6 and 7 show TEM micrographs of a variety of vermiculite which displays this phenomenon, and a typical scroll after separation from the plate is shown in Figure 8. The EDXA spectrum from the scroll is shown in Figure 9, illustrating that the

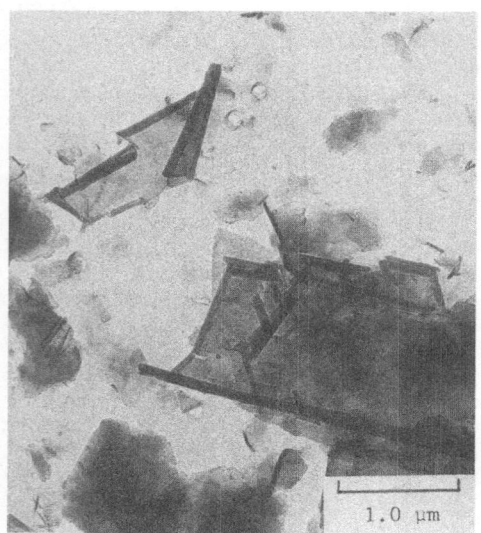

Fig. 6. TEM Micrograph of Vermiculite Sample Which Exhibits Scrolling Effect

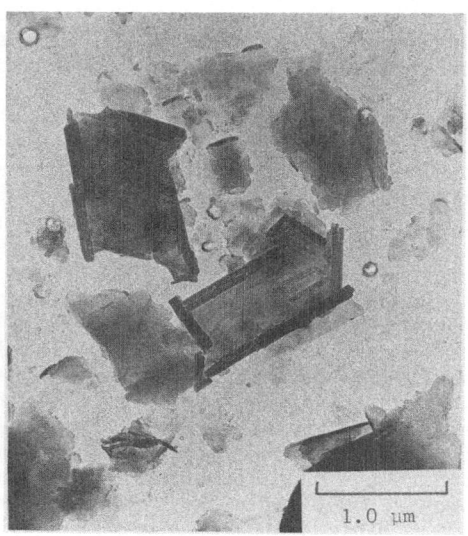

Fig. 7. TEM Micrograph of Vermiculite Sample Which Exhibits Scrolling Effect

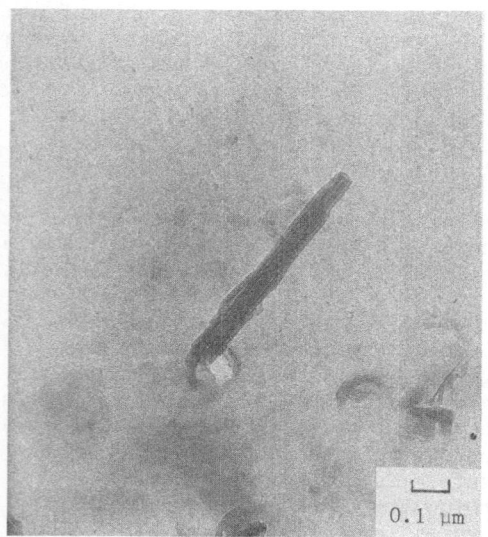

Fig. 8. TEM Micrograph of Detached Vermiculite Scroll

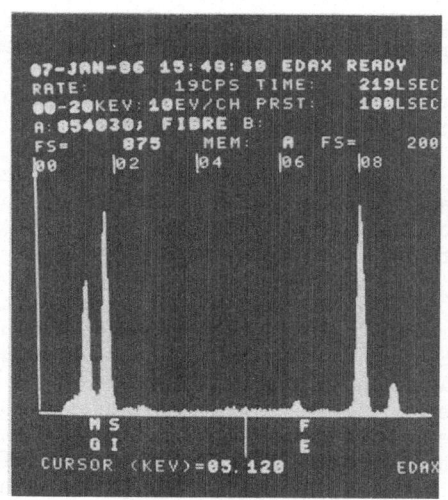

Fig. 9. EDXA Spectrum of Detached Scroll

composition is close to that of chrysotile. The SAED pattern from the scrolls usually does not contain the 0.73 nm (002) reflections typical of the SAED pattern from chrysotile, but for material from some deposits reflections occur which are very close to this value.

The correct classification of amphibole fibers is particularly important, because most of the amphiboles are neither considered mineralogically to be asbestos nor are they the subject of any legislative action. Accurate quantitative analyses permit amphibole fibers to be placed into their correct classifications, and the procedure for doing this has been published (International Mineralogical Association, 1978). The amphiboles have a wide range of composition, and even within a species of amphibole, there are compositional variations which do not exclude a fiber from being that particular amphibole variety. For example, the class-ification of a fiber as tremolite or actinolite may be totally incorrect if there is more than a specific concentration of aluminum in its composit-ion. At particular values of aluminum concentration, such a fiber should be reported as tremolitic hornblende, magnesio-hornblende or tschermakite.

Figure 10 shows the results of the quantitative EDXA analysis for a small richterite fiber. In recognition of the fact that the EDXA analyses are not absolutely precise, a range is assigned to the concentration of each element, and this results in selection of several minerals which have compositions consistent with this range. Routinely, the analysis would be carried no further because none of these minerals would be classified as asbestos, although some fibrous forms of richterite are known. If some of the minerals reported were asbestos-forming amphiboles, and others were similar in composition, but not amphiboles, zone-axis work would be performed. Whether a particular fiber is an amphibole variety or not can only be demonstrated by the use of zone-axis SAED methods. Since the lattice parameters of the amphiboles are all very similar, SAED is not normally capable of discriminating between the amphibole varieties, except for the general distinction between orthorhombic and monoclinic amphiboles. Even for this distinction, the fiber must present an appropriate orientation within the range of the TEM stage for definitive zone-axis SAED patterns to be obtained. In order to obtain a zone-axis SAED pattern, the fiber is tilted and rotated so that a principal zone-axis is aligned parallel to the electron beam. A simple SAED pattern is then obtained which is amenable to interpretation (Lee, 1978). The fiber can then be tilted to another zone-axis, and the angle between these can also be measured. The two zone-axis SAED patterns obtained in this way for the richterite fiber, at an angular separation of 59 degrees, are shown in Figure 11. These SAED patterns can be analyzed by a computer program (Rhoades, 1976) to examine their consistency with the structures of possible minerals, and the inter-zone-axis angle can also be calculated and compared for consistency.

Once a fiber has been determined to be an amphibole variety, it is still not necessarily asbestos. Only a few of the amphibole varieties display an asbestos habit. It is not currently possible routinely in the TEM to classify an individual small amphibole fiber either as asbestos or as a cleavage fragment. It has, however, been shown that populations of fibers that originate from the asbestos and non-asbestos varieties of the same mineral can be distinguished from each other on the basis of fiber aspect ratio distributions (Wylie, 1979; Chatfield, 1983b).

ANALYTICAL PRECISION, ACCURACY AND STATISTICAL CONSIDERATIONS

When airborne particulate samples are collected, the random deposition of fibers on the collection filter, in the absence of effects such as

PARTICLE IDENTIFICATION

PARTICLE: Suspected Amphibole Fiber

WIDTH OF PARTICLE: 0.200 Micrometers

X-ray Spectrum:	Element	Peak Area	Element	Peak Area
	Na	41.42	Mg	74.43
	Al	13.92	Si	235.98
	K	7.94	Fe	39.69

Calculated Atomic Ratios:	Element	Ratio	Element	Ratio
	Na	0.478	Mg	0.516
	Al	0.069	Si	1.000
	K	0.046	Fe	0.126

MINERALS WITH COMPOSITIONS CONSISTENT WITH X-RAY SPECTRUM

Fe-Eckermannite   Mg-Eckermannite
Fe-Katophorite   Mg-Katophorite
Richterite

Fig. 10. Results of Quantitative EDXA Analysis of Richterite Fiber

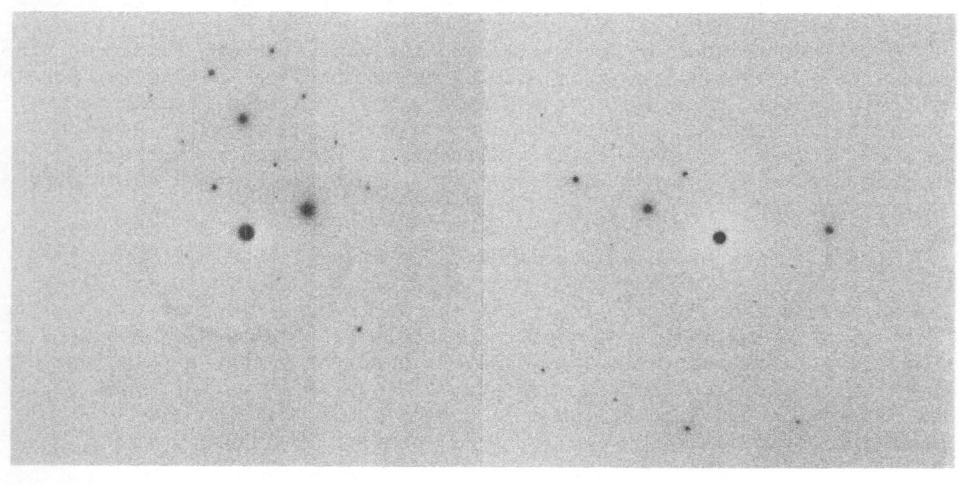

Fig. 11. Zone-Axis SAED Patterns Obtained From Fiber
of Richterite, at Angular Separation of 59°

aggregation or localized non-uniformities in the airflow through the filter, should be according to a Poisson distribution. This is the condition of the sampling filter which represents the optimum situation, and the precision of fiber counting is limited by Poisson statistics. However, the situation is not always ideal and many samples deviate from this. For example, the supporting structures behind the sampling filter cause localized variations in the airflow, which result in non-uniform deposits of particulate on the filter surface. The airborne fibers may also be non-uniformly dispersed in the atmosphere and may be aggregated with other fibers or with non-fibrous particulate. The relative proportion of fibers to other particulate may prevent sampling to yield an ideally-loaded filter, and then microscope limitations may affect the ability to count all relevant fibers. It has also been suggested that electrical charging of the filter surface and the filter cassette itself can deflect particulate away from the filter, or give rise to non-uniformity of the collected deposit. Although many of these effects can be minimized, they do cause to some extent a deviation in the fiber counts from the Poisson distribution. This is the case for any fiber counting technique, whether by PCM, SEM or TEM. Whether this deviation is significant in terms of the measurement is another question, but in any analysis its magnitude should be known. Where less than about 30 fibers are counted in a TEM analysis, it is not possible to measure the degree of deviation from Poisson behavior, although in some cases the magnitude of the deviation will be sufficiently large that it can be detected by the use of a simple statistical test. For example, in a count of 30 fibers on 20 grid openings, if 15 of those fibers were found on one grid opening, it is most unlikely that the fiber deposit was random, and such a sample would not be capable of providing a satisfactory analysis. In those cases where more than about 30 fibers have been detected, it is possible to calculate a standard deviation from the individual grid opening fiber counts, from which a confidence interval for the fiber concentration can be derived.

It has been found in practice that many ambient air and building atmosphere samples display very low fiber counts, and the interpretation of these fiber counts in terms of fiber concentrations must be done very carefully. In early work, it was common to see fiber concentrations reported as a single number, perhaps with more than one significant digit, on the basis of detection of one fiber. This has led to a great deal of confusion, because such measurements can appear incompatible with others if the data regarding precision for all measurements are not available. In other studies reported in the literature, the sampling conditions were such that a single fiber detected in the fiber count represented a very high fiber concentration. Quoted without specification of the precision, such a result is open to mis-interpretation, because the single fiber found may have originated from contamination of the unused filter or have been introduced as contamination during the analysis. In this type of analysis, it is entirely possible that a blank filter analyzed concurrently with the actual sample would have produced a zero fiber count, lending even more credibility to the high concentration reported. If, on the other hand, the Poisson 95% confidence interval were quoted for both the measurement and the blank, the fact that there is no statistically-valid difference would be made clear to the individual who interprets the data.

Because of the wide range of fiber loadings, filter uniformities and fiber counts, it is essential that all parameters used in each TEM analysis for asbestos be specified. Firstly, a chi-squared test should be carried out on the individual grid opening fiber counts to determine if there is any reason to suspect that the fibers are distributed on the filter in a non-uniform manner. Assuming that the uniformity is satisfactory, a 95% confidence interval can be specified for the fiber concentration, based on the Poisson distribution if there are only a few fibers, and based on the normal distribution if there are sufficient numbers of fibers in the grid openings to allow calculation of a standard deviation. Where no fibers are found, the mean fiber concentration cannot be specified, but a 95% upper confidence limit, equal to 3.69 times the analytical sensitivity, represents the analytical detection limit. For detection of 1 to 4 fibers, the lower 95% confidence limit for a Poisson distribution corresponds to less than one fiber, and accordingly, this is defaulted to zero and the result is simply stated as less than the corresponding upper 95% confidence limit. For 5 - 30 fibers, the lower 95% confidence limit corresponds to more than one fiber, and both a mean fiber concentration and a 95% confidence interval about the mean are quoted, based on Poisson statistics. For fiber counts exceeding 30 fibers, it is possible to calculate a standard deviation from the individual grid opening fiber counts. Here, 95% confidence intervals can be calculated on the basis of both Poisson and Gaussian statistics, and the larger interval about the mean is then reported. This approach has been specified in the analytical method for drinking water analysis published by the EPA (Chatfield and Dillon, 1983), and the same general approach can be applied directly to air sample analysis. This concept has been accepted by Working Group ISO/TC147/SC2/WG18-ISO/TC146/SC3/WG1 of the International Organization for Standardization, for reporting of asbestos fiber concentrations in both air and water samples.

The statistical procedures outlined above deal with the simple case of individual fibers, positioned randomly on a filter. The procedures assume that the position of any fiber is an independent event, and not affected by the proximity of other fibers. Chrysotile, in particular, dispersed from the solid material by agitation, yields a highly aggregated aerosol. The aggregates themselves probably behave in a Poissonian manner so far as the sample collection is concerned. However, the imposition of the normal fiber counting rules during examination of such a filter will produce an apparent fiber distribution which is far from Poissonian, and the procedures to be used for evaluation of such samples have not been developed. When an aerosol of this type is very dilute in terms of the number of aggregates, fiber counts will generally be correspondingly low, except that occasionally a grid opening will be encountered which contains one of the aggregates. If a fiber concentration is reported based on inclusion of data from one such grid opening, the value reported will depend strongly on how many of the essentially zero grid openings are incorporated in the calculation. Essentially, to obtain a reliable fiber concentration determination when a low concentration of aggregates is present, a statistically-valid number of aggregates must be included in the fiber count, just as for the individual fibers in the more simple situation where only single fibers are involved. Where the mechanism of aerosol dispersal is by disturbance of the solid material, it appears that this fiber aggregate problem may affect many of the measurements made in ambient air, and in the atmospheres of buildings, particularly when airborne fiber determinations are made in association with asbestos removal or renovation work.

The aggregate effect has been demonstrated during the measurements made on comparison of the direct and indirect preparation techniques for TEM specimens. Table 6 shows the results of applying the chi-squared uniformity test to directly and indirectly prepared TEM specimens of single fibril and aggregated aerosols. Although satisfactory uniformity can be demonstrated for direct preparation of filters exposed to the single fibril aerosol, the directly-prepared specimens from the aerosol generator would not pass the uniformity test at 0.1% significance. The chi-squared test for uniformity was passed easily by all of the indirectly-prepared samples.

Table 6. Uniformity of Chrysotile Fiber Deposition on
Membrane Filters

(Significance Level of Chi-Squared Test)

Aerosol Type	Direct Preparation		Indirect Preparation	
	Nuclepore	Millipore	Nuclepore	Millipore
Single Fibril	5%	10%	50%	99.5%
Aggregated	<0.1%	<0.1%	50%	25%

MEASUREMENT OF ASBESTOS FIBER CONCENTRATIONS IN BUILDING ATMOSPHERES

The specific problem of measurement of airborne asbestos fiber
concentrations in the atmospheres of general occupancy buildings, such as
schools and office buildings, arose when it was recognized that there could
be emissions of fibers from asbestos-containing building materials used
inside the building. Air sampling is carried out in the same manner as for
sampling of the ambient atmosphere, using a filter which is prepared for
TEM analysis. Early measurements were made by the indirect method of
analysis, and because of this, results were reported in terms of mass
concentration (Nicholson, 1980; Sebastien et al., 1979). At the time of
some of this early work, some sources of membrane filter had been found to
be contaminated by chrysotile asbestos (Chatfield, 1975; Chatfield et al.,
1978). This filter contamination was not at a constant level; some filters
were not affected, whereas others were contaminated to a degree that
interfered with interpretation of measurement results. It is possible,
therefore, that some of the measurements of that time were affected to some
extent, and this should be kept in mind when data are being interpreted.
The indirect method of TEM specimen preparation is particularly sensitive
to the presence of contamination in the collection filter material, since
the ashing procedure results in inclusion of the contamination fibers from
the whole thickness of the portion of the filter which is ashed.

There is a great deal of controversy as to the utility of air sampling
in general occupancy buildings. It has often been stated that air sampling
should not be used to define whether there are reasons to remove asbestos
from a building. Air sampling is, however, generally accepted as
appropriate to determine whether, after asbestos has been removed from a
building, the post-removal cleaning was adequate. "Aggressive" air
sampling is sometimes advocated for this purpose, the intention being to
re-suspend any dust on the floor or interior surfaces of the building, and
to collect an air sample while this is being done (Stewart, 1985; U.S. EPA,
1985). Some have advocated that "aggressive" air sampling should be used
in buildings which contain asbestos, in order to define whether asbestos
should be removed. It is not the purpose of this paper to enter into this
controversy, and comment will be confined to what information the various
approaches can provide.

One situation which frequently arises concerns the reassurance of
building occupants that under normal operating conditions there is no
excessive risk to their health from asbestos fibers originating within the
building. Some will argue that analysis of a limited number of air samples
is inadequate for this reassurance, since the measurement pertains only to

the conditions existing at the time of sampling; but this limitation also applies to practically every other measurement one might make in similar circumstances, even to measurements made in the general environment outside. Few question on these grounds the validity of formaldehyde measurements, or the measurement of occupational exposure to asbestos. Nevertheless, measurements of asbestos fiber concentrations are frequently made to show the levels of airborne asbestos in buildings under normal operating conditions. Sampling is arranged to be at breathing height for persons in their regular activity, and the air handling in the building is maintained at its usual level. The analysis is conducted in the same way as for an ambient air sample, aiming at an analytical sensitivity of about 0.001 fibre/mL. In the absence of any guideline for acceptable fiber concentrations, a comparison is made with the concentrations outside the building, measured concurrently. In urban locations, it is often found that the outdoor sample is heavily loaded by other particulate, and this may limit the analytical sensitivity which can be achieved. Usually, it is known what type or types of asbestos are present in the building, and the analysis can be arranged to look specifically for these. On occasion, the analyses indoors yield an unexpected result, either in the type of asbestos found or an unusually high concentration of asbestos. Such a result can lead the building manager to locate an unknown source of airborne asbestos.

By far the most frequent use of electron microscopy in building atmosphere measurement is for post-abatement clearance monitoring. This involves measurement of the airborne asbestos fiber concentration in a building after asbestos-containing materials have been removed, in order to specify that the removal contractor has satisfactorily completed the work. Although this determination is frequently performed using PCM methods, the fact that the PCM technique only accounts for a small minority of the number of fibers which may be present, and does not discriminate asbestos from other fibers, means that clearances based on these methods can be challenged by the building occupants or by the contractor as ineffective. PCM is capable of detecting only a very small proportion of the fibers in a typical chrysotile aerosol. A technique which detects all of the size range of fibers, particularly those fibers longer than 5 μm but too narrow to be seen by PCM, is often seen to be required. In addition, the analytical sensitivity of the PCM methods is inadequate for measuring asbestos concentrations which are expected to be comparable with acceptable ambient airborne asbestos levels. Sometimes, the SEM has been applied to this task, because it is more commonly available than the TEM, and to some the instrument performance seemed adequate. The practical limitations of the SEM for this purpose are now well recognized, and the TEM method has emerged as the method of choice for effective post-abatement monitoring, in spite of the higher cost of such analysis. The EPA has specified the sampling procedures to be used for post-abatement monitoring (U.S. EPA, 1985), and although a precise numerical clearance criterion was established for comparison with the PCM measurements when this technique is used, there was apparently some difficulty in specifying such a criterion for TEM measurements. The TEM clearance release criterion is specified in terms of a statistical comparison between 5 interior samples and 5 other samples collected concurrently just outside the work area. The problem arises because there is usually a detectable background asbestos level which is quantifiable when the TEM method is used. Nearly all of the TEM measurements will yield some value, either from filter contamination, or in most cases from the air which was sampled. Someone then must make a decision regarding what is a suitable level for clearance, and to date few legislators have been prepared to do this. In the case of PCM measurements, what is perceived as the analytical detection limit for large fibers of all species has been specified, this being the best value that could possibly be achieved by this measurement method. TEM measurements, however, are reporting identified <u>asbestos</u>, and since there is always some background level, specification of an acceptable numerical upper limit would undoubtedly be seen as condoning some asbestos exposure of the building occupants, regardless of how low that limit was. Some jurisdictions have, however, established either guidelines or target levels for asbestos fibers in ambient air, and using the criterion that a building owner has no obligation to provide air any cleaner than that which exists outside, these target levels can be applied to the air inside a building. Ontario, Canada, has established such a guideline of 0.04 fiber/mL, for asbestos fibers longer than 5 μm, measured by TEM. Germany has a target of 0.001 fiber/mL, for asbestos fibers longer than 5 μm, measured by the VDI SEM method. In France, the Conseil Superieur d'Hygiene Publique has recommended remedial action in buildings if the airborne level of asbestos exceeds 50 nanograms/cubic meter (Conseil Superieur d'Hygiene Publique, 1978). This latter recommendation appears to acknowledge the use of air sampling in order to define that remedial action is necessary, rather than the inspection approach used in North America.

The specification of clearance criteria based on asbestos fibers longer than 5 μm was discussed recently at a Workshop on Asbestos Measurement in Buildings (Ontario Research Foundation, 1986). One of the conclusions of this meeting was that the observation of long fibers in an air sample was in some way indicative of a release of asbestos into the building atmosphere, whereas some small asbestos fibers were almost always found in every sample analyzed by TEM (Chatfield, 1986). The other observation made by several of the participants was that the results of measurements in buildings were generally "spotty", in that a large number of very low results occur, and occasionally one value significantly higher is found. This is a characteristic of even some of the very early measurements (Sebastien et al., 1980). This observation seems to indicate that either the aerosol itself is of too "grainy" a character for the measurement techniques being applied, or that release of fibers into the interior atmosphere is only sporadic. Alternatively, in some cases, occasional contamination in the analysis may very well be occuring, and giving rise to these sporadic results. It is well known that extreme precautions must be taken in order to avoid contamination during analysis, but even using the best practice it is doubtful whether all of a group of perhaps 100 blank analyses would yield satisfactory values. Some recent supplies of filters have been quite unacceptable for this type of analysis on account of their contamination levels, and no information is likely to be available on whether this problem has existed all the time in a rather more sporadic manner. When a limited series of analyses have been performed to evaluate a particular batch of filters, the level of asbestos contamination has been found to fluctuate from one filter to the next, even within a single batch. The cost of extensive analysis certainly would have inhibited accumulation of adequate data to determine whether the sporadic high values observed during sampling were actually a consequence of sporadic filter contamination.

If the reason for the sporadic results in building atmosphere determinations is simply that there is a sporadic release of fibers within the building, or that the analysis method is statistically not adequate for the type of aerosol encountered, what are the implications for those measurements already made, and the buildings cleared using these measurements? In terms of occupant exposure, the fact that there are so many very low measurements gives confidence that the overall exposure is low. When a sporadic high concentration is found, after incorporation in a time-weighted average as is done in the asbestos workplace measurement, the average is also likely to reflect a low level of exposure. However, because of the high cost of analysis, in many cases the high measurement observed will not be accompanied by a statistically-appropriate number of low values, and there will be an over-estimate of the average exposure. In other cases, a limited series of essentially zero results will be obtained, leading to a possible under-estimate of the exposure levels. Assuming that the high values are, as is usually observed, infrequent, there will be a tendency for a slight under-estimate for most of the samples, with an occasional extreme over-estimate. The only way in which this whole situation can be resolved would be for a complete study to be made, using both long-term and short-term sampling, with analyses being made to establish the nature and frequency of fiber emission from various asbestos-containing materials. Whether the sampling and analytical techniques now being used are optimum for the purpose of post-abatement clearance could then be defined.

DETERMINATION OF RESPIRABLE FIBER COMPONENTS IN MINERAL PRODUCTS

When mineral products such as vermiculite are used, the dust which is created during handling of the material may contain respirable fibers of any asbestos which is sometimes associated with the crude ore. Amphibole minerals are frequently associated with vermiculite, and since any amphibole particle with an aspect ratio exceeding 3:1 is considered by some regulatory bodies to be an asbestos fiber, there is a need to characterize such particles very carefully. The regulators do not generally consider common amphiboles such as hornblende to be asbestos, and so a need arises to discriminate between different varieties of amphibole in the analysis of materials like vermiculite. Indeed, when airborne dust levels are measured using the PCM technique, which does not discriminate fiber types, even acicular fragments of the vermiculite will be included in the determination. Grinding of the product to particle sizes appropriate for microscopic analysis would present a totally false picture of what the potential for exposure of the user might be, because the material is not used in this finely divided state. The exposure potential from the dust is represented by a measurement of fibers within the respirable size range, present in the product as normally used. A procedure for obtaining this measurement has been developed (Chatfield and Lewis, 1980). In the case of vermiculite, ores from some sources display the scrolling effect described earlier, leading to the development of scrolls which can be mistaken for chrysotile. In other sources, a range of amphibole varieties may also be present along with the vermiculite. Only a small proportion of the total number of amphibole fibers found may be classified as possible asbestos fibers on the basis of their compositions. Even these may be acicular cleavage fragments of the corresponding non-fibrous amphibole, although in some cases the aspect ratio distribution may clearly demonstrate that the fibers originate from material of an asbestos habit. In the procedure of Chatfield and Lewis, the material is first examined optically to determine whether gross quantities of obviously fibrous material are present, in which case there is little point in performing any quantitative analysis, except to identify the fibrous species. If the vermiculite appears not to be seriously contaminated, a known weight is suspended in a liquid, and samples of suspended particulate are taken under conditions which ensure that all particles of respirable size would be included. These sub-samples of the liquid are filtered, TEM grids are prepared by the direct method, and the fiber counts are performed. The identification of each fiber includes morphology, SAED and quantitative EDXA. Using the sophisticated identification techniques available on a modern analytical electron microscope, chrysotile fibers can be discriminated from vermiculite scrolls, acicular fragments of vermiculite can be discriminated, and amphibole fibers can be correctly assigned to their specific amphibole variety, using the International Mineralogical Association Amphibole Nomenclature. The type, and range of amphiboles found in a typical vermiculite sample are shown in Table 7, in which it can be seen that in this particular sample, only 13% by number, and 3.5% by weight of the total

Table 7. Amphibole Fiber Classification in a Vermiculite Sample
Proportion of Total Amphibole

Amphibole Species	Numerical %	Mass %	Highest Aspect Ratio Observed
Tremolite	2	0.2	4.0
Actinolite	9	3	6.8
Actinolitic Hornblende	8	3	6.7
Mg-Hornblende	15	9	9.0
Fe-Hornblende	2	0.7	3.6
Tschermakitic Hornblende	6	2	4.0
Tschermakite	6	9	3.8
Richterite	9	11	13.0
Mg-Katophorite	27	55	25.0
Mg-Taramite	9	4	33.5
Taramite	6	3	4.0
Anthophyllite/Cummingtonite	2	0.3	6.3

amphibole fibers found could possibly be classified as asbestos. The question as to whether these fibers were actually asbestos, rather than cleavage fragments, could only be answered by a further analysis in which more than 100 fibers of the particular compositional classification would be measured in order to obtain a reliable aspect ratio distribution.

In the case of talc ore, the problem of determining the levels of asbestos is compounded by the fact that the ore frequently contains a high concentration of acicular talc fragments, and the composition of talc is very similar to that of chrysotile or anthophyllite. Both chrysotile and amphiboles have been found associated with talc ores. As with vermiculite analyses, determinations of asbestos levels in talc require the identification capabilities of an analytical electron microscope.

The results of an analysis as described above allow estimates to be made of the potential exposure of users of the product to be made, even before the material comes into production. This is a very useful approach when new ore bodies are under investigation: before the ore bodies are developed, the data permit decisions to be made concerning possible health risks, special production conditions can be defined, and potential markets or market restrictions for the product can be identified.

CONCLUSION

Analysis of air samples by electron microscopy for the presence of asbestos is a task which can vary widely in its requirements, and before measurements are made, there should be a clear understanding of what information is required. If the intention is to obtain further information on a fiber count obtained in an occupational setting, to determine the nature of fibers comprising a PCM fiber count, it is likely that a simple SEM/EDXA examination may provide the information required. In this case there is no interest in the smaller fibers, and attention can be concentrated on just that size range of fibers which were optically-visible. For measurements in building atmospheres, the material present or being disturbed may be known, in which case a TEM analysis can often be made using simple morphological classification, particularly if the only variety of asbestos present is chrysotile. In other situations such as analyses of ambient air, or analyses where litigation or product liability is involved, only the most stringent fiber identification will be sufficient. Analysis for asbestos is a complex mixture of mineralogy, optical microscopy, electron microscopy, X-ray analysis, statistics and aerosol physics, and a detailed understanding of all of these topics is required in order to perform the more sophisticated analyses. The methods now used have been under development concurrently with the generation of a significant amount of data. Some of the earlier measurements, therefore, should be treated with caution, having regard for possible defects in the methods used. There is still some need for air sampling methods to be optimized, and there is a pressing need for the medical community and legislative bodies to define the relevant parameters which should be measured.

REFERENCES

Asbestos International Association, 1979, Reference method for the determination of asbestos fibre concentrations at workplaces by light microscopy (membrane filter method), AIA Health and Safety Publication, Recommended Technical Method No. 1 (RTM1), Asbestos International Association, 68 Gloucester Place, London, W1H 3HL, England.

Asbestos International Association, 1984, Method for the determination of airborne asbestos fibres and other inorganic fibres by scanning electron microscopy, AIA Health and Safety Publication, Recommended Technical Method No. 2 (RTM2), Asbestos International Association, 68 Gloucester Place, London, W1H 3HL, England.

Asbestosis Research Council, 1971, The measurement of airborne asbestos dust by the membrane filter method, Technical Note 1, Asbestosis Research Council, Rochdale, Lancashire, England.

Baris, Y, 1980, The clinical and radiological aspects of 185 cases of malignant pleural mesothelioma, in: "Biological Effects of Mineral Fibres, Vol. 2," J. C. Wagner, ed., IARC Scientific Publications No. 30, International Agency for Research on Cancer, Lyon, France.

Beckett, S. T. and Attfield, M. D., 1974, Inter-laboratory comparisons of the counting of asbestos fibres sampled on membrane filters, Ann. Occup. Hyg. 17:85.

Berry, E. E., 1971, Thermal analysis of various chrysotiles using evolved water analysis techniques, 2nd International Conference on the Physics and Chemistry of Asbestos Minerals, Louvain, Belgium, Paper 2:7.

Burdett, G. J. and Rood, A. P., 1982, A membrane-filter, direct transfer technique for the analysis of asbestos fibres or other inorganic particles by transmission electron microscopy, Environ. Sci. and Tech., 17:643.

Chatfield, E. J., 1975, Asbestos background levels in three filter media used for environmental monitoring, in: "Proceedings, 33rd Annual Meeting, Electron Microscopy Society of America, Las Vegas, Nevada," C. J. Arceneaux, ed., Claitor's Publishing Division, 3165 S. Acadian at 1-10, P.O. Box 239, Baton Rouge, Louisiana 70821.

Chatfield, E. J., Glass, R. W. and Dillon, M. J., 1978, Preparation of water samples for asbestos fiber counting by electron microscopy, EPA Report EPA-600/4-78-011, Environmental Research Laboratory, Office of Research and Development, U.S. Environmental Protection Agency, Athens, Georgia 30613.

Chatfield, E. J., 1979, Preparation and analysis of particulate samples by electron microscopy, with special reference to asbestos, in: "Scanning Electron Microscopy/1979/I," O. Johari, ed., SEM Inc., AMF O'Hare, Chicago, IL 60666.

Chatfield, E. J. and Lewis, G. M., 1980, Development and application of an analytical technique for measurement of asbestos fibers in vermiculite, in: "Scanning Electron Microscopy/1980/I," O. Johari, ed., SEM Inc., AMF O'Hare, Chicago, IL 60666.

Chatfield, E. J., 1982, Measurement of asbestos fibre concentrations in workplace atmospheres, Study No. 9, Royal Commission on Matters of Health and Safety Arising from the Use of Asbestos in Ontario, Publications Mail Order Service, 880 Bay Street, 5th Floor, Toronto, Ontario, Canada, M7A 1N8.

Chatfield, E. J., 1983a, Short mineral fibres in airborne dust, in: "Short and Thin Mineral Fibres. Identification, Exposure and Health Effects," National Board of Occupational Safety and Health, Research Department, Solna, Sweden.

Chatfield, E. J., 1983b, Methods of fibre measurement in ambient air, in: "Fibrous Dusts," Verein Deutscher Ingenieure - Kommission Reinhaltung der Luft, Berichte Nr. 475.

Chatfield, E. J., 1983c, Measurement of asbestos fibre concentrations in ambient atmospheres, Study No. 10, Royal Commission on Matters of Health and Safety Arising from the Use of Asbestos in Ontario, Publications Mail Order Service, 880 Bay Street, 5th Floor, Toronto, Ontario, Canada, M7A 1N8.

Chatfield, E. J. and Dillon, M. J., 1983, Analytical method for determination of asbestos fibers in water, U.S. Environmental Research Laboratory, Athens, Georgia, Contract 68-03-2717, Available through: National Technical Information Service, 5285 Port Royal Road, Springfield, VA 22161, Order No. PB 83-260-471.

Chatfield, E. J., Dillon, M. J. and Stott, W. R., 1983, Development of improved analytical techniques for determination of asbestos in water samples, U.S. Environmental Research Laboratory, Athens, Georgia, Contract 68-03-2717, Available through: National Technical Information Service, 5285 Port Royal Road, Springfield, VA 22161, Order No. PB 83-261-471.

Chatfield, E. J., 1984, Fiber definition in occupational and environmental asbestos measurements, in: "Definitions for Asbestos and Other Health-Related Silicates, ASTM STP 834," B. Levadie, ed., American Society for Testing and Materials, Philadelphia.

Chatfield, E. J., 1986, Limitations of precision in analytical techniques based on fibre counting, in: "Asbestos Fibre Measurements in Building Atmospheres," E. J. Chatfield, ed., Ontario Research Foundation, Sheridan Park Research Community, Mississauga, Ontario, Canada, L5K 1B3.

Cliff, G. and Lorimer, G. W., 1975, The quantitative analysis of thin specimens, J. of Microsc., 103:203.

Conseil Superieur d'Hygiene Publique de France, 1978, Annual Report, Paris, France.

Crawford, N. P., 1986, Fibre assessment standards of UK laboratories engaged in asbestos monitoring, in: "Asbestos Fibre Measurements in Building Atmospheres," E. J. Chatfield, ed., Ontario Research Foundation, Sheridan Park Research Community, Mississauga, Ontario, Canada, L5K 1B3.

Hodgson, A. A., 1979, Chemistry and physics of asbestos, in: "Asbestos Properties, Applications and Hazards," L. Michaels and S. S. Chissick, eds., John Wiley and Sons, Chichester, United Kingdom.

International Mineralogical Association, 1978, Nomenclature of amphiboles (compiled by B. E. Leake), Can. Mineralogist, 16:501.

International Organization for Standardization, 1981, Determination of airborne inorganic fibre concentrations in workplaces by light microscopy (membrane filter method), Draft Proposal for International Standard, ISO/TC146/SC2/WG5, Convenor U. Teichert, GSA Gesellschaft fur Staubmesstechnik und Arbeitsschutz, Am Rottgen 126, D-4040, Neuss 16, West Germany.

International Organization for Standardization, 1982, Working Group
    ISO/TC147/SC2/WG18 - ISO/TC146/SC3/WG1 Meeting, Convenor,
    E. J. Chatfield, Ontario Research Foundation, Sheridan Park Research
    Community, Mississauga, Ontario, Canada, L5K 1B3.
Jaffe, M. S., 1948, Handling and washing fragile replicas,
    J. Applied Physics, 19:1187.
Jones, A. D. and Johnstone, A. M., 1985, Counting bias relating to sample
    density, in: "Fifth International Colloquium on Dust Measuring
    Technique and Strategy," F. Baunach, ed., Asbestos International
    Association, 68 Gloucester Place, London, W1H 3HL, England.
Konig, R., 1980, A new sample preparation technique for the measurement of
    the asbestos fibre concentration in ambient air, in: "Proceedings of
    the 3rd International Colloquium on Dust Measuring Technique and
    Strategy," D. Bouige, ed., Asbestos International Association,
    68 Gloucester Place, London, W1H 3HL, England.
Lee, R. J., 1978, Basic concepts of electron diffraction and asbestos
    identification using SAD. Part I: current methods of asbestos
    identification using SAD; Part II: single crystal and SAD, in:
    "Scanning Electron Microscopy/1978/I," O. Johari, ed., SEM Inc., AMF
    O'Hare, Chicago, IL. 60666.
LeGuen, J. M. M. and Galvin, S., 1981, Clearing and mounting techniques for
    the evaluation of asbestos fibres by the membrane filter method,
    Ann. Occup. Hyg. 24:273.
LeGuen, J. M. M., Rooker, S. J. and Vaughan, N. P., 1980, A new technique
    for the scanning electron microscopy of particles collected on membrane
    filters, Environ. Sci. and Tech., 14:1008.
Leidel, N. A., Bayer, S. G., Zumwalde, R. D. and Busch, K. A., 1979,
    USPHS/NIOSH membrane filter method for evaluating airborne asbestos
    fibers, U.S. Department of Health, Education and Welfare, Public Health
    Service, Center for Disease Control, National Institute for
    Occupational Safety and Health, 4676 Columbia Parkway, Cincinnati,
    OH 45226.
McCrone, W. C. and Stewart, I. M., 1974, Asbestos, American Laboratory,
    6:13.
National Institute for Occupational Safety and Health, 1984, NIOSH
    method 7400, U.S. Department of Health, Education and Welfare, Public
    Health Service, Center for Disease Control, National Institute for
    Occupational Safety and Health, 4676 Columbia Parkway, Cincinnati,
    OH 45226.
Nicholson, W. J., Rohl, A. N., Weisman, I. and Selikoff, I. J., 1980,
    Environmental asbestos concentrations in the United States, in:
    "Biological Effects of Mineral Fibres, Vol. 2," J. C. Wagner, ed., IARC
    Scientific Publications No. 30, International Agency for Research on
    Cancer, Lyon, France.
Ogden, T., 1981, Unpublished Data. Health and Safety Executive, 403 Edgware
    Road, London, NW2 6LN, England.
Ontario Research Foundation, 1986, "Asbestos Measurements in Building
    Atmospheres," E. J. Chatfield, ed., Ontario Research Foundation,
    Sheridan Park Research Community, Mississauga, Ontario, Canada,
    L5K 1B3.
Ortiz, L. W. and Isom, B. L., 1974, Transfer technique for electron
    microscopy of membrane filter samples, Am. Ind. Hyg. Ass. J., 35:423.
Pearson, E. S. and Hartley, H. O., 1958, "Biometrica Tables for
    Statisticians, Volume I," Cambridge University Press, 32 East 57th
    Street, New York, NY.
Pott, F., Dolgner, R., Friedrichs, K. H. and Huth, F., 1976, L'effet
    oncogene des poussieres fibreuses, Ann. Anat. Pathol., 21:237.
Pott, F., 1978, Some aspects on the dosimetry of the carcinogenic potency
    of asbestos and other fibrous dusts, Staub-Reinhalt., Luft, 38:486.

Rhoades, B. L., 1976, Xident – A computer technique for the direct indexing of electron diffraction spot patterns, Research Report 70/76, Department of Mechanical Engineering, University of Canterbury, Christchurch, New Zealand.

Ross, M., 1978, The "asbestos" minerals; definitions, description, modes of formation, physical and chemical properties, and health risk to the mining community, in: "Workshop on Asbestos: Definitions and Measurement Methods, NBS Special Publication 506," C. C. Gravatt, P. D. LaFleur and K. F. J. Heinrich, eds., National Measurements Laboratory, National Bureau of Standards, Washington, D.C. 20234.

Royal Commission on Matters of Health and Safety Arising From the Use of Asbestos in Ontario, 1984, Report, Ontario Ministry of Government Services, Publications Service Branch, 880 Bay Street, Toronto, Ontario, Canada, M7A 1N8.

Sebastien, P., Bignon, J., Gaudichet, A., Dufour, G. and Bonnaud, G., 1976, Les pollutions atmospheriques urbaines par l'asbeste, Rev. Franc. Mal. Resp.,4:51, (Supp. 2).

Sebastien, P., Billon, M. A., Dufour, G., Gaudichet, A., Bonnaud, G. and Bignon, J., 1979, Levels of asbestos air pollution in some environmental situations, in: "Health Hazards of Asbestos Exposure," I. J. Selikoff and E. C. Hammond, eds., Annals of the New York Academy of Sciences, 330:401.

Sebastien, P., Billon-Galland, M. A., Dufour, G. and Bignon, J., 1980, Measurement of asbestos air pollution inside buildings sprayed with asbestos, Report No. EPA-560/13-80-026, Office of Pesticides and Toxic Substances, U.S. Environmental Protection Agency, Washington, D.C. 20460.

Selikoff, I. J. and Lee, D. H. K., 1978, "Asbestos and Disease", Environmental Sciences, An Interdisciplinary Monograph Series, Academic Press, New York.

Small, J. A., Heinrich, K. F. J., Newbury, D. E. and Myklebust, R. L., 1979, Progress in the development of the peak-to-background method for the quantitative analysis of single particles with the electron probe, in: "Scanning Electron Microscopy/1979/II," O. Johari, ed., SEM Inc., AMF O'Hare, Chicago, IL 60666.

Speil, S. and Leineweber, J. P., 1969, Asbestos minerals in modern technology, Environ. Res., 2:166.

Spurny, K. R., Boose, G., Hochrainer, D. and Monig, F. J., 1976, A vibrating bed aerosol generator of fibrous and powder dust particles, Institut fur Aerobiolgie der Fraunhofer-Gesellschaft Grafschaft, Zbl. Bakt. Hyg., I Abt. Orig. B 161:326.

Stanton, M. F. and Layard, M., 1978, The carcinogenicity of fibrous minerals, in: "Workshop on Asbestos: Definitions and Measurement Methods, NBS Special Publication No. 506," C. C. Gravatt, P. D. LaFleur and K. F. J. Heinrich, eds., National Measurements Laboratory, National Bureau of Standards, Washington, D.C. 20234.

Steel, E. B. and Small, J. A., 1985, Microanalysis of asbestos in air for nonoccupational settings, in: "Workshop on the Monitoring and Evaluation of Airborne Asbestos Levels Following an Abatement Program," D. E. Lentzen, coordinator, Research Triangle Institute, Research Triangle Park, North Carolina.

Stewart, I. M., 1985, Aggressive sampling techniques for contamination and detection, in: "Workshop on the Monitoring and Evaluation of Airborne Asbestos Levels Following an Abatement Program," D. E. Lentzen, coordinator, Research Triangle Institute, Research Triangle Park, North Carolina.

Teichert, U., 1980, Influence of some parameters on the detection limit by phase contrast microscopy, in: "Proceedings of the 3rd International Colloquium on Dust Measuring Technique and Strategy," D. Bouige, ed., Asbestos International Association, 68 Gloucester Place, London, W1H 3HL, England.

U. S. Environmental Protection Agency, 1985, Guidance for controlling asbestos-containing materials in buildings, Report EPA 560/5-85-024, Office of Pesticides and Toxic Substances, Washington, D.C. 20460.

Verein Deutscher Ingenieure, 1983, Measurement of inorganic fibrous particulates in ambient air, Guideline 3492, VDI Publishers, P.I. Box 1139, D 4000, Dusseldorf 1, FRG.

Whittaker, E. J. W., 1979, Mineralogy, chemistry and crystallography of amphibole asbestos, in: "Short Course in Mineralogical Techniques of Asbestos Determination," R. L. Ledoux, ed., Mineralogical Association of Canada, Department of Mineralogy, Royal Ontario Museum, 100 Queen's Park, Toronto, Ontario, Canada M5S 2C6.

Wicks, F. J., 1979, Mineralogy, chemistry and crystallography of chrysotile asbestos, in: "Short Course in Mineralogical Techniques of Asbestos Determination," R. L. Ledoux, ed., Mineralogical Association of Canada, Department of Mineralogy, Royal Ontario Museum, 100 Queen's Park, Toronto, Ontario, Canada M5S 2C6.

Wylie, A. G., 1979, Fiber length and aspect ratio of some selected asbestos samples, in: "Health Hazards of Asbestos Exposure," I. J. Selikoff and E. C. Hammond, eds., Annals of the New York Academy of Sciences, 330:605.

COMPARISON OF THE PARTICLE SIZE DISTRIBUTION OF ALPHA QUARTZ FOUND IN
RESPIRABLE DUST SAMPLES COLLECTED AT UNDERGROUND AND SURFACE COAL MINES
WITH FOUR REFERENCE STANDARDS

Charles W. Huggins, Suzanne, J. Johnson, Joe M. Segreti
and Janet G. Snyder

U.S. BuMines
4900 LaSalle Rd.
Avondale, Md. 20782-3393

ABSTRACT

The object of this research was to compare the size distribution of respirable quartz found in samples collected at nine underground and thirteen
surface coal mines with that of four reference standards used to quantify
quartz in respirable dust samples. Particle size measurements were made on
all samples using a scanning electron microscope in the backscatter electron
mode of operation, interfaced with an image analysis system. The Mine Safety
and Health Administration, Pittsburgh, Pa. provided the respirable dust
samples which were collected at coal mine sites in 7 states. The accumulated
particle size measurements indicate three reference standards, minus 5-μm
Min-U-Sil, minus 5-μm Supersil, and NBS 1878 would be preferred over the
fourth, Silver Bond B for quantitative quartz determinations in coal mine
dusts by x-ray diffraction and infrared spectrometry.

INTRODUCTION

The Federal Coal Mine Health and Safety Act of 1969 established a maximum level for exposure of coal mine workers to respirable dust and authoriz-
ed research on development of new or improved methods of reducing the concentration of respirable dust in the mine atmosphere of active coal mine operations. Effective December 30, 1972, this standard was set at two milligrams per
cubic meter total dust for each eight hour shift. One mission of the Bureau
of Mines is to conduct research on the total control and measurement of coal
mine dust. Enforcement of coal mine safety regulations since 1978 has been
the responsibility of the Mine Safety and Health Administration (MSHA).
Medical responsibility is appropriated to the National Health Institute of
Occupational Safety and Health.

Respirable dust is the portion of airborne dust that penetrates the
deepest portion of the lungs and is generally considered to be particles
less than 10 μm in size. Major sources of dust generation in coal mines
are roof bolting, continuous miner, and drilling (Figs.1-3). The respirable
quartz particles in coal dusts are the cause of "silicosis", the major contributor to "black lung" disease. Compensation cost for "black lung" is
approximately $1.9 billion per year (1).

Dust conditions in coal mines have improved owing to new methods of
dust control but still need to be monitored. Approximately 40,000 dust

FIGURE 1. - Dust being generated by roof bolting in underground coal mine.

samples are collected by inspectors annually, and the quartz content of the
samples is determined by either x-ray diffraction or infrared spectrometry.
Both methods require quartz standards for quantitative measurement of quartz
in coal mine dusts. The response of both methods, however, as shown by
Tuddenham (5), Huggins (2), and Klug (3) is particle size dependent. Joint
studies by BuMines and MSHA show that inaccuracy in quantitative quartz
values may be as much as 30% when the particle size distribution of the
quartz standard varies significantly from that of the coal mine dust. This
is illustrated in the quartz calibration curves (Fig.4) for three of the
four reference standards. It is essential that the particle size distribution of both the respirable quartz standard and the coal mine dust be established in order to select the appropriate quartz reference standard to
provide an accurate determination.

The object of this research was to determine the particle size distribution of quartz in respirable quartz standards and in coal mine dust
samples collected from underground and surface coal mines. Twenty-three
respirable dust samples collected at surface and underground coal mines
in seven states were provided by MSHA, Pittsburgh, PA, for this research.
The quartz reference standards used for this research were minus 5-μm
Min-U-Sil, minus 5-μm Supersil, Silver Bond B, and NBS 1878. From this

FIGURE 2. - Dust being generated by continuous miner in underground coal mine.

FIGURE 3. - Dust being generated by drilling operation at surface coal Mine.

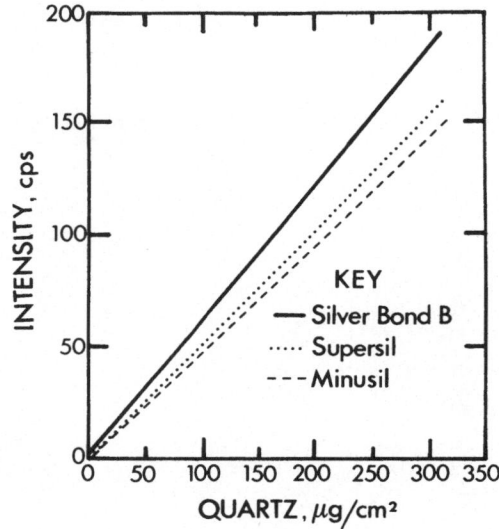

FIGURE 4. - Quartz calibration curves of three reference standards

research, a better match of standard to coal dust should be achieved.

Supersil was supplied by the Pennsylvania Glass and Sand Co., Berkeley Springs, WV, as minus 325-mesh material. It was wet-sieved at MSHA, Pittsburgh, PA, to obtain minus 5-μm material. Minus 5-μm Min-U-Sil, also from Pennsylvania Glass and Sand Co., was used as received. Since minus 5-μm Min-U-sil requires no laboratory preparation and is similar to quartz found in most respirable dust samples, it is preferred by many laboratories as their reference standard. The Silver Bond B, obtained from Tammsco Inc., Tamms, IL, was prepared by sedimentation at MSHA, Denver, CO, to obtain only particles smaller than 10 μm.

NBS 1878 has been available since late in 1983. In preparation of this standard, NBS started with commercially available minus 5-μm Min-U-Sil, and slightly improved the purity of the quartz. During purification, some of the very fine quartz particles may have been lost. The crystalline purity reported by NBS is 95.5 plus or minus 1 wt pct crystalline alpha quartz. The mass spherical diameter is 1.62 um

SAMPLE PREPARATION

The sampes were collected by the mine inspectors from cassettes similar to the one being worn by the coal miner in figure 5. The dust was collected on MSA filters similar to those shown in figure 6. All samples were taken from the mines for regulatory purposes and were low temperature ashed and redeposited by MSHA from the original monitoring collection filters onto Gelman DM-450 polyvinyl chloride membrane filters for quantitative quartz measurements by infrared spectroscopy. The samples were then hand carried to the BuMines laboratoy to prevent any loss of quartz particles. Preparation for image analysis was similar to that described by Snyder and Huggins (4). Wedge shaped pieces, approximately one eight of each filter, were cut and low temperature-ashed to remove the organic based filter. The remaining ash was subsequently suspended in isopropyl alcohol, ultrasonicated at 80 kHz for ten minutes, and deposited on a 0.2 μm Nuclepore filter. Two rectangular pieces aproximately 1 by 1.5 cm were cut from each Nuclepore filter sample and mounted with carbon paint on scanning electron microscope (SEM) stubs. The samples were carbon coated in a vacuum prior to measurement.

FIGURE 5. - Miner with collection filter cassette on jacket and pump unit on belt.

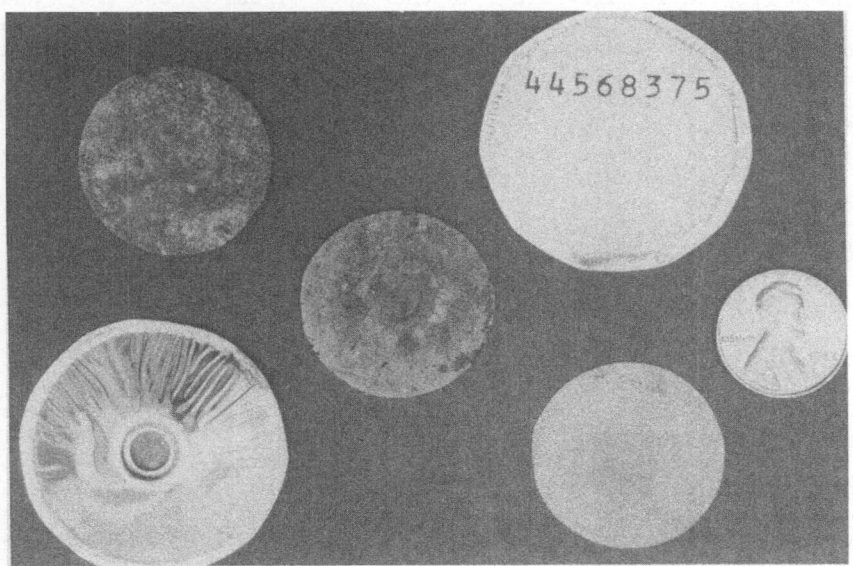

FIGURE 6. - Top and bottom of filter cassette, three collection filters, and penny used for filter size comparison.

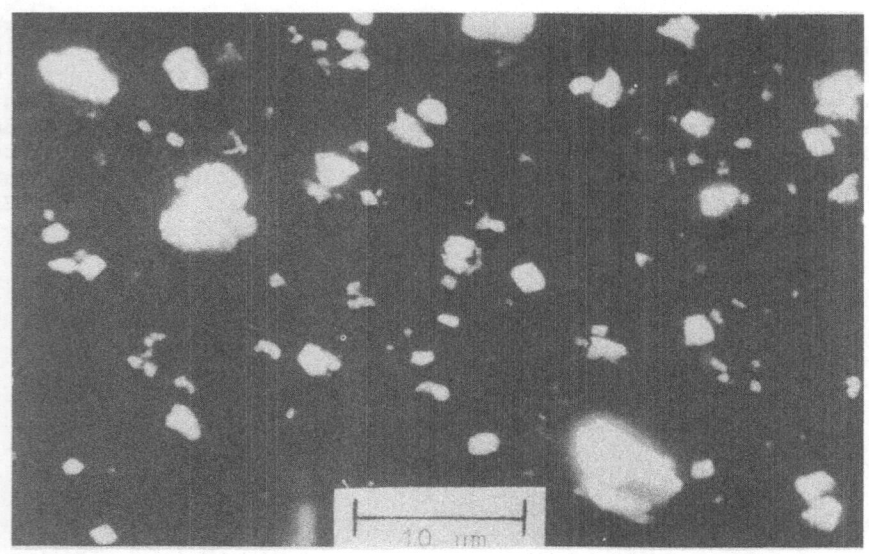

FIGURE 7. - Respirable dust particles dispersed on Nuclepore filter for
image analysis.

IMAGE ANALYSIS

Figure 7 shows particles dispersed on Nuclepore filter for sizing of
the quartz particles by image analysis. Instrumentation consisted of an Amray
model 1400 SEM equipped with a LeMont Scientific model DB-10 image analysis
system and a Kevex model 8000 energy dispersive x-ray analysis system (EDS).
The SEM was operated at 20 kV using a backscatter electron detector to provide the best contrast between particles and filter substrate and consequently enhanced gray level differences in the video signal of the image analyzer.
The contrast "threshold" level in the image analyzer was set just above
background to ensure measurement of all particles. Once a particle is detected in the binary image of the image analyzer, it is sized by deflection of
the electron beam in a series of horizontal, vertical, and diagonal movements.
The off-point density was set at 256 to ensure location of particles 0.20
µm or larger on the 10-cm cathode ray tube screen. The on point density was
set at 1,024, thus achieving a particle measurement precision of plus or
minus one "on point" spacing of 0.044 µm. All particles in the 45 square
micron field of view were measured. Sizes were measured at a magnification
of 2,000. Magnification calibration was performed with a Ladd Research Institute 15,240 line-per-inch grating using a magnification calibration program provided by LeMont Scientific.

Following each particle sizing, an x-ray spectrum was acquired at the
geometric center of the particle until a preset integral of 750 x-ray
counts was reached or a maximum of 10 seconds was met. Windows for detection
of elements commonly present in respirable coal mine dusts were set in the
EDS multichannel analyzer, thus monitoring the elements Al, Ca, Fe, K, Mg,
Na, S, Si, and Ti. A minimum of 30 x-ray counts was required for an element
to be considered present. The particles were categorized into two classes
by a chemistry definition file, and the information stored on diskettes. A
particle was classified as quartz if 80 pct or more of the total x-ray
count was due to silicon. All other particles were classified as miscellaneous. Approximately 400 quartz particles were measured in each sample.

RESULTS

Figure 8 shows the histograms of the four reference standards, and figures 9-11 show the histograms of the 9 underground coal mine samples taken
from mines in Virginia, West Virginia, and Kentucky; figures 12-16 show the
histograms for the 14 surface coal mine dust samples taken from mines in
Virginia, West Virginia, Pennsylvania, Ohio, Indiana, and Tennessee. Particles smaller than 0.3 µm were not measured as most of them were lost in
the MSHA sample preparation onto Gelman DM-450 filters which have a pore
size of 0.45 µm.

The surface coal mine dusts contained slightly more particles in the
larger size ranges than were found in underground coal mine dusts. This
can be seen in the particle size ranges of 4.2 to 9.6 µm. The average frequency percent of particles in this range for the underground coal mine
samples is 2.7 whereas 6.4 pct was found for the surface mine samples
examined in this study. Few of the 400 quartz particles measured in each
sample were longer than 9.6 µm.

An examination of the four reference standards in the particle size
range of 4.2 to 9.6 µm shows that Silver Bond B has approximately 66 pct
of its particles in this size range. In comparison, the minus 5-µm Min-
U-Sil, the minus 5-µm Supersil and NBS 1878 reference standards exhibit
respectively 2.8, 2.2, and 4.5 pct of their particles in the 4.2 to 9.6 µm
size range.

Based of these quartz particle size distribution conmparisons, the
minus 5-µm Supersil, the -5µm Min-U-Sil and NBS 1878 all are similar to
surface and underground coal dust samples. The Silver Bond B reference
standard has a significantly greater percentage of particles in the larger
size ranges than any of the dust samples analyzed and, therefore, would
not be a suitable reference standard for quartz in respirable coal mine
dusts.

CONCLUSION

Surface and underground coal mine dust samples show only slight differences in particle size distribution. Samples collected from surface coal
mines show a slightly larger quartz particle size distribution than those
collected from underground mines. Neither, however, on the basis of particle size, can be significantly differentiated for choice of reference standard for quantitative quartz determinations by x-ray diffraction and infrared spectrophotometry. Neither surface or underground coal dust samples is
significantly different from minus 5 µm Supersil, minus 5-µm Min-U-Sil, and
NBS 1878 all currently available as reference standards for quartz determinations in respirable coal mine dusts. Minus 5-µm Min-U-Sil and NBS 1878
would be preferred over minus 5-µm Supersil as they require no sample preparation before usage. Silver Bond B is unsuitable as a reference standard
for quantitative measurements of quartz in dusts collected from coal mines.
The particle size distribution in Silver Bond B does not adequately match
the size distribution found in the airborne coal mine dusts.

ACKNOWLEDGMENT

This chapter is an elaboration of a paper which was presented at the
Forensic, Occupational and Environmental Health Symposium at the 1985 Joint
National Meetings of the Electron Microscopy Society of America and the
Microbeam Analysis Society in Louisville, Kentucky. An extended abstract
of that presentation has been published by San Francisco Press.

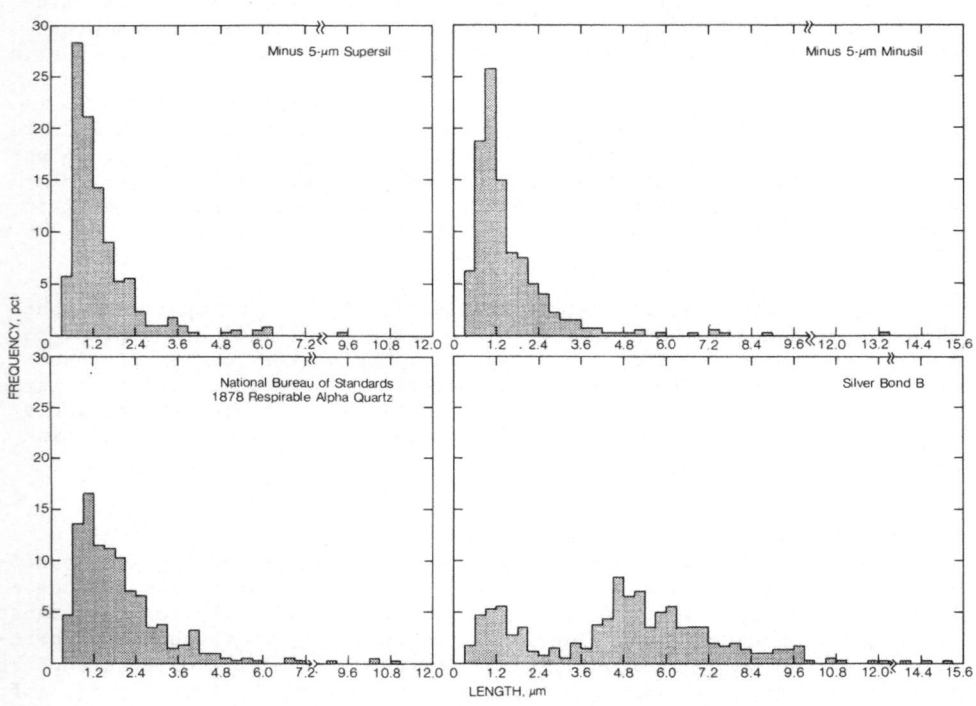

FIGURE 8. - Histograms of the four reference standards used for quantitative quartz determinations.

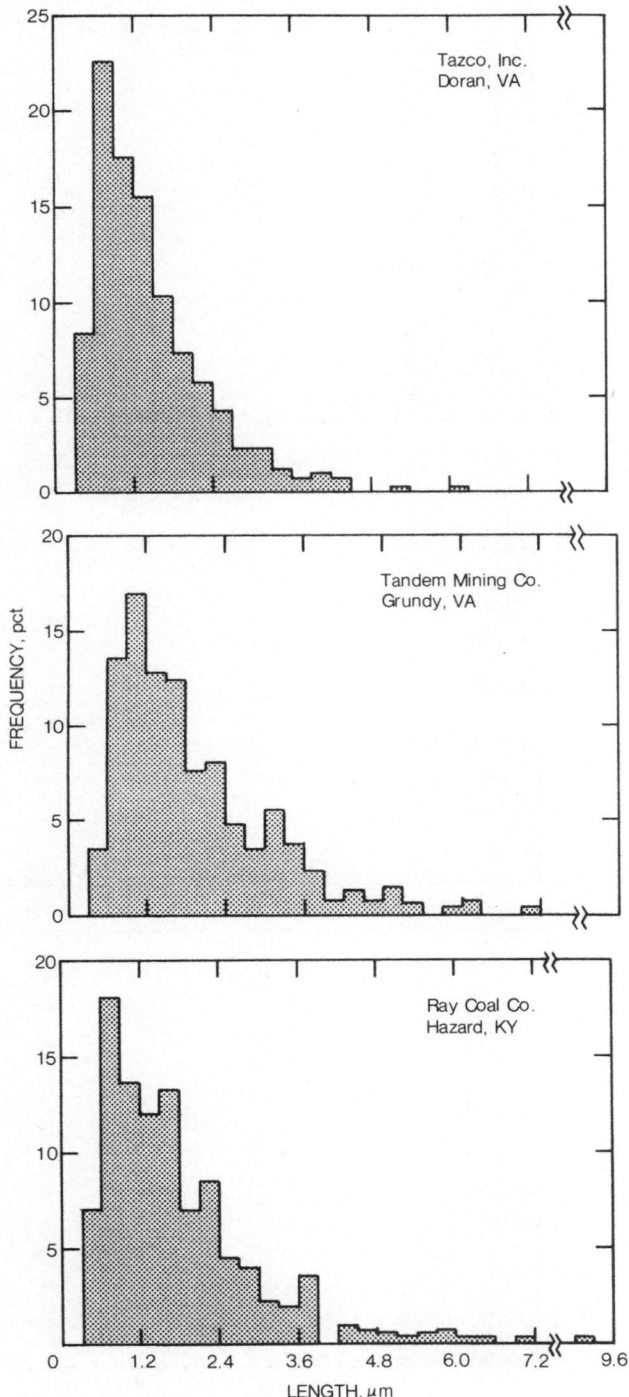

FIGURE 9. - Histograms of quartz particle size distribution found in respirable dust samples collected at underground coal mines in Virginia and Kentucky.

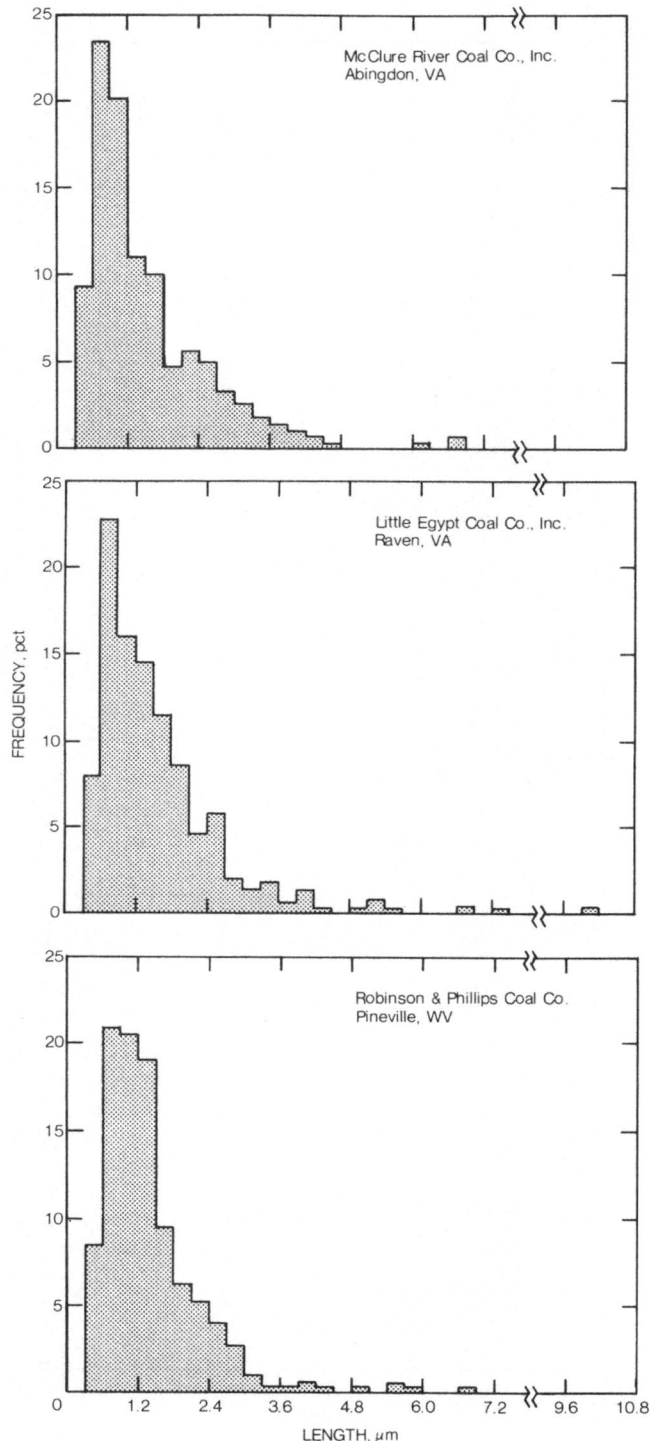

FIGURE 10. - Histograms of quartz particle size distributions found in respirable dust samples collected at underground coal mines in Virginia and West Virginia.

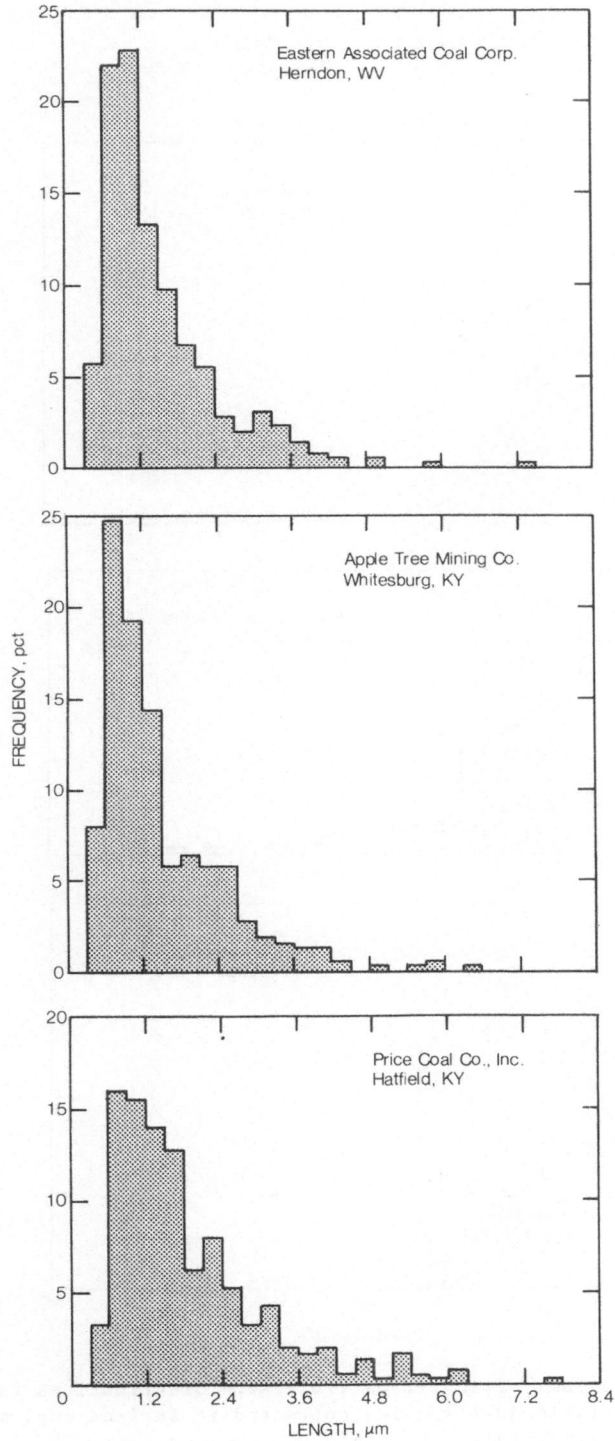

FIGURE 11. - Histograms of quartz particle size distributions found in respirable dust samples collected at underground coal mines in West Virginia and Kentucky.

197

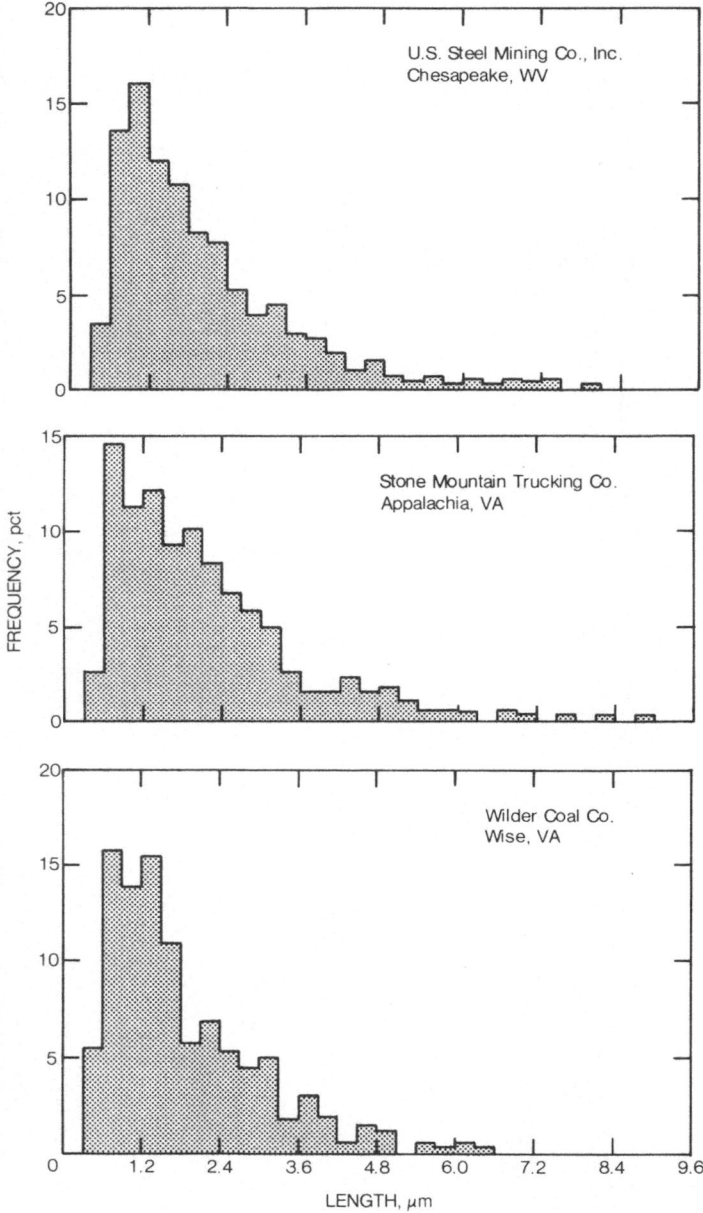

FIGURE 12. - Histograms of quartz particle size distributions found in respirable dust samples collected at surface coal mines in West Virginia and Virginia.

198

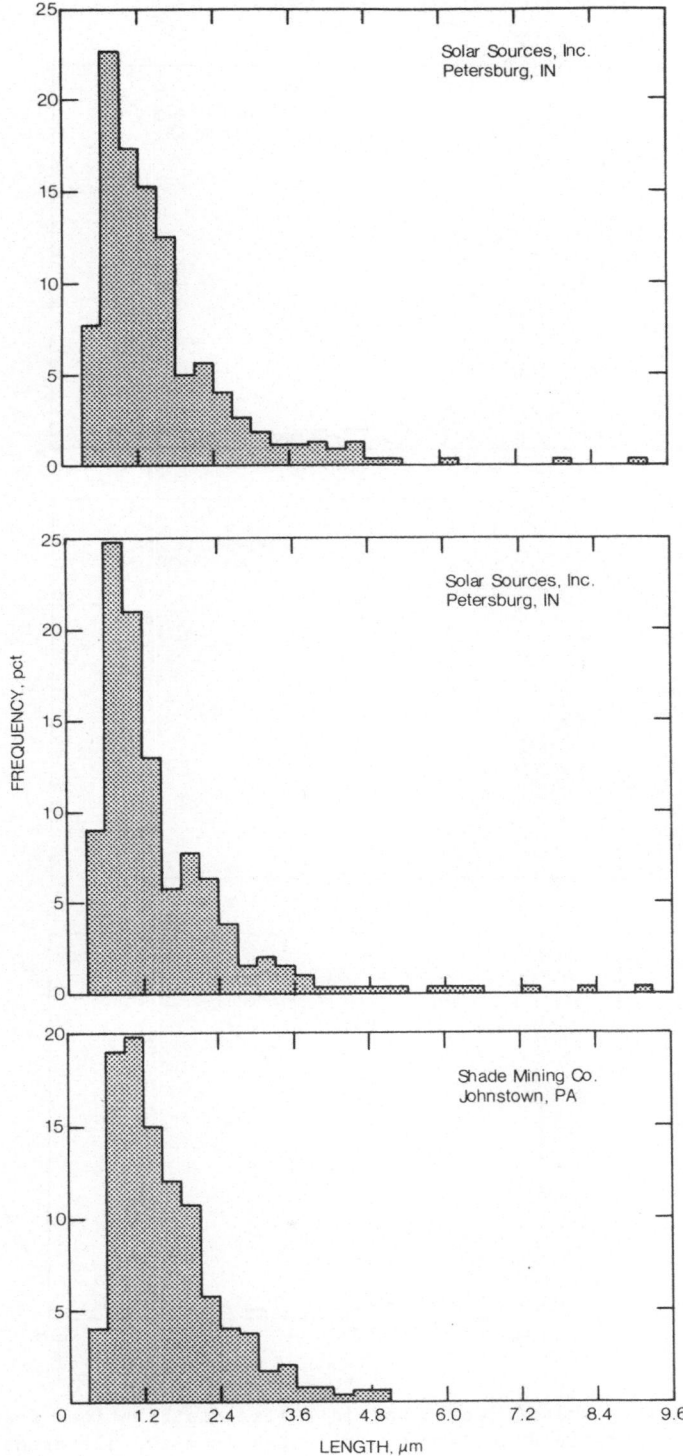

FIGURE 13. - Histograms of quartz particle size distributions found in respirable dust samples collected from two different locations in Indiana surface coal mine and one at Pennsylvania coal mine.

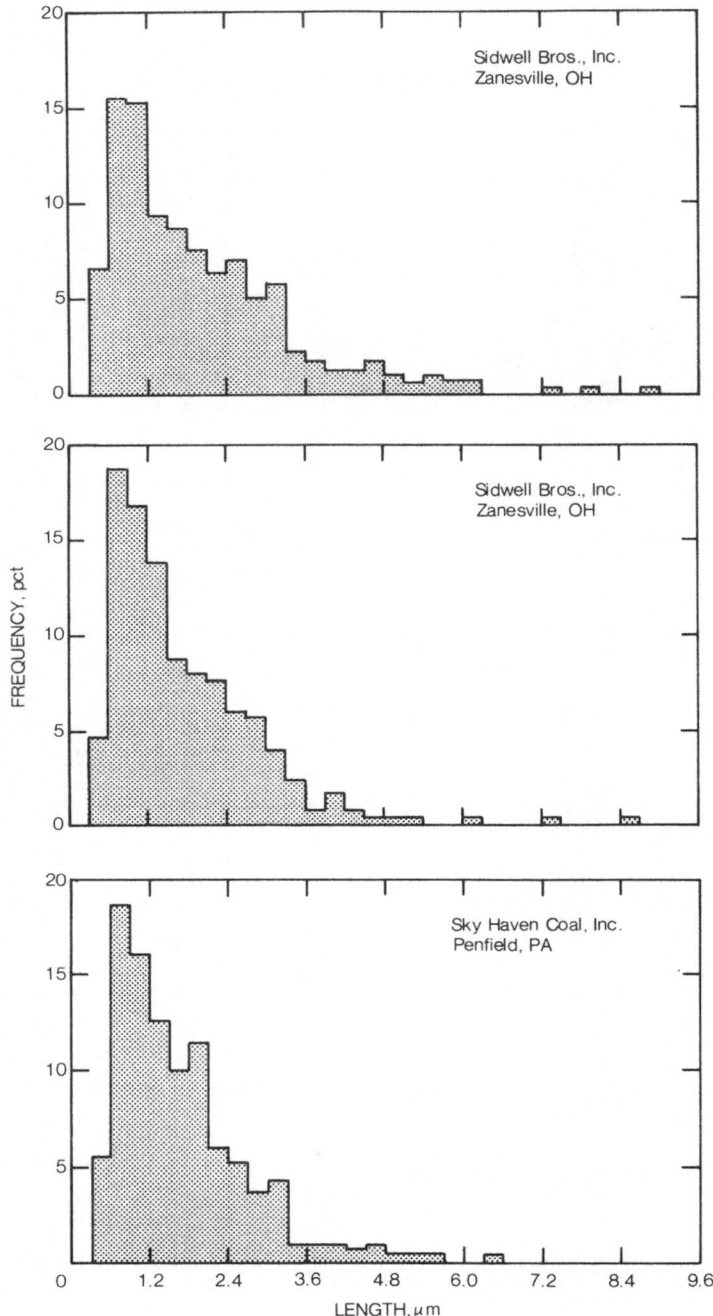

FIGURE 14. - Histograms of quartz particle size distributions found in respirable dust samples collected from two different locations in Ohio surface coal mine and one at Pennsylvania surface coal mine.

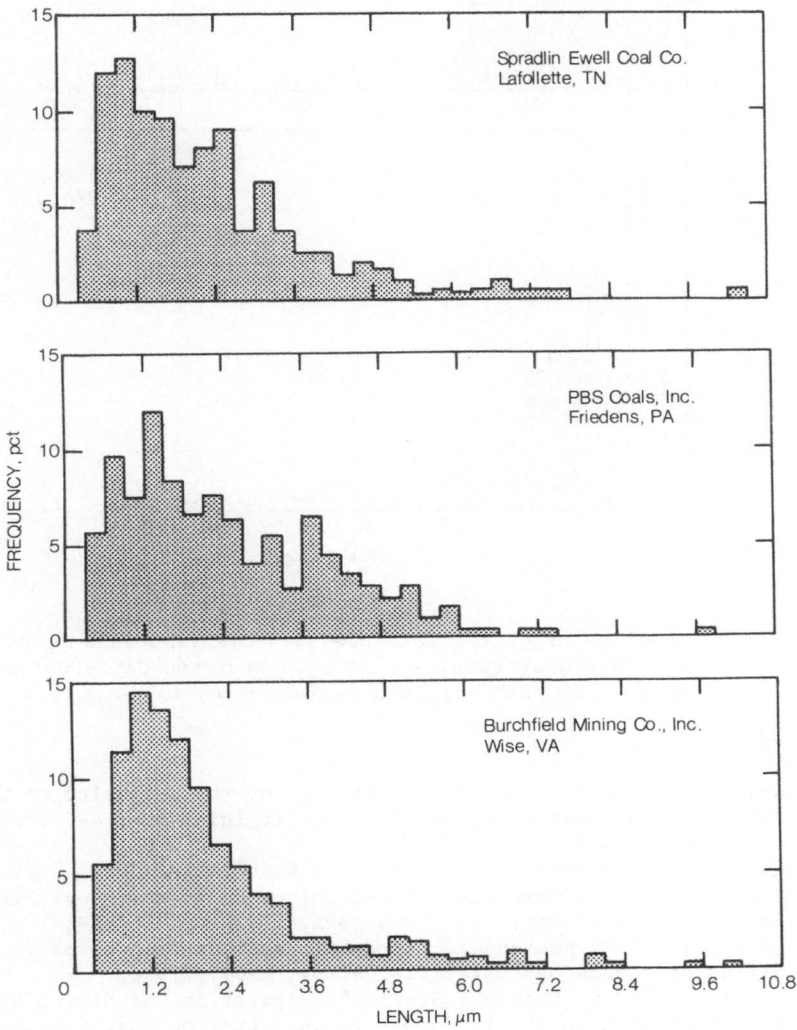

FIGURE 15. - Histograms of quartz particle size distributions found in
respirable dust samples collected at surface coal mines in
Tennessee, Pennsylvania, and Virginia.

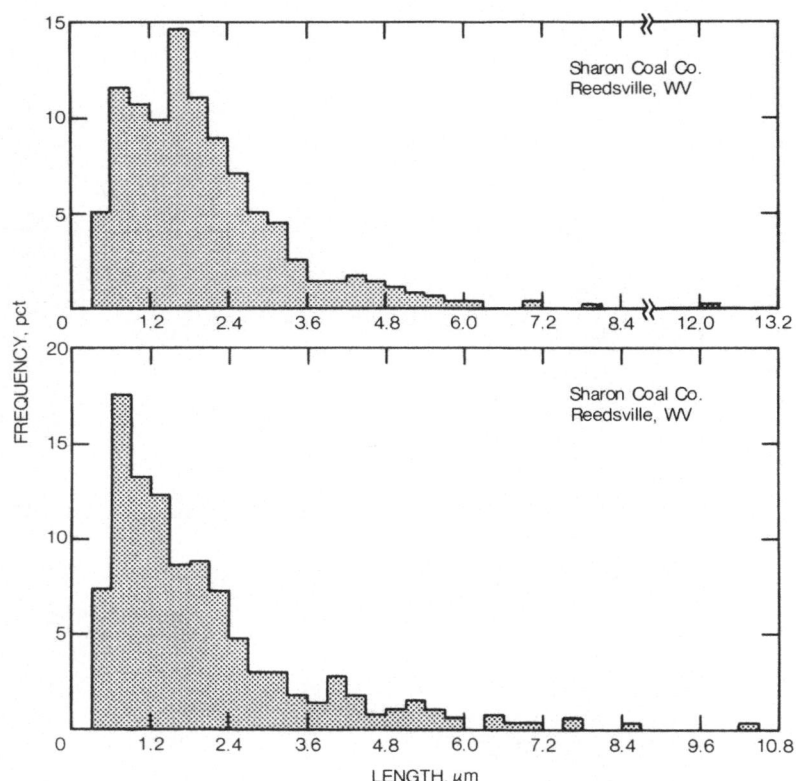

FIGURE 16. - Histograms of quartz particle size distributions found in respirable dust samples collected from two different locations at West Virginia surface coal mine.

REFERENCES

1. Brennan, J.P. Address. Paper in Proceedings of the Symposium on Control of Respirable Coal Mine Dust, Beckley, W. Virginia, Oct. 4-6 1983. MSHA, 1983, pp. 7-13.
2. Huggins, C.W. Roundrobin Investigation of Respirable Quartz Dust. Paper in Proceedings of the Symposium on Control of Respirable Coal Mine Dust, Beckley, W. Virginia, Oct. 4-6, 1983 MSHA, 1983, pp. 287-296.
3. Klug, H.P., and L.E. Alexandeer. X-ray Diffraction Procedures for Poly-crystalline and Amorphous Materials. Wiley, 1954, 716 pp.
4. Snyder, J.G. and C.W. Huggins. Specimen Preparation and Sizing by Image Analysis of Respirable Quartz Particles Collected on Coal Mine Air Monitoring Filters. Microbeam Analysis Soc. Proc., 1983, pp. 22-26.
5. Tuddenham, M.V., and R.P. Lyon. Infrared Techniques in the Identification and Measurements of Minerals. Anal. Chem., v. 32, 1960, pp.1630-1634.

SUBCELLULAR LOCALIZATION OF NICKEL IN THE MYOCARDIUM

Arisztid G.B. Kovach*, Istvan Balogh* and Gabor M. Rubanyi†

*Departments of Physiology and Forensic Medicine,
 Semmelweis Medical University, Budapest, and
†Department of Physiology, Mayo Clinic, Rochester, MN 55905

ABSTRACT

The aim of the study was to analyze uptake of exogenously administered nickel chloride ($NiCl_2$) and redistribution of endogenous nickel under various pathological conditions in myocardial cells by the newly developed dimethylglyoxim electron-cytochemical method. Electron-dense particles are observed in the capillary lumen, endothelial cells, in the cytoplasm and mitochondria of myocardial cells along with ultrastructural damages after perfusion of isolated rat hearts with 1 to 100 μM $NiCl_2$. The presence of nickel in these particles was verified by chloroform-extraction, electron-diffraction and electronprobe microanalysis. Without added $NiCl_2$, similar electron-dense particles and ultrastructural alterations are observed after carbon-monoxide poisoning, hemorrhagic shock and acute burn in rat, dog and human myocardium. Chronic feeding of rabbits with a nickel-rich diet caused nickel accumulation and ultrastructural damages in myocardial cells. It is concluded, that (1) the dimethylglyoxim-method can be used to detect subcellular distribution of nickel and (2) nickel (and other trace metals) may play an important role in cardiomyopathies caused by various pathological conditions.

INTRODUCTION

Earlier work has focused attention on the potentially pathogenetic role of the essential trace metal nickel in myocardial ischemia (Rubanyi, et al., 1984a,b). However, subcellular localization of nickel under normal conditions and its redistribution during pathological events have not been systematically analyzed, primarily because there were no selective electron-cytochemical methods available. We have developed a selective method (using dimethylglyoxim) for the subcellular localization of nickel (Balogh, et al., 1981, 1982; Rubanyi, et al., 1980a) and we studied its redistribution during various pathological conditions in the rat, rabbit, dog and human myocardium.

---

†To whom all communications must be addressed.

MATERIALS AND METHODS

Isolated perfused rat hearts

White Wistar rats of either sex weighing 200 to 300 grams were decapitated by a guillotine after intraperitoneal injection of heparin (5 IU per g body weight). Hearts were rapidly removed and the aortic stump cannulated to allow retrograde coronary perfusion (modified Langendorff technique) by Krebs-Henseleit bicarbonate buffer solution (KHB), equilibrated with 95% O_2-5% CO_2 gas mixture (pH 7.4) and containing 10 mM glucose and 10 mU/ml insulin (Rubanyi and Kovach, 1980). The hearts were perfused by a constant flow peristaltic pump (Watson-Marlow) via bubble-trap, thermostat (37°C) and filter. In some experiments the solution was equilibrated with a gas mixture of 45.1% CO-4.8% CO_2 and 50.1% N_2.

In situ dog heart

Mongrel dogs of either sex weighing 16 to 33 kg were anesthetized by glucochloralose (100 mg per kg body weight) with additional anesthetic given as needed to maintain a constant level of anesthesia. The animals were immobilized by flaxedyl (2 mg/kg) and pulmonary ventilation was accomplished by a positive pressure respirator (Harvard) with room air enriched by 100% oxygen. Blood gases and pH were monitored (Radiometer Copenhagen, Type ABL 1) (Rubanyi, et al., 1984a). In 8 dogs 5 to 7 vol % carbon monoxide (CO) was added to the inhaled gas mixture. Blood COHb concentration (determined by the Wolff-Maehly method from heparinized blood taken from the femoral artery) rose from 0 to 50 to 60 rel % in 4 animals, while in the other 4 dogs it remained below 30 rel %. Control animals (n = 4) were not treated with CO.

Experimental pathological models

Hemorrhagic hypotension in the rat. The rats (n = 8) received heparin after pentobarbital anesthesia (5 mg per 100 g body weight) and both left and right femoral arteries were cannulated (Rubanyi, et al., 1980b; Rubanyi, 1981). One of these was used for continuous monitoring of arterial blood pressure and the other for bleeding of the animal into a polyethylene reservoir until mean arterial pressure reached 35 mmHg. This low pressure was maintained for 3.5 hours by regulating the amount of blood given up by the animal. After 2.5 hours the blood was retransfused and 10 minutes later the heart was removed and perfused with the Langendorff-technique for 30 minutes prior to 10 min perfusion-fixation with 2.5% glutaraldehyde (buffered with 0.1 M cacodylate).

Standardized thermal injury in the rat. A standardized burn model was carried out as modified in our laboratory (Rubanyi, et al., 1983). White laboratory rats (n = 12) of either sex weighing 180-250 g were anesthetized by pentobarbital (50 mg/kg i.v.). Immersion of the shaved back of the animals in a water bath at 98°C for 15 sec caused full-thickness skin burn covering approximately 20-25% of the total body surface area. The rats received 5% body weight Ringer solution intraperitoneally after burn and were kept in individual cages at room temperature. They received water and food ad libitum. Skin lesions were not treated locally. Hearts excised from burned rats on the seventh postburn day were perfused with Krebs-Henseleit solution for 30 min, then perfusion-fixed with 2.5% glutaraldehyde.

Feeding of rabbits with a nickel-rich diet. Adult male New Zealand rabbits (2.5-3.5 kg body weight) were fed with a nickel-rich diet (12.5 mg/kg body weight/day) for 3 months. Control animals were kept under identical conditions except nickel chloride was not added to their diet. After 3 months they were anesthetized (sodium pentobarbital; 50 mg/kg) and the heart removed for ultrastructural analysis (see below).

Human hearts

One hour post mortem, a tissue sample was taken from the papillary muscle of the left ventricle by VIM-Silverman needle for ultrastructural analysis from human cadavers, who died as a consequence of acute carbon-monoxide poisoning (n = 40), of severe burn (n = 40) or of traffic accident (n = 10).

Ultrastructural analysis

Following fixation in glutaraldehyde, post fixation in OsO_4 and dehydration, multiple blocks were embedded in Durcupan (ACM Fluka), and ultrathin sections were cut by a Reichert OM U2 ultramicrotome. The sections were examined in a JEM 100B electron microscope at 60 kV acceleration voltage.

Subcellular localization of nickel

For intracellular localization of Ni the dimethylglyoxim cytochemical method was used as developed in our laboratory (Rubanyi, et al., 1980a; Balogh, et al., 1981, 1982). Following perfusion-fixation of the rat hearts, or fixation of the excised myocardium of the rabbit, dog and human hearts with 2.5% glutaraldehyde, the specimens were incubated in 0.1% dimethylglyoxim (Reanal; dissolved in 70% ethanol) for 5 min. One percent OsO_4 was used for afterfixation. Following this procedure, some specimens were treated with neutral chloroform to remove the metals from the metal-dimethylglyoxim complexes. Analytical electron microscopy was performed by an ORTEC energy dispersive electronprobe spectrometer in combination with a JEOL 100B electron microscope provided with a JEOL scanning attachment (Rubanyi, et al., 1981). The specimen, held by copper grids, was examined in the scanning mode at 40-80 kV. Total counting time was 200 sec. In some cases the analysis was performed by the electron diffraction method (Balogh, et al., 1981, 1982).

RESULTS

Effect of perfusion with $NiCl_2$ on the ultrastructure and subcellular Ni-localization of perfused rat hearts

The normal ultrastructure of the myocardium of isolated rat hearts fixed after 30 min control perfusion and treated with dimethylglyoxim is shown in Figure 1. Infusion of dimethylglyoxim does not result in the appearance of electrondense particles in myocardial cells, only the characteristic (metal-free) concentric dimethylglyoxim structure can be observed in the capillary (Fig. 1). No ultrastructural damage was observed following 10 min perfusion with a solution containing 1 µM $NiCl_2$, but electrondense nickel-dimethylglyoxim complexes can be seen in the capillary lumen, along the gap-junctions and pinocytotic vesicles in the capillary wall, in the pericapillary interstitial space and also in the cytoplasm of myocardial cells (Figs. 2-4), but never in the mitochondria or other intracellular structures. Perfusion of the hearts with 100 µM $NiCl_2$, caused intracellular edema and characteristic changes of the mitochondrial structure (intramitochondrial accumulation of glycogen-like particles; rupture and lysis of the cristae), and nickel-dimethylglyoxim complexes were found in the mitochondria as well (Figs. 5-7).

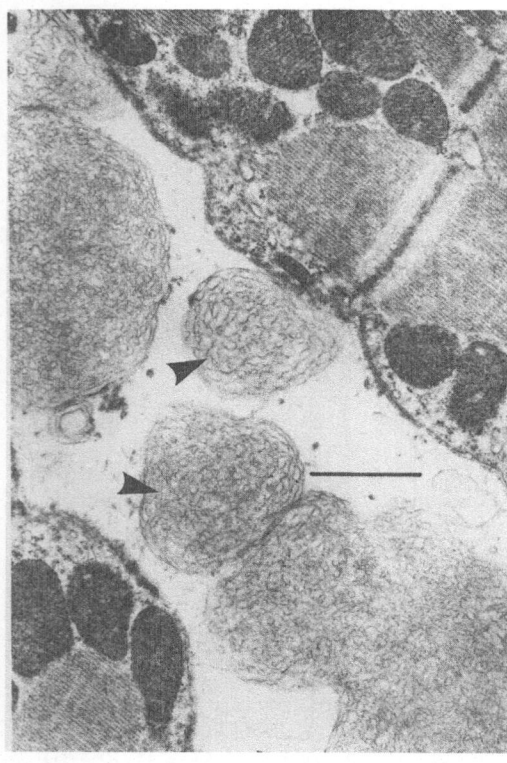

Fig. 1. Ultrastructure of rat myocardium after 30 min perfusion with control solution and treatment with dimethylglyoxim. The characteristic dimethylglyoxim structure (without electrondense metal particles) are seen in the capillary lumen. The solid bars in this and in all other figures represent 1 μm.
Arrows: Dimethyl glyoxim structures

Several tests were performed to verify the presence of nickel in the electrondense particles within the dimethylglyoxim complexes. Treatment of the specimen with neutral chloroform completely removed the electrondense particles, and only the "metal-free" dimethylglyoxim structure remained (not shown), as in the control hearts (see Fig. 1). Electron diffraction (inset in Fig. 6) and energy-dispersive x-ray microanalysis (inset in Fig. 7) of the electrondense particles (observed after perfusion of the rat hearts with 100 μM $NiCl_2$) verified the presence of nickel in the reaction complexes.

Effect of Ni-rich diet on the rabbit myocardium

Feeding the rabbits with a nickel-rich diet (12.5 mg/kg b.w./day $NiCl_2$) for 3 months induced characteristic changes in the myocardial ultrastructure: distension of the Z-line-zone, increased number of myofibrils and accumulation of collagen fibers in the interstitial space (Fig. 8). In some (but not in all) specimens nickel-dimethylglyoxim complexes can be observed in the mitochondria, which otherwise did not show pathological structure alterations (Fig. 8).

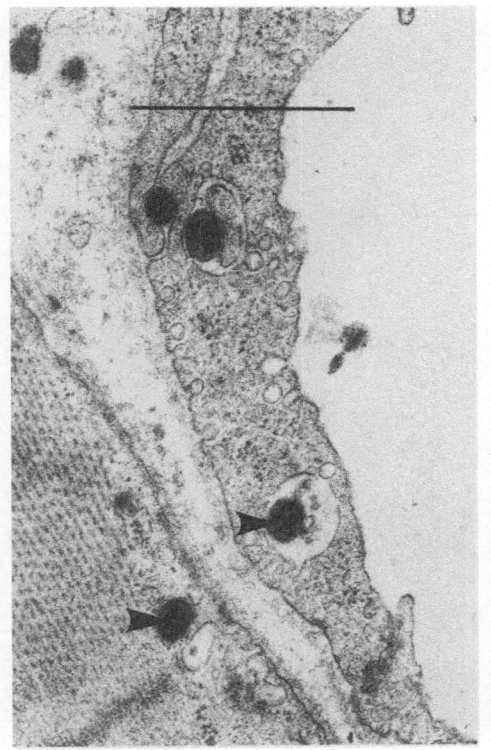
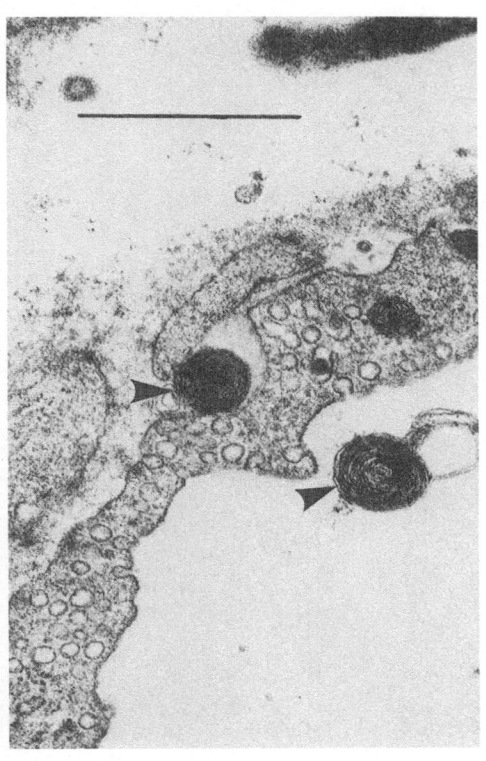
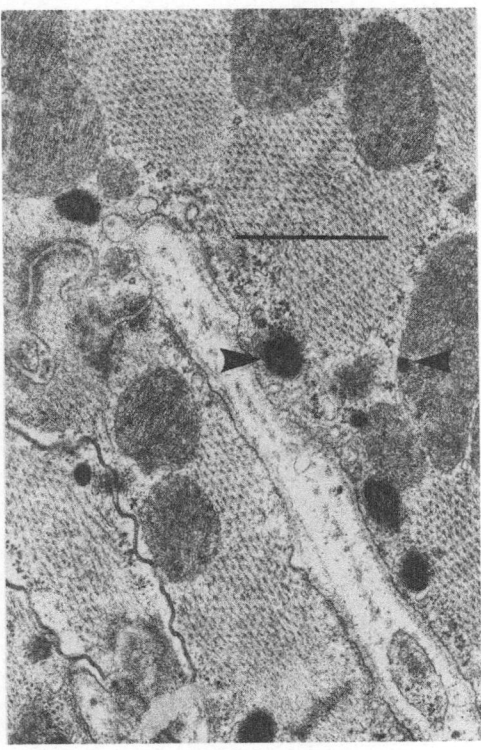

Figs. 2-4. Effect of exogenous NiCl$_2$ (1 μM) on the ultrastructure and sub-cellular Ni-distribution in isolated perfused rat hearts. With no alterations of the ultra-structure, electrondense Ni-dimethylglyoxim complexes are seen in the capillary lumen (Fig. 3), in the gap junctions (Fig. 2) and pinocytotic vesicles of the capillary endothelium (Figs. 2 and 3), in the pericapillary space (Fig. 3) and in the cytoplasm of myocardial cells (Figs. 2 and 4). Arrows: Ni-dimethyl-glyoxim particles

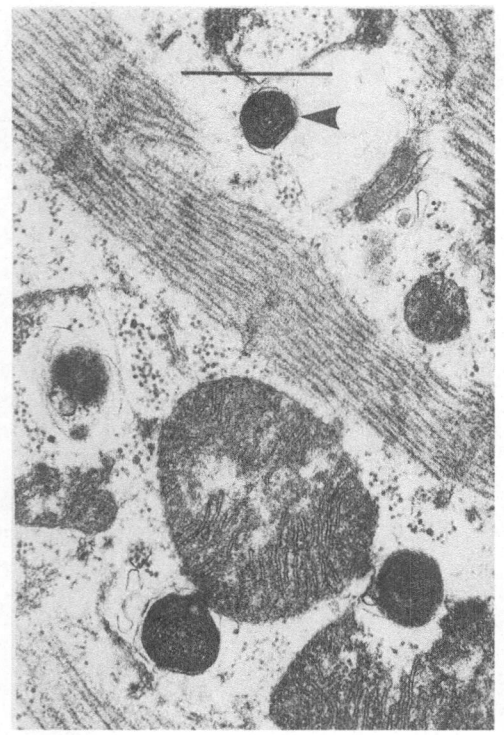

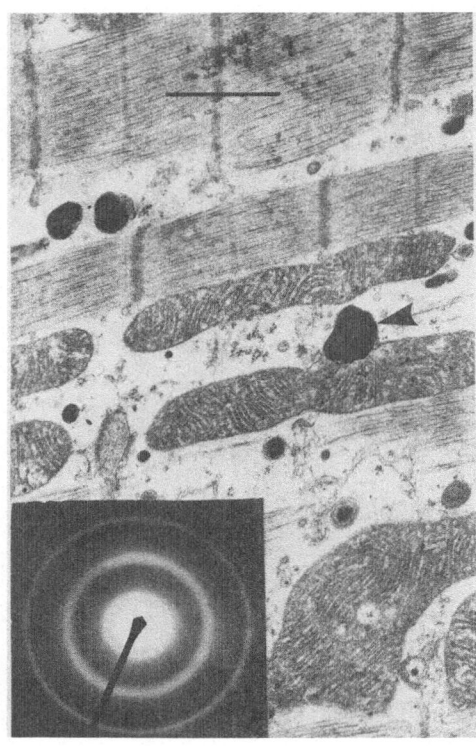

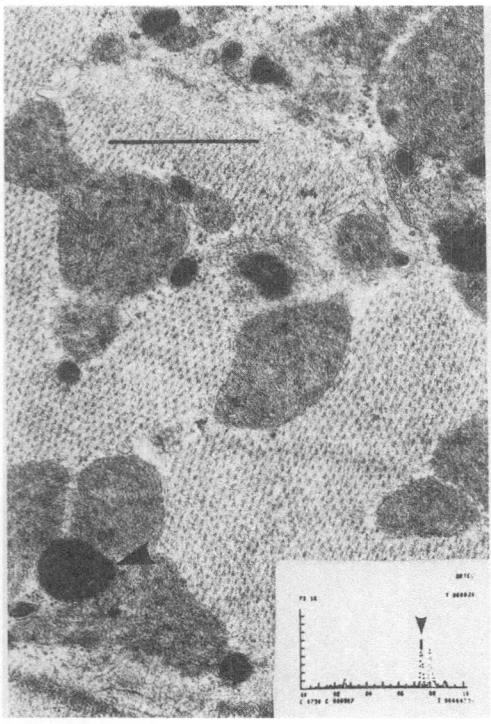

Figs. 5-7. Ultrastructural damage
and intracellular and
intramitochondrial
accumulation of Ni-
dimethylglyoxim particles
in isolated rat hearts
following perfusion with
100 μM $NiCl_2$. Electron
diffraction (inset in
Fig. 6) and electronprobe
(inset in Fig. 7) micro-
analysis of the electron-
dense particles verified
the presence of nickel
in the reaction. The
first large peak (indi-
cated by the arrow) on
the energy-dispersive
spectrum of the electron-
probe analysis represents
nickel, while the second
peak the grid-material,
copper (Fig. 7).
Arrows: Ni-dimethyl-
glyoxim particles

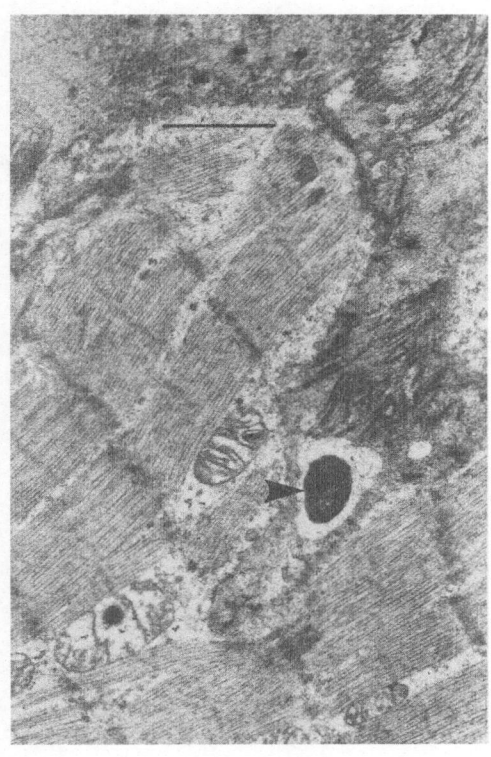

Fig. 8.   Ultrastructural alterations
and subcellular Ni-
detection (by dimethyl-
glyoxim) in the rabbit
myocardium following 3
months feeding of the
animals with a Ni-rich
diet (12.5 mg/kg body
weight/day).

Detection of endogenous nickel accumulation/redistribution

Carbon monoxide poisoning. Specific nickel-dimethylglyoxim complexes
can be observed in the myocardial cells of in situ dog hearts after
inhalation of 5 to 7 vol % carbon monoxide (COHb rel %: 50-60%; Fig. 9)
and in the heart of human cadavers died of acute CO-poisoning (Fig. 10).
Similar observations were made with isolated rat hearts perfused with
45.1% CO (data not shown). No nickel-dimethylglyoxim complexes could be
found in the heart of control dogs, or of CO-poisoned dogs, with less
than 30 rel % COHb in the blood and of persons who died in traffic
accidents (not shown).

Hemorrhagic shock. Following 2.5 h hemorrhagic hypotension
characteristic electrondense nickel-dimethylglyoxim complexes can be
detected in the myocardium of rat hearts (Fig. 11) in parallel with
swelling of the mitochondria.

Burn injury. Seven days after exposure of the rats to severe burn
injury, nickel-dimethylglyoxim complexes can be seen in the myocardium
(Fig. 12). Post mortem analysis of the biopsy material taken from the
heart of patients who died as a consequence of severe third or fourth
degree burn showed the presence of large number of nickel-dimethylglyoxim
complexes (Fig. 13), which exceeded the amount observed in the other
pathological conditions. As described earlier (Balogh, et al., 1981),
the presence of nickel in the electrondense material was always verified
by the tests, shown for the verification of nickel in perfused rat hearts
exposed to exogenous $NiCl_2$ (see Fig. 7).

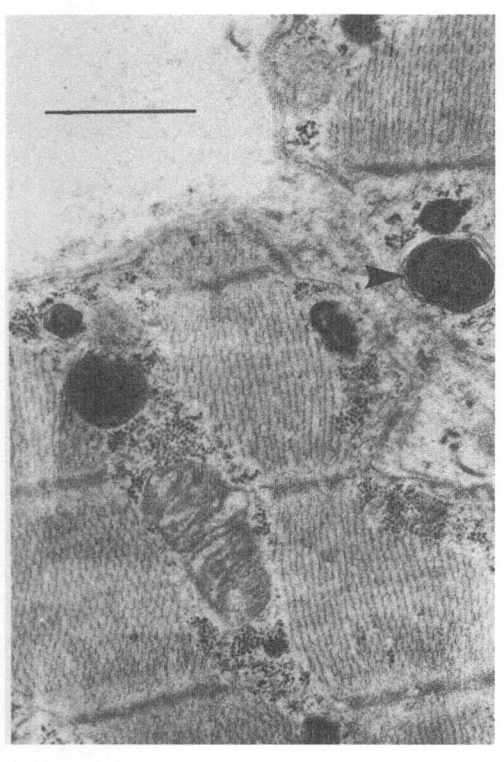

Fig. 9. Effect of carbon monoxide
inhalation (5 to 7 vol %)
on myocardial ultra-
structure and intracellular
localization of Ni-di-
methylglyoxim complexes
in the _in situ_ dog heart.
Electrondense reaction-
complexes can be observed
in the cytoplasm and in
the damaged mitochondria.
Arrows: Ni-dimethyl-
glyoxim particles

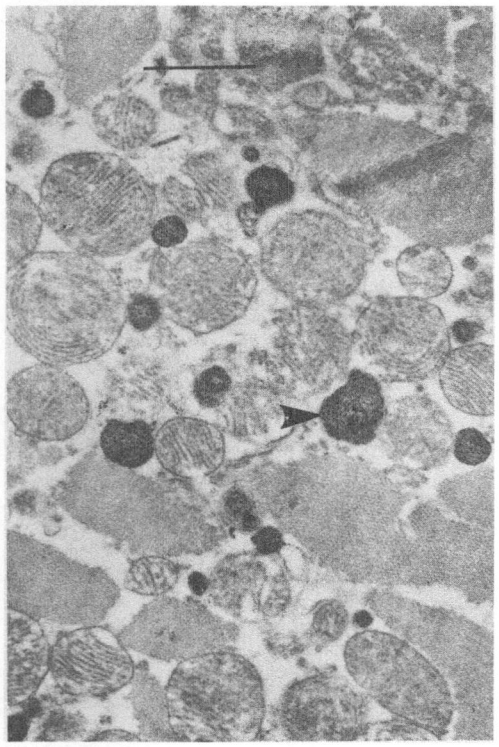

Fig. 10. Ni-dimethylglyoxim
reaction in the myo-
cardial cell of a
human heart one hour
after acute death due
to carbon monoxide
poisoning.

Fig. 11. Swelling of mitochondria and characteristic Ni-dimethylglyoxim complexes in myocardial cells of the rat heart following 2.5 h hemorrhagic hypotension. Inset: electron diffraction picture of the electron-dense reaction complex. Arrows: Ni-dimethylglyoxim particles

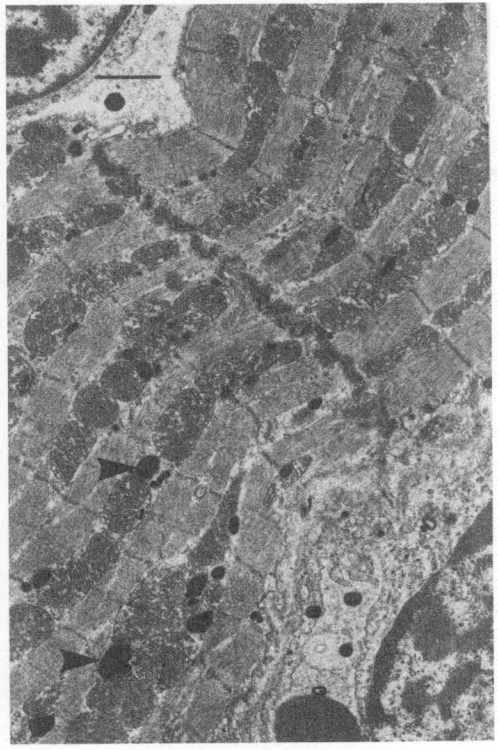

Fig. 12. Intracellular localization of Ni-dimethylglyoxim particles in the myocardium of rat hearts seven days after severe burn injury. Arrows: Ni-dimethylglyoxim complexes

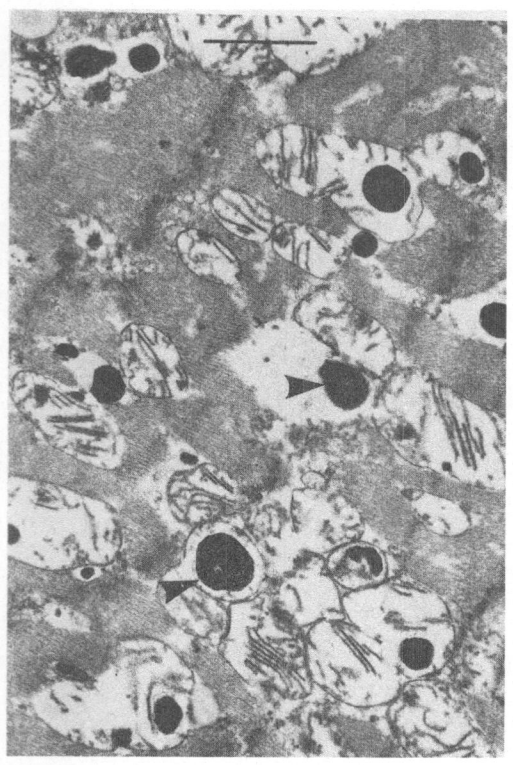

Fig. 13. Ni-dimethylglyoxim
reaction in the myo-
cardial cells of
human hearts one hour
after death following
severe burn. Note the
signs of autolysis.
Arrows: Ni-dimethyl-
glyoxim complexes

DISCUSSION

The present study demonstrates that the newly developed dimethyl-
glyoxim-electrocytochemical method is an adequate technique for the
subcellular detection of nickel. Its specificity is verified by chloroform-
extraction and by the more selective electron diffraction and electronprobe microanalyses. The technique can be successfully used for the
detection of (1) the transport and subcellular accumulation of exogenously
administered nickel in the normal myocardium; and (2) the accumulation/
release/redistribution of endogenous nickel under various pathological
conditions. However, this technique cannot be used in its present form
to detect endogenous nickel in the myocardium under normal conditions,
despite the fact that the myocardium contains relatively large amounts
of this trace metal (Rubanyi, et al., 1981b). These facts indicate that
under normal conditions nickel is in a (bound) form in the myocardium
which is not accessible for complex reaction with dimethylglyoxim. If,
however, free nickel is administered exogenously, or endogenous (bound)
nickel is liberated or accumulated in the myocardium, dimethylglyoxim
is able to react with the trace metal to form electrondense particles.
Although this electrocytochemical technique is not quantitative enough to
estimate accurate changes in endogenous nickel content (for that purpose
the atomic-absorption spectrophotometric method can be used; Rubanyi, et
al., 1981b, 1984a), it can serve as a valuable tool to detect liberation
of "free" nickel under various pathological conditions and to characterize
its subcellular distribution or transport across the capillary wall or
myocardial cell membrane.

These features of the technique may be of great importance in the analysis of the recently discovered pathophysiological significance of this trace metal. It was shown that low concentrations of exogenous $NiCl_2$ induce coronary arterial vasoconstriction in isolated perfused rat hearts (Rubanyi and Kovach, 1980), in the in situ dog heart (Rubanyi, et al., 1981a,b; 1984a,b) and in isolated canine coronary artery strips (Rubanyi, et al., 1982). Exogenous $NiCl_2$ inhibits the coronary vasodilator action of arterial hypoxemia, coronary occlusion and exogenous adenosine in the in situ dog heart (Rubanyi, et al., 1984a,b) and in isolated rat hearts (Rubanyi, et al., 1981b). Adenosine, hypoxia and hypoperfusion augment the coronary vasoconstrictor action of exogenous $NiCl_2$ in the in situ dog heart (Rubanyi, et al., 1984a,b). The pathological significance of these observations with exogenously administered $NiCl_2$ is emphasized by the findings that serum nickel level is significantly elevated in patients with acute myocardial infarction (D'Alonzo and Pell, 1963; Tarala, 1970; McNeely, et al., 1971) and in experimental animals during general or regional myocardial ischemia (Aranova, 1973; Rubanyi, et al., 1981a,b; 1984a,b). A significant rise of serum nickel concentration during myocardial ischemia in the pig (Anke, et al., 1980), during prolonged severe hemorrhagic hypotension in the rat (Rubanyi, et al., 1981b) and during the acute phase of burn injury in the rat (Rubanyi, et al., 1983) and in patients (McNeely, et al., 1971) is accompanied by significant decrease of total nickel content of the myocardium. The present demonstration of the appearance of nickel-dimethylglyoxim particles in myocardial cells of the rat heart after hemorrhagic hypotension, of the rat and human heart after burn injury and in the rat, dog and human heart following severe carbon monoxide poisoning is in good agreement with these previous observations and is a visual representation of nickel mobilization in the myocardial cells under these conditions. Although the presence of nickel in the electrondense particles was verified in these pathological conditions (Balogh, et al, 1981), the contribution of other metals cannot be excluded, as evidenced by our preliminary observations with vanadate and cobalt (Kovach, et al., 1985).

Ultrastructural alterations observed in the rat and rabbit myocardium in response to acute (rat) or chronic (rabbit) exposure to excessive amounts of exogenously administered $NiCl_2$ indicate that in addition to the pathophysiological significance of endogenous nickel-liberation, intake of excessive amounts of exogenous nickel chloride may lead to secondary cardiomyopathy. This finding may add to the list of already known toxicological actions (e.g. carcinogenesis; Sunderman, 1977) of this trace metal. However, recent observations that other trace metals (e.g. cobalt and vanadate) can also induce ultrastructural alterations of myocardial cells (Kovach, et al., 1985), may indicate the involvement of metals (other than nickel) in the pathogenesis of cardio-myopathies.

Finally, the present studies indicate that the dimethylglyoxim method can be useful in the determination of the cause of death. This conclusion is based on the finding that despite various degrees of post mortem autolysis, the nickel-dimethylglyoxim particles can be selectively detected in the myocardium of cadavers who died as a consequence of burn or carbon monoxide poisoning, but not in those who died in traffic accidents. After systematic elaboration of this feature of the newly developed technique, it may be of great importance in forensic medicine.

ACKNOWLEDGEMENTS

This chapter is largely an elaboration of an abstract published by San Francisco Press, Inc. (Kovach, et al., 1985).

REFERENCES

Anke, M., Schneider, N.J., Bruckner, C., 1980, "3rd Spurenelement Symposium on Nickel", University Press, Jena.

Aranova, A.V., 1973, Certain trace elements in experimental myocardial infarction with hypertension and atherosclerosis in the background, Kardiologia 13: 43-47.

Balogh, I., Rubanyi, G., Kovach, A.G.B., Sotonyi, P., Somogyi, E., 1981, Electron cytochemical detection of nickel ions under pathological conditions in the rat heart, Acta Morph. Acad. Sci. Hung. 29: 87-90.

Balogh, I., Somogyi, E., Rubanyi, G., 1982, Nickel cytochemistry: applicability of a new cytochemical technique in forensic medicine, Acta Med. Leg. Soc. (Liege) 32: 459-464.

D'Alonzo, C.A., Pell, S.A., 1963, A study of trace metals in myocardial infarction, Arch. Envir. Hlth. 6: 381-389.

Kovach, A.G.B., Balogh, I., Rubanyi, G., 1985, Effects of the elements Ni, V, Co on ultrastructure of the heart muscle, in: Proc. 43rd Annual Meeting of the Electron Microscopy Society of America, G.W. Bailey, ed., San Francisco Press, Inc., San Francisco, pp. 94-97.

McNeely, M.D., Sunderman, F.W., Jr., Nechay, M.W., Levine, H., 1971, Abnormal concentrations of nickel in serum in cases of myocardial infarction, stroke, burns, hepatic cirrhosis and uremia, Clin. Chem. 17: 1123-1128.

Rubanyi, G., Kovach, A.G.B., 1980, Cardiovascular actions of nickel ions, Acta Physiol. Hung. 55: 345-353.

Rubanyi, G., Balogh, I., Somogyi, E., Kovach, A.G.B., Sotonyi, P., 1980a, Effect of nickel ions on ultrastructure of isolated perfused rat heart, J. Mol. Cell. Cardiol. 12: 609-618.

Rubanyi, G., Koltay, E., Dora, T., Balogh, I., Kovach, A., Somogyi, E., 1980b, Effect of hemorrhagic shock on mechanical activity, O_2-consumption and ultrastructure of isolated rat heart, Circ. Shock, 7:59-71.

Rubanyi, G., 1981, Control of coronary vascular tone in hemorrhagic shock, in: "Injury and shock", Biro, S., Kovach, A.G.B., Spitzer, J.J., Stoner, H.B., Adv. Physiol. Sci., vol. 26; Pergamon Press, Oxford/Akademiai Kiado, Budapest, pp. 99-108.

Rubanyi, G., Ligeti, L., Koller A., 1981a, Nickel is released from the ischemic myocardium and contracts coronary vessels by a Ca-dependent mechanism, J. Mol. Cell. Cardiol. 13:1023-1026.

Rubanyi, G., Ligeti, L., Koller, A., Bakos, M., Gergely, A., Kovach, A.G.B., 1981b, Physiological and pathological significance of nickel ions in the regulation of coronary vascular tone, Adv. Physiol. Sci., vol. 27, Szentivanyi, I., Juhasz-Nagy, S., eds., Pergamon Press, Oxford/Akademiai Kiado, Budapest, pp. 133-154.

Rubanyi, G., Kalabay, L., Pataki, T., Hajdu, K., 1982, Nickel stimulates isolated canine coronary artery contraction by a tonic calcium activation mechanism, Acta Physiol. Hung. 59: 155-160.

Rubanyi, G., Szabo, K., Balogh, I., Bakos, M., Gergely, A., 1983, Endogenous nickel release as a possible cause of coronary vaso-constriction and myocardial injury in acute burn of rats, Circ. Shock, 10: 361-370.

Rubanyi, G., Ligeti, L., Koller, A., 1984a, Possible role of nickel ions in ischemic coronary vasospasm in the dog heart, J. Mol. Cell. Cardiol., in press.

Rubanyi, G., Ligeti, L., Koller, A., Kovach, A.G.B., 1984b, Nickel ions and ischemic coronary vasoconstriction, in: "Vasodilator Mechanisms"; Vanhoutte, P.M., Vatner, S., eds., Karger, Basel, pp. 200-208.

Sunderman, F.W., Jr., 1977, A review of the metabolism and toxicology of nickel. Ann. Clin. Lab. Sci. 7: 377-398.

Tarala, G.T., 1970, The content of zinc, copper, nickel, manganese, lead and silver in the blood of patients with angina pectoris at different stages of the disease, _Kardiologia_, 10: 146–147.

PARTICLE CONTENTS OF HUMAN LUNGS

L. E. Stettler,* S. F. Platek,* D. H. Groth*
F. H. Y. Green** and V. Vallyathan**

*Division of Biomedical and Behavioral Science
National Institute for Occupational Safety and Health
Cincinnati, Ohio 45226
**Division of Respiratory Disease Studies
National Institute for Occupational Safety and Health
Morgantown, Virginia 26505

ABSTRACT

An automated analysis technique utilizing scanning electron microscopy, energy dispersive x-ray analysis, and image analysis was used to determine the exogenous particulate burdens for 48 urban lung specimens from the Cincinnati, Ohio area. Over 60,000 individual particle analyses were performed for the sample group. The mean exogenous particle concentration was $508 \pm 417 \times 10^6$ particles per gram of dry lung, with a range of $107 - 1844 \times 10^6$ particles/g. The particle compositions found in all lungs were very similar. Major particle types included various aluminum silicates, silica, rutile, iron oxide, talc, and alumina. This data will serve as baseline data for lungs with no overt pneumoconioses against which the particle contents of diseased lungs can be compared.

INTRODUCTION

Scanning electron microscopy (SEM) and energy dispersive x-ray analysis (EDXA) have been used by many investigators (e.g., Abraham, 1979; Brody, et al, 1976; DeNee, 1976; Stettler, et al., 1977; and Vallyathan, et al., 1980) to investigate the particle content of human lungs. Most of these analyses have been performed on lung specimens with known or suspected lung disease. The object of this work is to relate the composition and concentration of particles present in the lungs to the disease. With few exceptions (Abraham and Burnett, 1983; Churg and Wiggs, 1985), the major factor missing from lung microanalysis work has been data from typical "normal" lungs, with which the analyses of diseased lungs could be compared. Currently in progress at NIOSH, is a study in which the particle contents of 96 urban lungs from the Cincinnati, Ohio area are being determined. Preliminary summary reports for the analysis of 33 (Stettler, et al., in press) and 35 of these lungs (Stettler, et al., 1985) have been given elsewhere. A total of 48 of the 96 lungs have now been analyzed and will be summarized and discussed in this report. When completed, the analysis data from these

96 specimens will serve as baseline data, with which comparisons of the particle contents of diseased lungs can be made.

MATERIALS AND METHODS

Lung Specimens

The left lungs from 96 deceased residents of the Cincinnati, Ohio urban area make up the sample group for this study. These lungs have been extensively analyzed for trace metal contents using bulk analytical procedures. Complete descriptions of the sample group, as well as these bulk chemical analyses, have been reported elsewhere (Sweet, et al., 1978). The sample group was about evenly divided between men and women, blacks and whites. Ages ranged from 23 to 96 years, with a median age of 66 years. Thirty-six of the subjects were non-smokers; the remaining 60 were light to heavy users of tobacco. The range of occupations represented included foundry workers, machinists, farmers, gardeners, salesmen, domestics, waitresses, clerical workers and homemakers. The subjects' deaths were from natural causes, mostly chronic degenerative conditions of advancing age. While there were some acute disease conditions, there were no traumatic deaths. Autopsy reports for each of the subjects were available for analysis. Hematoxylin and eosin (H&E) stained sections from the right lungs of approximately 80% of the subjects were also available for evaluation.

Tissue Processing

The tissue specimens received for microanalysis consisted of freeze-dried homogenates of the left lungs. The procedure used to prepare the freeze-dried homogenates has been reported elsewhere (Sweet, et al., 1978). Small portions of the homogenates of each lung specimen weighing from 0.1 - 0.2 g were processed separately for microanalysis. In this procedure, a specimen was placed in a new, chromic acid-washed scintillation vial and then ashed for seven hours in a low temperature asher at 90 watts, using an oxygen pressure of 2 torr. A sample blank for each lung specimen, starting with an empty crucible in the asher, was carried through each step in the preparation procedure.

A suspension of the ash from the sample crucible was then made using 50 ml of a 0.05% solution of Aerosol OT® in filtered, de-ionized water. The blank crucible was similarly treated. New acid-cleaned glassware were used throughout this and following procedures. The resulting suspensions from the lung and blank were sonicated in an ultrasonic bath for 15 minutes. The suspensions were then made up to a final volume of 100 ml with filtered, de-ionized water, to which 1 ml of glacial acetic acid had been added and then allowed to sit overnight. It was found that this acetic acid treatment successfully removed most of the endogenous calcium and phosphorus-containing particles from the ash suspensions.

Aliquots of the suspensions for sample and blank were then filtered onto 0.1 µm pore size, 25 mm diameter Nuclepore® polycarbonate filters. These filters were then attached to carbon planchets using colloidal graphite and examined in the SEM to determine whether particle loading was suitable for subsequent analysis. Based upon this examination, a final filter preparation was made for each sample. Final aliquots for the samples analyzed in this report ranged from 1 to 20 ml.

Filter Analyses

The final filter preparations for each specimen, including the blank filter, were analyzed in the SEM, which was equipped with a Kevex 7000 EDXA system and a LeMont Scientific DA-10 image analyzer using the backscattered electron image. Complete descriptions of SEM-based image analysis instrumentation and its application to particle analysis have been reported elsewhere (Kelly, et al., 1980; Lee, et al., 1978, 1980).

For each filter, a minimum of 1000 exogenous particles in a minimum of 20 randomly selected fields of view at a magnification of 1000X were sized and analyzed for 31 elements using an x-ray spectrum acquire time of five seconds. A list of the elements analyzed is given in Table 1. Unless otherwise noted, the K_α x-ray lines were used in the analyses. Each particle is classified by the image analyzer using an operator-defined chemistry definition file, which classifies the particles based upon their major elemental components and their net fractional x-ray intensities. A single chemistry file consisting of 14 major chemical classes, shown in Table 2, was used for all of the lungs in this study. The first four classes in this chemistry file are for particles which contain only one major peak; i.e., silicon, iron, titanium, or aluminum. All particles which are placed into these categories are presumed to be oxides.

The Aluminum Silicate class allows for a broad range of aluminum, silicon, and other element x-ray intensities. This Aluminum Silicate class has been further divided into subclasses with more specific elemental definitions, which allowed these aluminum silicates to be further classified on the basis of the major cations present such as potassium, sodium, calcium, iron, etc. The Magnesium Silicate class allows for a broad range of magnesium, iron, and silicon intensities. A subclass, which was based upon the analysis of a known talc sample, was also written for this category.

The Exogenous-Endogenous Combination class contains particles which are considered as endogenous (formed within the lungs from biological products) or combinations of endogenous and exogenous (from the environment) particles. The major characteristic of these particles is that they all contain phosphorus. The class was further subclassified

Table 1 -- Elements Analyzed

Na	Cl	Ti	Ni	Se	Mo
Mg	Cd*	V	Cu	Br	
Al	K	Cr	W*	Sr	
Si	Sn*	Mn	Zn	Y	
P	Ca	Fe	Hg*	Zr	
S	Sc	Co	Pb*	Nb	

*L_α peak used for analysis

Table 2 -- Chemistry Definition File - Major Classes

Silica	Silicon Rich
Iron Oxide-like	Iron Rich
Rutile-like (TiO$_2$)	Titanium Rich
Alumina-like	Aluminum Rich
Aluminum Silicate	Other Aluminum Silicates
Magnesium Silicate	No X-ray
Exogenous-Endogenous Combination	Miscellaneous

to separate particles into endogenous and endogenous-exogenous
categories. Particles considered as completely endogenous contained
various combinations of phosphorus, iron, calcium, sodium, potassium,
chlorine, and sulfur. The majority of these particles contain only iron
and phosphorus. For the purpose of endogenous-exogenous classification,
particles which contain a net fractional exogenous element intensity of
0.15 or more were considered as exogenous. These particles are thought
to be either aggregates or exogenous particles which are coated with
endogenous components, probably hemosiderin, since the major endogenous
components are usually iron and phosphorus.

Particles placed in the Silicon Rich, Iron Rich, Titanium Rich, and
Aluminum Rich categories may be either individual particles containing
major amounts of the specified elements or particle aggregates.
Particles in these categories were further subclassified based upon
their other major element components. The particles in the Other
Aluminum Silicate class are probably particle aggregates. Particles in
the No X-ray category contain no peaks, and are presumably tissue
components which were not completely ashed. These No X-ray particles
are considered as endogenous. All remaining particles are placed in the
Miscellaneous class.

It should be emphasized that the particle data from all analyses are
stored on floppy disks. The analyst has the opportunity to create
chemistry definition files after the analyses have been performed.
Consequently, the particle data may be reprocessed using new or modified
chemistry files at any time. Following completion of the x-ray analyses
for each lung, the image analyzer calculates the net total exogenous
particle concentration and concentration for each particle category
using the particle densities on the final filter preparation and the dry
lung weight.

RESULTS

The results for over 60,000 individual particle analyses performed
for the 48 lungs are summarized in Table 3. The exogenous particle
concentrations for these lungs ranged from 107 to 1844 x 10^6 particles

Table 3 -- Summary of Exogenous Particle Data
from 48 Urban Lungs

Particle Type	Mean	Standard Deviation	Min.	Max.	Median
Total Exogenous (millions/g dry lung)	508	417	107	1844	370
% Silica	18.2	5.9	8.9	39.1	18.0
% Al Silicate	38.3	10.3	21.9	60.8	36.6
% Mg Silicate	2.7	3.2	0.0	17.1	1.7
% Iron Oxide-like	6.0	3.9	0.7	19.3	5.0
% Rutile-like	9.9	9.1	1.9	38.6	7.0
% Alumina-like	0.6	0.8	0.0	4.9	0.4
% Exog.-Endog. Comb.	6.5	6.7	0.6	30.4	4.0
% Silicon Rich	3.0	1.8	0.8	8.2	2.6
% Iron Rich	4.0	2.0	0.9	9.6	3.4
% Titanium Rich	2.7	0.9	1.1	4.5	2.6
% Other Al Silicates	1.5	0.6	0.6	3.1	1.5
% Alum. Rich	4.1	2.3	0.6	10.4	3.8
% Miscellaneous	2.6	1.2	0.6	7.7	2.4

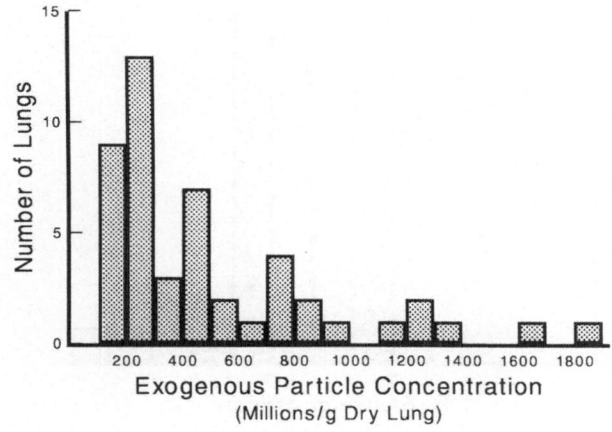

Figure 1.   Total exogenous particle concentration
distribution for the 48 lungs.

per gram of dry lung. The exogenous particle concentrations for the 48
lungs are shown in Figure 1.

   The median circular area equivalent diameters for the analyzed
exogenous particles ranged from 0.38 μm to 0.99 μm. The distribution of
these median diameters is shown in Figure 2. In general, the median
diameters for the major particle classes were similar to the overall
total exogenous median diameter for each lung, with two major
exceptions. Particles classified as Rutile-like had median diameters
consistently smaller than the overall exogenous median diameter for each
lung. Particles classified as Magnesium Silicate had median diameters
consistently larger than the overall exogenous median diameter.

   In terms of particle composition, all of the analyzed lungs
contained the same major particle types. The most frequently
encountered particle types were Aluminum Silicate, Silica, and
Rutile-like. The percentage distributions for these three particle
types, plus Iron Oxide-like and Magnesium Silicate, are shown in
Figures 3-7.

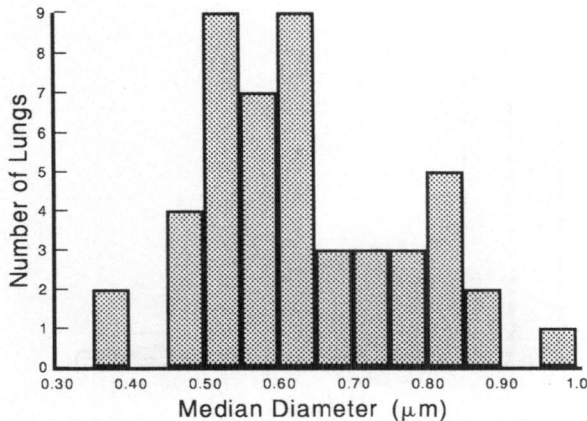

Figure 2.   The median circular area equivalent
diameter distribution for exogenous
particles for the 48 lungs.

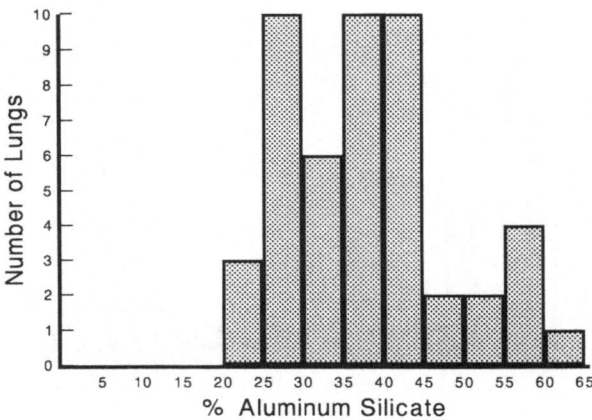

Figure 3.   Aluminum Silicate percentage
distribution for the 48 lungs.

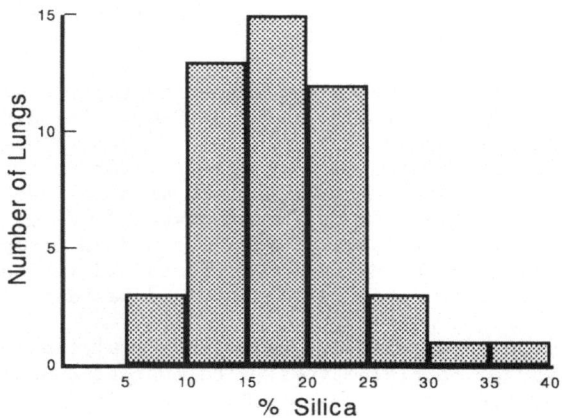

Figure 4.   Silica percentage distribution for
the 48 lungs.

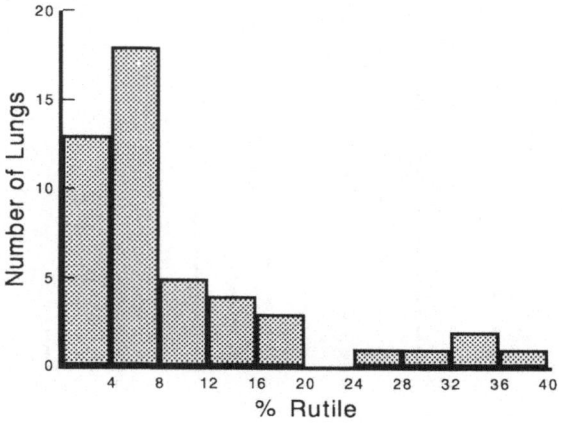

Figure 5.   Rutile-like percentage distribution
for the 48 lungs.

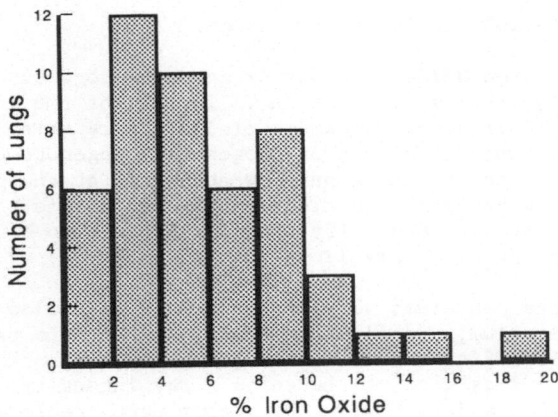

Figure 6. Iron Oxide-like percentage
distribution for the 48 lungs.

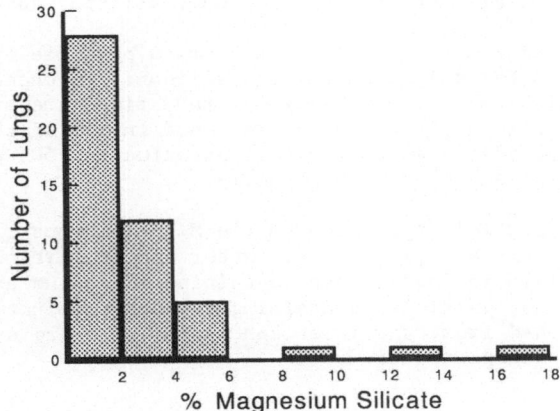

Figure 7. Magnesium Silicate percentage
distribution for the 48 lungs.

In 42 of the 48 lungs analyzed, Aluminum Silicates were the most
frequently encountered species. In the remaining six lungs, this
category was the second most often encountered species. The Aluminum
Silicate particles typically fell into two major subclasses; i.e., those
which also contained a major peak for potassium, and those which
contained no peak other than aluminum and silicon. The
potassium-containing particles are most likely either mica or a
feldspar. The "pure" aluminum silicate particles are presumably
kaolinite. Other common subclasses of the Aluminum Silicate category
included particles which contained peaks for sodium and/or calcium.
These particles which were seen in most lungs, although at very low
concentrations, are most likely plagioclase feldspars.

Particles classified as Silica were the most often encountered
species in only one of the 48 lungs. However, silica was the second
most often encountered species in 31 of the 48 lungs, and the third
ranked particle in 13 lungs. Particles classified as Rutile-like
(TiO_2) were the first ranked in four lungs, the second ranked species
in three lungs, and the third ranked particle in 21 lungs. Iron
oxide-like particles accounted for greater than 10% of the exogenous
particles in six of the 48 lungs. Magnesium Silicates, presumably talc,

were major species (> 10%) in two of the lungs.

The data for the Iron Oxide-like and Exogenous-Endogenous Combination categories is very interesting. In most of the lungs having high concentrations of Iron oxide-like particles, there were correspondingly large concentrations of Exogenous-Endogenous Combination particles. This leads to the question of whether all of the particles in the Iron Oxide-like category are truly exogenous. It is possible that the acetic acid used to decalcify the lung ashes also cleaned up hemosiderin to the point that only iron is detectable.

Each lung analyzed contained varying amounts of particles classified as Silicon, Iron, Titanium, and Aluminum Rich. Most of the particles in these classes contained from two to four major peaks, with the category peak accounting for 50% or more of the total x-ray intensity. Binary combinations of silicon and iron were the predominant species for both the Silicon and Iron Rich categories. For the Titanium Rich category, three types of particles were commonly found. These included particles with peaks for iron and titanium, particles containing iron, silicon, and titanium, and particles containing aluminum, silicon, and titanium.

In the Aluminum Rich category, the most common particle type was one which contained peaks for only aluminum and silicon. In these particles, the aluminum peak was usually 1.5 to 3 times the size of the silicon peak. Particles such as these were found in all of the lungs analyzed, and in most cases, accounted for approximately 50% of the total particles in the Aluminum Rich category.

The most common particle type seen in the Miscellaneous class was one which contained only a peak for tin. Other common particle types in the Miscellaneous class included those containing zirconium and silicon, presumably zircon sand; particles containing various combinations of aluminum, titanium, and iron; particles containing zinc and sulfur; and various lead containing particles.

DISCUSSION

The lungs thus far analyzed in this study are typical for subjects from an urban area which have terminal illnesses. It must be emphasized, however, that there is a distinction between these urban lungs and normal lungs. These lungs are not normal, since there are disease processes present in most of them. However, in most cases, these disease processes are typical considering the age and health of the subjects in question. It should be emphasized that in none of these lungs was there evidence of an overt pneumoconiosis.

A complete statistical analysis of the data obtained from these analyses has not been completed at this time. Correlations, if any, between total exogenous particle concentrations and individual particle class concentrations with parameters such as age, smoking history, and occupation will be determined when the entire 96 lung sample group has been analyzed. Preliminary statistical evaluation performed after completion of the first 33 lung analyses found no correlations between total exogenous particle concentration and sex or smoking history (Stettler, et al., in press).

Very little data concerning the exogenous particle concentrations in lungs with no pneumoconioses has been published. Recently, Churg and Wiggs (1985) reported data on the mineral contents of 28 lungs, 14 with lung cancer and 14 controls. There was no history of occupational or other dust exposure for any of the 28 subjects. The average lung particle content determined by Churg and Wiggs for these lungs was $525 \pm 369 \times 10^6$ particles/g of dry lung for the cancer subjects, and $261 \mp 175 \times 10^6$ particle/gram for the controls. The average lung particle content determined in this work, $508 \pm 417 \times 10^6$ particles/gram compares very favorably with that of Churg and Wiggs. The major particle types seen in both studies were also very similar; i.e., various aluminum silicates and silica.

The lung particle data generated in this study will serve as baseline data for typical non-diseased lungs, with which comparisons of the particle content of diseased lungs can be made. We have found that the total particle contents of diseased lungs are, in general, much higher than seen in our control lungs. For example, the total particle contents of four silicotic lungs analyzed ranged from 5 to 32 times that of the average of the control lungs, as shown in Table 4. The silica content of these lungs ranged from 17 to 75 times that seen in the control lungs. In future work, we will investigate dose-response relationships for toxic particles such as silica, as well as the relationships of other particles present in the lung such as aluminum silicates on the overall disease processes.

Table 4 -- Exogenous Particle Contents
for Some Silicotic Lungs
(millions/gram of dry lung)

Case	Total Exogenous	Silica	Aluminum Silicate
1	15,840	7,480	3,790
2	16,300	5,100	8,890
3	4,350	2,320	1,720
4	2,420	1,760	165

REFERENCES

1. Abraham, J. L., 1979, Documentation of environmental particulate exposures in humans using SEM and EDXA, Scanning Electron Microsc., II:751.
2. Abraham, J. L. and Burnett, B. R., 1983, Quantitative analysis of inorganic particulate burden in situ in tissue sections, Scanning Electron Microsc., II:681.
3. Brody, A. R., Vallyathan, N. V., and Craighead, J. E., 1976, Distribution and elemental analysis of inorganic particulates in pulmonary tissue, Scanning Electron Microsc., I:477.
4. Churg, A. and Wiggs, B., 1985, Mineral particles, mineral fibers, and lung cancer, Environ. Res., 37:364.
5. DeNee, P. B., 1976, Identification and analysis of particles in biological tissue using SEM and related techniques, Scanning Electron Microsc., I:461.
6. Kelly, J. F., Lee, R. J., and Lentz, S., 1980, Automated characterization of five particles, Scanning Electron Microsc., I:311.
7. Lee, R. J., Huggins, F. E., and Huffman, G. P., 1978, Correlated Mosbauer - SEM studies of coal minerology, Scanning Electron Microsc., I:561.
8. Lee, R. J. and Kelly, J. F., 1980, Overview of SEM-based automated image analysis, Scanning Electron Microsc., I:303.
9. Stettler, L. E., Groth, D. H., and Mackay, G. R., 1977, Identification of stainless steel welding fume particles in human

lung and environmental samples using electron probe microanalysis, Am. Ind. Hyg. Assoc. J., 38:76.

10. Stettler, L. E., Platek, S. F., Groth, D. H., Green, F. H. Y., and Vallyathan, V., 1985, Particle contents of human lungs, in: "Proceedings of the 43rd Annual Meeting of the Electron Microscopy Society of America," G.W. Bailey, ed., San Francisco Press, San Francisco, page 116.

11. Stettler, L. E., Groth, D. H., Platek, S. F. and Burg, J. E., in press, Particulate concentrations in urban lungs, in: "Microprobe Analysis in Medicine," J. Shelburne and P. Ingram, eds., Hemisphere Publishing Corp., New York (in press).

12. Sweet, D. V., Grouse, W. E., and Crable, J. E., 1978, Chemical and statistical studies of contaminants in urban lungs, Am. Ind. Hyg. Assoc. J., 39:515.

13. Vallyathan, N. V., Green, F. H. Y., and Craighead, J. E., 1980, Recent advances in the study of mineral pneumoconiosis, Path. Ann., 15:77.

LIVER TISSUE PREPARATION USING A MODIFIED CRYOULTRAMICROTOMY KIT

Patrick J. Clark, James R. Millette, Allan L. Allenspach†,
Paul T. McCauley, and Isaac S. Washington

Toxicology and Microbiology Division
Health Effects Research Laboratory
U.S. Environmental Protection Agency
Cincinnati, OH   45268

†Department of Zoology
Miami University
Oxford, OH   45056

ABSTRACT

The great potential of cryoultramicrotomy is that it is possible to observe the morphology of a sample in the electron microscope while at the same time analyze for its elemental composition using x-ray microanalysis. There are other methods using indirect means of studying the chemical composition of cell organelles, such as digestion and centrifugation, but cryoultramicrotomy is the only direct method. This unique ability will make cryoultramicrotomy a vital tool in the field of cell biology and pathology in the near future. Our interest is the quantitative analyses of diffusible elements in the mitochondria of rat liver before and after exposure to toxins both singularly and in mixtures. To obtain reliable and reproducible data it is critical that each step in the technique be carried out correctly. Any deviation in any of the steps will leave the final results in doubt.

INTRODUCTION

In spite of the potential of cryoultramicrotomy for x-ray microanalysis the technique is not widely employed for several reasons: (1) equipment is expensive to purchase and maintain, (2) the methodology is still in the developmental stage and hence requires specialized training in equipment operation and data analysis, and (3) there is only limited agreement among researchers about acceptable methods of tissue preparation for use with X-ray microanalytic methodology (Moore et al., 1984). In addition a high degree of precision and patience is needed to carry out the procedure.

Recently studies using cryoultramicrotomy and X-ray analysis have been published in biological journals (Hardgest et al., 1985; Hanaichi et al., 1984). We have found it necessary to modify these techniques to increase the efficiency and reproducibility of cryosectioning. Since methodology actually consists of several techniques associated with their own

227

inherent problems, each step will be considered individually. The focus of this paper will be on: (1) how the tissue was prepared for elemental analysis in the electron microscope, (2) how the problems inherent in the techniques were overcome, and (3) how the cryokit was modified to facilitate the cutting of sections.

METHODS AND MATERIALS

Female Sprague-Dawley rats of approximately 300 grams were used for testing. Both control and test animals were sacrificed by guillotine after set intervals (1-40 hours). The excised liver was placed on dental wax where a strip of tissue was cut from the central lobe using acetone cleaned razor blades. This tissue was then cut into pyramid shaped pieces (1 mm) and placed onto a silver pin using a wood applicator stick. A small amount of 30% glycerol was used for adhesion between the sample and the pin. The pins were next plunged into an isopentane slush which had been supercooled to -135°C using liquid nitrogen. The whole process from excision to plunging took approximately 90 seconds. Removing the pins from the isopentane left a thin residual film adhering to the tissue which was removed by rolling the pin on a Whatman filter paper which was elevated slightly above LN_2 in a styrofoam bucket. The pins were then stored in vials in a LN_2 dewar.

A modified LKB IV cryokit (Figure 1) was used for sectioning. The kit was enlarged in both width (13 inch) and height (4 inch) to add more room and facilitate a more stable temperature environment. To prevent cracking, all corners were sealed with a silicone, non-corrosive adhesive (Dow Corning 738 RTV) good to -70°C. The walls were insulated both inside and outside using 1-inch thick styrofoam (Somlyo and Silcox, 1979 and the inside was also lined with 1/8 inch Teflon sheets. Teflon sheets were also used beneath the cryokit and below the styrofoam on

Figure 1. Modified Cryobox on LKB Microtome. There are several modifications shown here such as (1) increase in size (2) insulation both inside and out, (3) swingout binoculars, (4) multiple thermocouples with digital readout, (5) optic fiber lighting system, and (6) nitrogen gas line for cooling chamber.

which the kit rests, insuring a much tighter fit of the styrofoam around the base of the knifeholder, thus preventing any nitrogen leaks. It is imperative that this styrofoam be checked for cracks caused by the combination of cold temperature and advancement of the knife because (1) a small leak can cause the micro advance mechanism to freeze up, stopping the whole operation and (2) if enough nitrogen leaks, a negative pressure of the gas ensues, causing "snow" to fall into the kit and onto the knife and tissue. Temperature equilibrium was maintained by feeding LN_2 from a large Dewar (1 lb./sq. in.) into an aluminum foil boat filled with fine steel wool. To offset any turbulence caused by the constant flow of nitrogen gas an aluminum air screen was placed around the knife holder.

The binoculars were modified to swing away from the kit to enlarge the operating area for knife and sample positioning. To increase the brightness and hue of the interference colors, a Ehrenreich MKII fiber optic lighting system was employed. An anti-static bar (Hagler et al., 1979) was used to neutralize any charge buildup on the tissue sections. A Fluke multiple thermocouple was used to monitor temperatures at various points in the cryobox. This includes one both 5 mm above and below the knife edge (-100°); one at the press block (-110°); and the final two in the nitrogen well (-160°) and at the top of the box (-90°).

The tissue was cut at a thickness of approximately 900 A on dry glass knives (Figure 2) set at a 5° cutting angle. Only transparent sections with good interference colors were transferred by means of an eyelash hair on an applicator stick to a formvar- and carbon-coated grid. If the coated grids were not used immediately, they were glow discharged in a Denton vacuum evaporater to remove oils or other contamination. This was accomplished with a high voltage transformer which ionizes the air to neutralize and clean the grids. The tissue on the grid (Figure 3) was then sandwiched between another grid. The grids were then separated and

Figure 2. Tissue slices collected on a dry glass knife.
The knife is set at a 5° cutting angle and is kept at -110°C. The tissue is transferred from here to a grid by means of an eyelash hair on an applicator stick. Note that this knife is kept frost free.

placed in an EM specimen holder which had been LN$_2$ cooled in the cryo-chamber. The specimen holder was then transferred in a styrofoam and glass nitrogen chamber to the EM where it was placed under vacuum and the tissue was freeze dried. A special silicon-rubber O-ring, which remains pliable at extremely low temperatures, enabled the cold specimen holder to enter the scope.

Adjacent tissue samples were fixed and plastic embedded using conventional methods. These were then cut in a LKB microtome at a 3° cutting angle, collected on a 200-mesh copper grid, and stained with lead citrate. Analysis was performed with an ORTEC energy-dispersive spectrometer in combination with a JEOL 100CX electron microscope provided with a JEOL ASID scanning attachment as described in Millette et al. (1985).

RESULTS AND DISCUSSION

The modifications of the LKB cryokit had to do with: (1) enlargement for more work area, (2) use of styrofoam, silicon sealant, teflon, and thermocouples for more temperature stability and control, and (3) manuverable binoculars, fiber optic light, and air foils for more convenience and efficiency. Also filling an aluminum foil boat for both the specimen and chamber cooling facilitated the cutting of sections. The original specimen coolant chamber was not unlike the one described by Curtis (1982), in which the chamber is periodically refilled with LN$_2$ in response to a level sensor. This variation in liquid nitrogen level can lead to small changes in dimensions due to temperature fluctuations and can affect the uniformity of tissue thickness. The liquid nitrogen also tends to splash over the guard and onto the knife, endangering any sections present. By placing steel wool in the front of the kit away from the specimen holder, we were able to maintain a more constant temperature and at the same time were able to disconnect the specimen cooling chamber for more consistent section thickness.

New microtomes on the market utilize these and additional modifications of the cryokit. We have tried two of the recent models, the DuPont Sorval MT-500 and the Reichert Ultracut FC4, and found both are

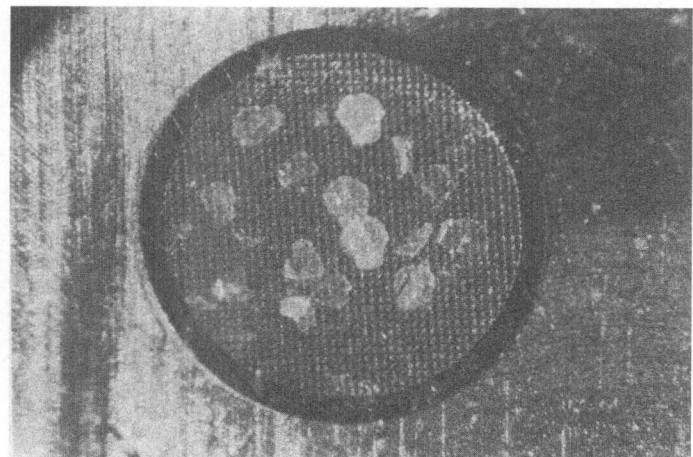

Figure 3. Optical microscope picture of tissue slices on a formvar, carbon-coated grid. Note that the tissue is flat and transparent which is indicative of good freezing.

230

able to cut good sections quickly and consistently. The cryo chambers are cooled separately and specimen and knife temperatures are accurate and quick in response. Sections can be cut continuously for at least 8 hours with no freeze up and little frost. Despite the increased cost of these units, these accure savings in time, liquid nitrogen use, consistency, and reproducibility.

There are many hypotheses about the best way to freeze tissue. Sub-zero temperature processes such as ice nucleation, vitrification, recrystalization, and elemental migration must be considered (Franks, 1980). Another aspect of importance in the freezing of biological tissue is the heterogeneous nature of the substrate. It is known that water contained in thin capillaries does not freeze at the same temperature as bulk water (Vigness and Disrema, 1974). The analogy that tissue is similar to a block of ice may be misleading. It is more appropriate to consider frozen tissue as a heterogeneous material having a variable degree of temperature dependent ductility (Saüberman, 1980). Because the basic composition of tissue is variable, the properties of the frozen block of two dissimilar tissues at the same temperature may vary considerably. Like metals, it appears that frozen tissue has a variable ductility depending upon its temperature (Appleton, 1974). It is apparent that for different tissues not only the minimum freezing temperature but also the optimum cutting temperatures will vary.

For freezing of the tissue we originally chose isopentane because of its low freezing point, supercooling capacity, relative safety, and ease of use. It is of critical importance that the isopentane be of the right temperature and, therefore, viscosity for proper freezing of the tissue. A syrupy viscosity at -135° allows for quick penetration by the pin and also quick freezing of the tissue. Only tissue which retains its original pinkish color was used for analytical studies. If the tissue turned white it was considered ice crystal damaged and was discarded. Efficiency was also enhanced by the pyramid shape of the tissue which relinquished the need for trimming and thus retains the best freeze areas. There are other techniques that utilize the ultrarapid freezing methods of Heuser et al. (1979), in which tissues are rapidly plunged onto a liquid helium cooled copper block (Parsons et al., 1984; Tormey 1983). There are several modifications of this procedure including one in which polished copper is placed on both ends of a small pair of pliers which is subsequently cooled and used to press a wafer of frozen tissue. This tissue can then be trimmed and placed in a modified chuck (Tormey, 1983) for cutting in the microtome. This method is extremely efficient in that (1) tissue is frozen very quickly after removal from the animal, (2) it does not come into contact with any liquid medium, (3) temperature can be consistent and extremely cold, and (4) tissue can be trimmed to the most convenient size while still retaining its best freeze areas. For a comparison of several methods and a survey of each one's advantages see Elder et al. (1982).

According to Moore et al. (1984), all cryoultramicrotomes must have certain common fundamental characteristics. They must be able (1) to hold a temperature below 100 C for at least several hours, (2) to control and display specimen and knife temperatures independently (3) to control changes in temperature quickly, (4) to permit complete control of sectioning thickness (from 30 to 250 nm), and (5) to allow the operator sufficient room to pickup and transfer cryosections quickly and conveniently. Precautions should also be taken to (6) to prevent buildup of frost inside of the chamber to eliminate ice contamination of the knife samples, (7) to eliminate static charge, (8) to minimize turbulence, and (9) to provide a way for the microtome to dry itself after use so that no moisture will freeze in subsequent sectioning.

The temperatures in the cryokit and storage dewar are very important. Stability for centuries or millenia requires temperatures below -130°C. Many cells stored above -80°C are not stable, probably because traces of unfrozen solution still exist (Mazur, 1970). No thermally driven reactions occur in aqueous systems at liquid N_2 temperatures (-196°C), the refrigerent commonly used for low temperature storage. One reason is that liquid water does not exist below -130°C. The only physical states that do exist are crystalline or glassy, and in both states viscosity is so high (10^{-3} poises) that diffusion is insignificant over less than geological time spans (Mazur, 1984). Moreover, at -196°C, there is insufficient thermal energy for chemical reactions (McGee, 1962). For the cryosectioning itself the temperature is critical for several reasons. First it must be cold enough to prevent melting of sections due to friction from the knife. Secondly, it must be warm enough to prevent fracturing of the tissue. The work of Frederick and Busing (1981) indicates that when cutting at -100°C the absence of the growth of ice crystals along with the lack of melting artifacts shows that no substantial through-section melting takes place. We found that cutting at -100°C gives consistent gold interference colors with no visible fracture lines.

The temperature is also important for freeze-drying of the tissue, whether it is done in the cryokit, under vacuum in the microscope, or elsewhere. In principal freeze drying can occur under atmospheric pressure as long as the vapor pressure of water in the atmosphere surrounding the object is kept lower than the saturated vapor pressure over ice at the temperature of the object (Frederick, 1982). However, this process is much slower than that under vacuum. Frederick calculated that thin sections obtained at -80°C will be dry by staying 5-10 minutes in the cold chamber of a microtome, a normal time in the routine preparation of the tissue. So sections normally cut at this temperature do not need further drying, but only protection from the environment before putting them in the microscope. This is where our glass LN_2 holder enables us to transfer our sections from the cryokit to the scope without exposure to the surroundings. Commercial freeze-drying units are now available, such as the Gatan, which increase the control over the freeze drying.

The transfer of the tissue is important in that exposure to warm air causes thermal expansion damage and prevents freeze drying. This damage is easy to observe in that the resolution of the image becomes soupy. If the tissue is too warm initially, or warms too quickly, then "melting" will take place and poor contrast will be apparent. The specimen stage was modified so that LN_2 could be added to the stage to keep it cold during transfer and freeze drying. Once the stage is put into the EM column the tissue slowly warms in the vacuum. Here, due to the different moisture contents of the cell organelles, different organelles appear as different contrasts. The mitochondria are very dark electron dense bodies as opposed to the lighter nuclei (Figures 4 and 5). With the development of better freeze drying techniques, the resolution of cell organelles should compare favorably to plastic sections (Figure 6). At this point the cells are able to be analyzed using X-ray analysis for changes in elemental composition due to exposure to the toxin.

Ideally this data should be very accurate because (1) this tissue has never been exposed to chemicals as it is in other preparation methods, (2) the tissue has been kept frozen to prevent any elmental redistribution, and (3) the tissue is frozen almost immediately after death and, therefore, should closely represent life-like conditions.

232

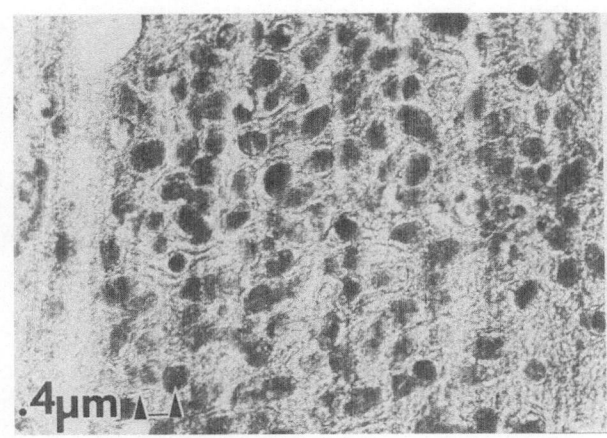

Figure 4. High mag picture of freeze dried
liver tissue. Present are mitochondria and endoplasmic reticulum.

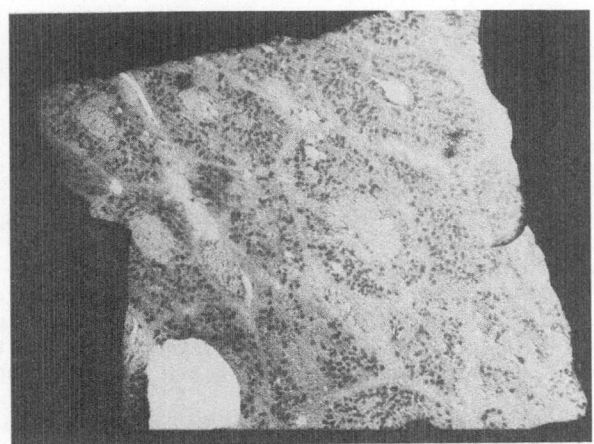

Figure 5. Low mag picture of freeze dried
liver tissue. Due to differences
in free water of the organelles
there are differences in contrast.
The nuclei and cell membranes are
lighter in color while the mitochondria are more electron dense.
Also present are sinusoids, bile
canaliculi, and Kupfler cells.

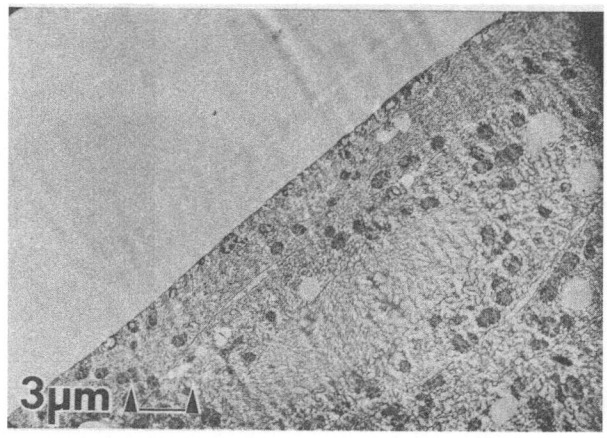

Figure 6. A plastic section of freeze substituted tissue. Note that the freeze
is best at the surface with ice
crystal damage more apparent with
prenetration into the tissue.

CONCLUSIONS

    Rapid freezing of liver tissue is critical for good cryosectioning
and elemental analysis. There are several methods for freezing and ours
in which tissue is cut and placed on silver pins to be plunged into
isopentane was consistent. Also, the use of an isopentane-propane mixture
enables the temperature to get much colder (-185°C) and does not leave a
film on the tissue. However, freezing of tissue with cooled polished
copper tipped pliers has the advantage of (1) little or no cutting of the
tissue to prevent damaged cells, (2) rapid freezing resulting in life-like
tissue, and (3) rapid set due to the simplicity of the aparatus.

    The modified cryokit worked well and produced sections with gold
interference colors on a fairly routine basis. The problem was the
cryokit often had to be taken apart and reinsulated and resealed to
prevent freeze up in the advance mechanism of the microtome and frost
buildup in the cryokit. Also there was not much manuverability of the
knife. The new cryoultramicrotomes take care of these problems and are
indespensible for efficient and routine sectioning.

    Modified transfer devices work in that they can keep the tissue from
being exposed to the ambient environment and thereby prevent thermal
damage. Of all the factors involved in providing quality thin sections
of fresh frozen tissue, none is more important than the maintenance of a
liquid nitrogen environment surrounding the specimen (Hagler et al.,
1984). Also important is the ability to control the temperature of the
stage and specimen while the tissue freeze dries. The new cryotransfer
stages are best suited for the purpose of increasing the efficiency and
reproducibility of cryosection transfer and freeze drying.

ACKNOWLEDGMENTS

We are grateful to John Silcox of Pennsylvania University and Ron Warner and Mark Meyer of Proctor and Gamble for their many tips on modified cryokits; we thank Bearings, Inc. for their help in finding the specialized O-rings; and we express our thanks to Pat Underwood for her aid in preparing this manuscript. The research described in this article has been reviewed by the Health Effects Research Laboratory and approved for publication. Mention of trade names or commercial products does not constitute endorsement or recommendation for use.

REFERENCES

Moore, P. L., Simson, J. A. V., Bank, H. L., and Balentine, J. D., 1984, X-Ray microanalysis: Identification and quantification of elements in normal and pathologically altered cells, Meth. Achiev. Exp. Pathol., Vol. 11, 138.

Hargest, T. E., Gay, C. V., Schraer, H., and Wasserman, A. J., 1985, Vertical distribution of elements in cells and matrix of epiphyseal growth plate cartilage determined by quantitative electron probe analysis, J. Histochemistry and Cryochemistry, Vol. 33, 275.

Hanaichi, T., Kidokoro, R., Hayashi, H., and Sakamoto, N., 1984, Electron probe x-ray analysis of human hepatocellular lysosomes with copper deposits: Copper binding to a thiol-protein in lysosomes, Lab. Invest., 51(5):592.

Somlyo, A. P., and Silcox, J., 1979, Cryoultramicrotomy for electron microprobe analysis, in: Lechene, Warner, "Microbeam Analysis in Biology," Lechene, Warner, eds., Academic Press, New York.

Hagler, H. K., Burton, K. P., Sherwin, L, Greica, C., Siler, A., Lopex, L. and Buja, L. M., 1979, Analytical electron microscopic studies of ishemic and mypoxic myocardial injury, Scanning Electron Microscopy, II, 723.

Millette, J. M., Allenspach, A., Clark, P. J., Washington, I., and McCauley, P. T., 1985, X-Ray microanalysis of calcium, potassium, and phosphorus in liver mitochondria stressed by carbon tetrachloride, J. Analy. Tox., Vol. 9, 145.

Curtis, R. M., 1982, A modification of the LKB cryokit, Science Tools, Vol. 29, No. 2.

Franks, F. L., 1980, Physical, biochemical and physiological effects of low temperatures and freezing--Their modification by water soluble polymers, Scanning Electron Microscopy, II, 349.

Vigness, M., and Disrema, K. M., 1974, A model for the freezing of water in a dispersed medium, J. Colloid Interface Science: 49:165.

Saüberman, A. J, 1980, Application of cryosectioning to x-ray microanalysis of biological tissue, Scanning Electron Microscopy, II, 4121.

Appleton, T. C., 1974, A cryostat approach to ultrathin "dry" frozen sections for electron microscopy: A morphological and x-ray analytical study, J. Microsc., 100:49.

Heuser, J. E., Reese, T. S., Dennis, M. J., Jan, Y., Jan, L., and Evans, L., 1979, Synaptic vesicle exocytosis captured by quick freezing and correlated with quantal transmitter rrlease, J. Cell Biol., 81:275.

Parsons, D., Bellotto, D. J., Schulz, W. W., Buja, M., and Hagler, H. K., 1985, Towards routine cryoultramicrotomy, EMSA Bulletin, 14(2):49.

Tormey, J. McD., 1983, Improved methods for x-ray microanalysis of cardiac muscle. Microbeam Analysis, 221.

Elder, H. Y., Gray, C. C., Jardine, A. G., Chapman, J. N., Biddlecombe, W. H., 1982, Optium conditions for cryoquenching of small tissue blocks in liquid coolants, J. Microsc., 126:45.

Mazur, P., 1970, Cryobiology: The freezing of biological systems, Science, 168: 939.

Frederick, P. M., and Busing, W. M., 1981, Strong evidence against section thawing whilst cutting on the cryo-ultratome, J. Microsc., 122: 217.

MICROANALYSIS OF AIRBORNE LEAD PARTICULATES IN AN URBAN

INDUSTRIAL ENVIRONMENT

F. E. Doern* and D. L. Wotton**

* Whiteshell Nuclear Research Establishment
Pinawa, Manitoba, Canada
** Province of Manitoba
Environmental Management Division
Winnipeg, Manitoba, Canada

SUMMARY

Automated SEM-EDX of particulates is a technically viable and econom-
ically feasible analytical technique in environmental modelling. A case
study involving the microanalysis of lead particulates on aerosol filters
from the Weston community of northwest Winnipeg is detailed. The sampling
requirements and analytical restrictions inherent in the technique provide
a complementary rather than competitive method to conventional chemical
analysis techniques. Additional software enhancements to assist in database management are described for use with existing commercial particle
analysis hardware/software packages.

INTRODUCTION

Environmental impact assessment models have developed along two phi-
losophical paths (see Figure 1). The dispersion model determines a source
inventory, calculates emission rates and emission parameters, factors in
meteorological variables and subsequently estimates ambient concentrations
at a variety of receptor sites. Conversely, the receptor model starts
with an ambient aerosol or water sample (usually a filter containing particulates) and uses chemical trace elements or "fingerprints", along with
particulate morphology and sample variability to derive an estimate of the
chemical contributions from a number of localized sources.[1,2]

The analytical basis for receptor modelling is often divided into
chemical methods (such as X-ray fluorescence, atomic absorption, and neutron activation techniques) or microscopic methods (including light optical microscopy, ion microscopy, and electron microscopy). Automated electron microscopy uses computer-controlled movement of the electron beam for
rapid X-ray microanalysis of the individual particles or features of interest in the microscope's field of view.

This article examines the applicability of automated electron microscopy for the analysis of airborne lead particulates in an urban environment, the Weston community of Winnipeg, Manitoba; discusses some of the
limitations of the technique as it is applied to environmental, forensic,

or occupational health problems; and outlines some recent efforts to integrate particulate analysis with digital imaging procedures and computer graphics to improve the analytical method.

Lead in the Weston Community[3]

Weston is a small residential neighborhood in the northwest quadrant of the City of Winnipeg (see Figure 2). The community is surrounded with light industrial activity including many warehouse facilities, and secondary manufacturing industries. Weston is unusual because it has a secondary lead smelter (Canadian Bronze Co. Ltd. - labelled Smelter A in Figure 2) within its residential area, and within two city blocks of Weston Elementary School (W in Figure 2). There are two other secondary lead smelters in Winnipeg, both of which are also located in the northwest of the city (Smelters B and C in Figure 2). Weston School is located along a major traffic artery, and the contribution of lead from automobile exhaust was unknown.

During the period 1979-81, the Environmental Management Division of the Province of Manitoba conducted extensive soil surveys to monitor lead levels around the Weston community. Following the guidelines of the Ontario Ministry of the Environment, the Environmental Management Division adopted the concentration level of 2600 µg/g of lead in soil as the criterion for unacceptable lead levels in sod and soil of residential areas or areas frequented by children. Based on the results of the soil surveys[4], remedial action, which included a physical clean-up of the Weston schoolyard and a sod/soil removal program of twenty-six residential properties in the school and smelter vicinity, was carried out in 1981-82.

The need to verify the lead source(s) became apparent when the lead in soil at Weston School was approaching the maximum permissible levels within twelve months after the clean-up operations. Consequently, in 1982 the Whiteshell Nuclear Research Establishment (WNRE), the Province of Manitoba Environmental Management Division (EMD), and Canadian Bronze Co. Ltd. implemented a joint research project with two central objectives:

(1) To characterize the lead species originating from automobile exhaust and various secondary lead smelters in northwest Winnipeg.
(2) To identify the primary sources of lead contamination within the Weston Elementary School playground.

EXPERIMENTAL PROCEDURES

The scanning electron microscope, with integrated X-ray microanalysis hardware and software for automated particle characterization, was chosen as the primary analytical instrument to complement the existing programs of soil bulk analysis for lead and other metals by conventional atomic absorption methods.[5,6]

Environmental Sampling

The Environmental Management Division developed an experimental design for sampling locations, and conducted the air-sampling program. Air filters were analyzed from eighteen locations throughout the city. The samples included reference filters of automobile exhaust from a vehicle burning leaded gasoline, and stack samples from the three secondary lead smelters; control samples taken during periods when Canadian Bronze was not conducting smelter operations; and filters from selected sites near the various smelter locations, at schools near major traffic arteries, and at quiet residential locations, during periods when all smelters were operational.

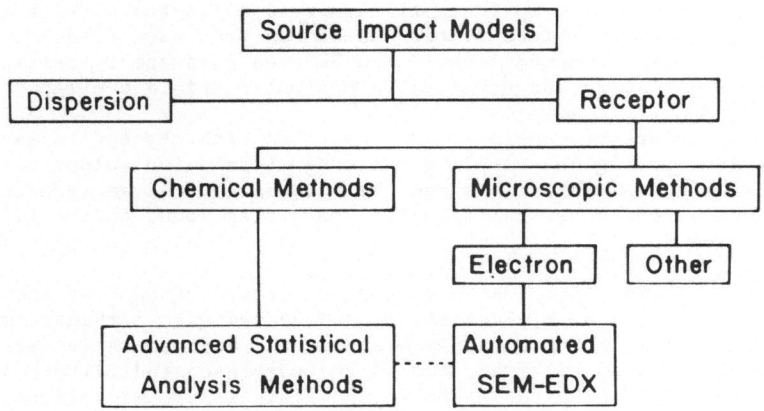

Figure 1.   Environmental modelling strategies. The dotted
            line joining Automated SEM/EDX and Advanced
            Statistical Analysis Methods is indicative of
            future effort.

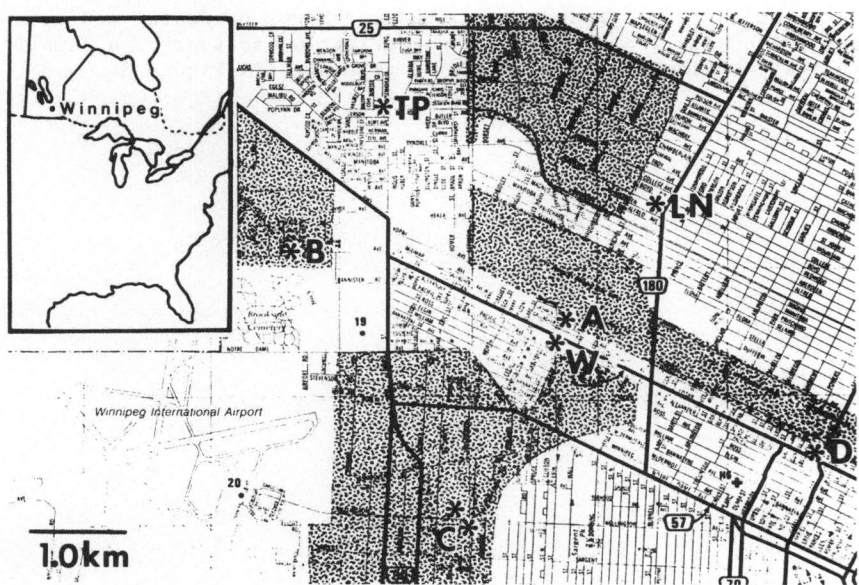

Figure 2.   Map of the northwest quadrant of Winnipeg showing
            the location of the Weston community and selected
            sampling sites. Shaded areas indicate industrial
            activity.

The air-particulate samples were collected on 47 mm diameter polycarbonate (Nucleopore) filters having a 0.4 µm pore size. Sample collection periods ranged from 2 seconds to 145 minutes, depending upon site location. Duplicate or triplicate filters were collected at each sampling site, and the appropriate filter was chosen from each site such that the collection time produced less than a 5% area coverage by particulates on each of the filters and resulted in minimal particle overlap.

Appropriate precautions were exercised with respect to meteorological conditions during the sampling periods. Local wind velocities were less than 25 km/h, and this minimized the frequency of large (greater than 10 µm equivalent spherical diameter) particles found on the filters due to resuspended dust.

A 10 mm wide strip was cut from the centre of each of the 22 filters analyzed, mounted on a glass slide, and coated with a thin film of carbon (15–20 nm thick) using a vacuum evaporator. Representative areas (approximately 10 mm x 10 mm) from each of the carbon-coated filters were subsequently mounted on carbon stubs for analysis in the microscope.

Microanalysis Parameters

Microanalysis was conducted at WNRE using an International Scientific Instruments DS130 scanning electron microscope (SEM) equipped with a lanthanum hexaboride electron gun and a high efficiency scintillator (Robinson) backscattered electron detector. The attached X-ray spectrometer system was a Tracor Northern TN2000 energy dispersive X-ray (EDX) spectrometer with digital electron beam control hardware, and both digital image processing and particle characterization software. A simplified schematic drawing of the hardware is presented in Figure 3.

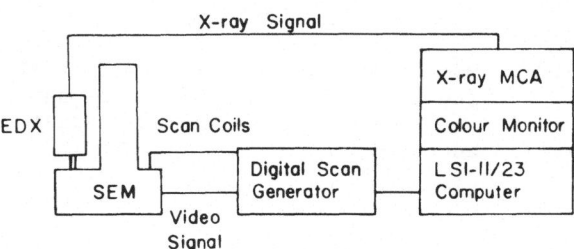

Figure 3. Diagram of the SEM–EDX configuration for automated particle characterization.

High resolution secondary electron images were collected on the top stage of the SEM, while X-ray microanalysis of individual particulates was conducted on the bottom stage of the SEM under the control of the Particle Recognition and Characterization (PRC) computer program developed by Tracor Northern and U. S. Steel.

A detailed description of the PRC software may be found elsewhere.[7] A simplified explanation of the PRC approach is required, however, to appreciate some of the limitations and constraints in using the software. PRC rasters the electron beam across the field of view until it detects a sharp change in video contrast indicative of a particle edge. The electron beam is then moved in smaller increments to form a horizontal chord across the particle (see Figure 4a). A vertical chord is "drawn" at the midpoint of the horizontal chord; if the two chords are perpendicular bisectors of each other, the program assumes the centre of the particle has been located. From the centre of the particle, eight rotating diameters (see Figure 4b) are determined, and based on the values of the eight diameters the size and shape of the particle are calculated. The electron beam is fixed at the centre of the particle and the EDX microanalysis data is collected for a specified period of time or until sufficient X-ray counts are received to satisfy statistical counting criteria. The procedure is repeated for each feature (particle) within the selected field of view.

The principal constraints to successful PRC operation are:

(1) good video contrast between the feature of interest and the surrounding matrix or background
(2) simple feature geometry or shape to ensure accurate size and area calculations
(3) features must exhibit minimal overlap.

The main experimental parameters of PRC were set to characterize particulates with an equivalent spherical diameter of 0.25 – 10 μm (as viewed in a backscattered electron image). The 0.25 μm minimum particle size was chosen based on the signal/noise characteristics of the EDX spectra, and the 10 μm maximum particle diameter was selected to restrict analysis to particles of respirable dimension.

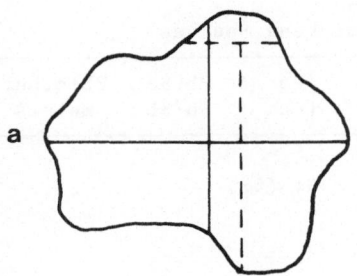
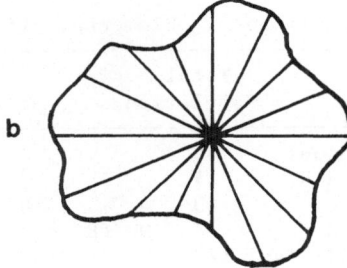

a   b

Figure 4.   (a) Bisecting chords used to locate the particle centre.
            (b) Rotating diameters used to determine particle size
                and shape characteristics.

All EDX analysis were performed using a 25 keV electron beam to facilitate both lead L X-ray and lead M X-ray identifications and minimize confusion from sulphur and molybdenum spectral interferences, and at an optimum specimen geometry (20 degree tilt, 22–24 mm working distance) to maximize minor element X-ray intensities. A maximum X-ray analysis time of 20 seconds/particle was used.

RESULTS AND DISCUSSION

From the 22 filters examined, 3389 particles were analyzed. All particles observed were of respirable dimensions; the largest particle was 6 μm equivalent spherical diameter, and the majority were less than 2.5 μm in diameter. Since a detailed discussion of the particle analysis from each of the 22 filters and 18 sampling locations is beyond the scope of this article, the filters will be grouped into appropriate categories.

The Particle Database

Reference Samples – The key to successful receptor modelling is to have definitive chemical or morphological "fingerprints" for each of the local sources. Figure 5 illustrates the various particulate shapes and their corresponding EDX spectral patterns for the auto exhaust and smelter stack emissions. It is important to note that all of the particles from these various sources are typically less than 1.0 μm in diameter, and are often found as particle agglomerates. Table 1 summarizes the PRC data for the reference filters. Automotive exhaust was always found with a lead-bromine (Pb-Br) or lead-bromine-chlorine (Pb-Br-Cl) spectral pattern. The prismatic crystals observed in the smelter A stack sample were rich in zinc, probably present as zinc oxide or zinc oxyhydroxide; the lead was always clustered with the zinc crystals. Thus the spectral fingerprint for smelter A was lead-zinc (Pb-Zn). Since the primary feedstock to smelter A was leaded brass, this Pb-Zn pattern was consistent with the chemistry of the smelter. The emission from the stack of smelter B was almost pure lead, and the particles were typically small spheroids or clusters of spheroids; the spectral fingerprint was Pb-only. Stack emissions from smelter C were not detectable using PRC software. The individual particles shown in the secondary electron micrograph of Figure 5 were only visible when the SEM accelerating voltage was reduced to 10 kV and substantial differential amplification was applied to increase the contrast of the video signal. The low material density of the particles was inferred from their electron transparency near the filter pores. Based on the crystal habit of the particles visible in the micrograph, the faint yellow appearance of the original filter to the naked eye, and the prior knowledge that a source of feedstock to the smelter was old car batteries, the conclusion was that smelter C was emitting a sulphur plume from its

Table 1. Elemental "Fingerprints" of Lead Sources

Location	Total Population	Pb only	Pb + Br/Cl	Pb + Zn	Pb+Sn Sn/Sb	Pb+other metals
Auto Exhaust	40		40			
Smelter A	124			64 (46)		
Smelter B	63	63				
Smelter C	No PRC data available					
1476 Well. Ave	173	6 (1)	(3)		4 (5)	11 (3)

Pb confirmed: both lead L and M X-rays present
(#) Pb suspected: only lead M X-rays detected

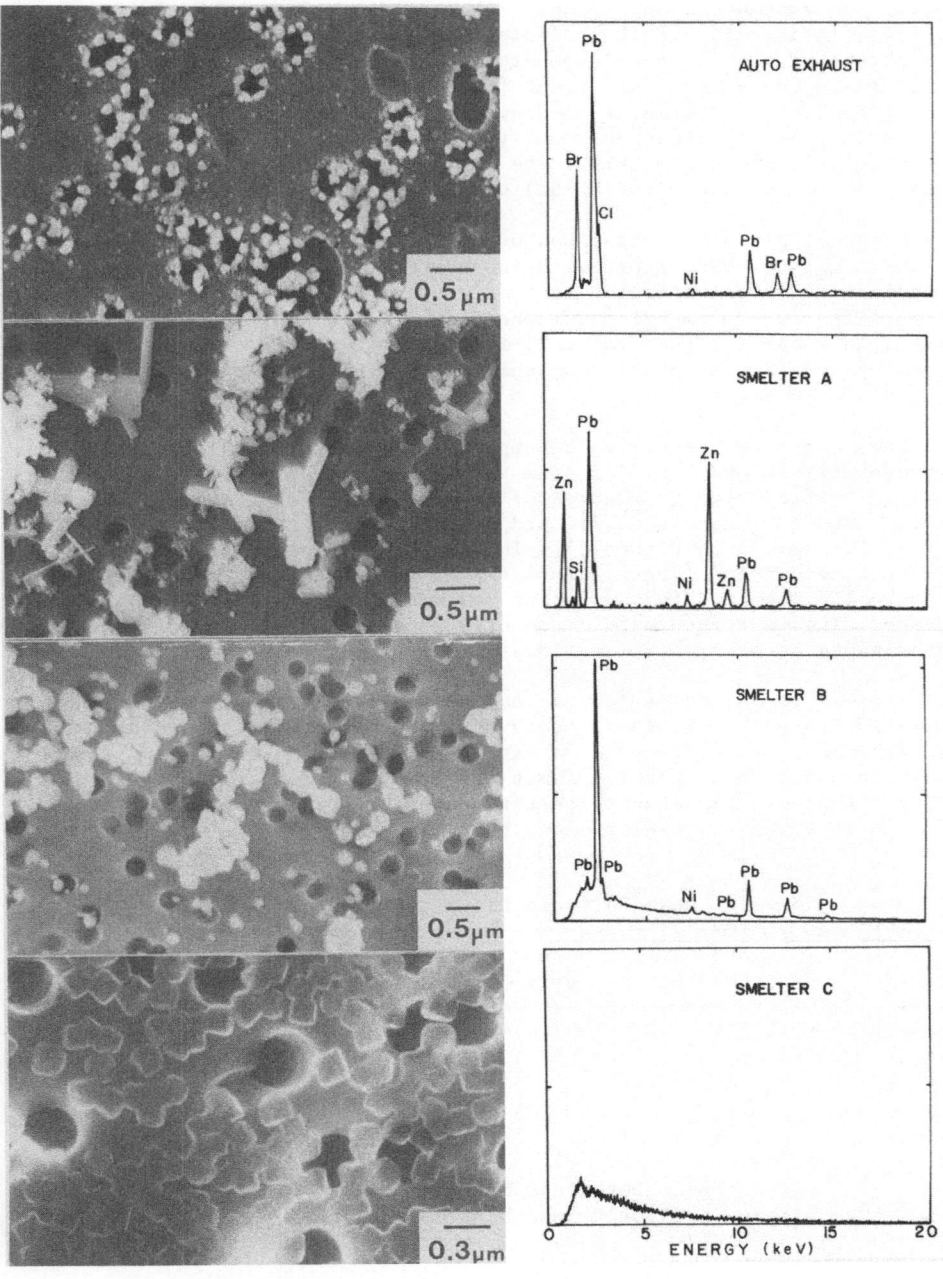

Figure 5. Secondary electron images and corresponding X-ray
"fingerprints" for lead source characterizations
in northwest Winnipeg. (Reproduced with permission
from Doern and Wotton, "Proc. EMSA 1985", San Francisco
Press, pp. 112-115.)

stack. The absence of any white, high secondary electron emitting particles in the micrograph indicates the fraction of any high-density lead particles must be small compared to the amount of sulphur. A filter sample collected across the street from smelter C at 1476 Wellington Avenue did show a substantial amount of lead, along with trace amounts of the elements tin (Sn) and antimony (Sb). These latter elements are characteristic of lead babbit. The smelter C spectral fingerprint was considered to be lead- tin (Pb-Sn) or lead-tin-antimony (Pb-Sn-Sb).

Control Samples – Comparison of control samples collected when smelter A was not operating, with those when all smelters were operating, showed a consistent variation in the frequency of zinc particles observed on the filters. A comparison of the Weston School control and surveillance samples (see Table 2) implied that the major source of lead contamination at the school was vehicular exhaust rather than smelter A emissions.

Environmental Surveillance Samples – The data in Table 2 are representative of PRC results from surveillance samples. The particulate study identified high incidences of lead at other schools along major traffic arteries; specifically Lord Nelson School (LN in Figure 2) and Dufferin School (D in Figure 2). The low lead values for samples collected near Tyndall Park School (TP in Figure 2) were consistent with the school's location in an area of lower traffic density. The study also provided clues for future environmental monitoring programs of trace metal contaminants at other sampling locations.

Table 3 is an estimation of lead accumulation rates based on the PRC data. The values in Table 3 should not be construed as absolute concentrations of lead, since no effort has been made to convert lead X-ray intensity to an absolute mass value. The values suggest a trend, however, that placing schools next to major traffic arteries may expose children to a higher level of lead contamination.

Table 2. Lead Particle Distributions Near Schools

Location	Total Population	Pb only	Pb + Br/Cl	Pb + Zn	Pb+Sn Sn/Sb	Pb+other metals
Weston (control)	151	2 (6)	25 (19)			(7)
Weston	161	2 (1)	9 (8)	2 (2)	1	(10)
Dufferin	161		25 (21)			(6)
Lord Nelson	122	(2)	22 (16)	1		(1)
Tyndall Park	155		4 (15)	1		

Pb confirmed: both lead L and M X-rays present
(#) Pb suspected: only lead M X-rays detected

Table 3. Lead Accumulation Rate – Confirmed Lead Species Only
(particles/mm^2 filter/hour)

Lord Nelson	132
Dufferin	79
Weston	40
1476 Wellington	32
Weston (control)	18

Table 4. Sampling Strategies

1. Aerosol filters for microscopy
   - very source specific
   - instantaneous "snapshots"
   - many sampling anomalies
     e.g. weather, traffic patterns

2. Soil/sod bulk chemical analysis
   - historic record
   - trace elements difficult to detect
   - does not reflect recent control efforts

and chlorine in the area around Weston School were consistent with the automated SEM-EDX conclusion that vehicular exhaust was a major contributor to the lead burden. Additional lead in sod and soil studies conducted by the Environmental Management Division found the lead levels at the other schools to be below the accepted guideline of 2600 µg/g.

Analytical Strategy

Table 4 summarizes some of the key considerations in developing an analytical approach to environmental problems, and in comparing microscopic methods to those of "bulk" chemical analysis. In most cases the automated SEM-EDX techniques provide valuable screening opportunities before conventional analytical chemical operations are undertaken. The two approaches must be considered as complementary rather than competitive analytical strategies, since many of the source specific "fingerprints" that are readily apparent in the SEM may be present as only trace elements in the bulk analysis.[8] Quantitative determinations by automated microscopic methods are very susceptible to "snapshot" sampling anomalies and require verification by comprehensive sampling procedures or other complementary analytical methods where possible. These restrictions are applicable to many other analytical methods which are plagued with the statistical uncertainties of obtaining small representative aliquots from large heterogeneous samples.

IMAGING, ANALYSIS, AND GRAPHICS

Technique Limitations

Table 5 lists some of the major limitations of the PRC software which is available for the TN2000 EDX spectrometer. PRC operation assumes that the analyst has considerable prior knowledge of the X-ray characteristics of features (particles) that are to be analyzed. This is not always possible. When setting up the chemical classifications or "types", rather rigid restrictions are placed on the presence or absence of a particular X-ray line (spectral region of interest), and on the relative intensity of the individual X-ray peaks in the total spectrum. When analyzing larger particles, topography tended to alter the relative X-ray yields and the particles would be categorized as unknowns. Similarly, if a smaller particle was attached or in close proximity to the feature undergoing analysis, or if an agglomerate particle was analyzed, the presence of an unexpected X-ray peak (even if it was relatively small) would also produce an unkown classification. Consequently, the chemical classification process tended to be iterative and somewhat unfriendly. The particle classification process was also restrictive since the software assumed only the 15 analyst-defined X-ray peaks could be present during the sample analysis. In the normal mode of operation any other spectral peaks would be lost,

Table 5. Method Limitations

1. Chemical "typing" – iterative, difficult
2. Large particles – topography alters X-ray ratios
3. Agglomerates distort X-ray analysis
4. Low density particles missed
5. X-ray intensities not elemental concentrations

and the probability for operator induced error was quite large. A number of the classification difficulties have been addressed in the latest hardware and software from commercial manufacturers, but in many cases retrofitting these options onto existing equipment is not possible.[9]

The example of stack emissions from smelter C illustrates the limitation on particle visibility/detectability as a function of size and density. Small particles (less than 0.2 μm diameter) of light elemental composition can not be analyzed by conventional automated SEM-EDX techniques.

Finally, the software produces data only on the basis of X-ray intensity and no effort is made to convert the peak intensities to elemental concentrations. Quantitative or even semi-quantitative estimates of analytical concentrations required careful manipulation of the particle database, and considerable additional computation.

Database Management

An alternative to the iterative chemical classification procedure of PRC is to collect the X-ray data for the individual features in the "search" mode. An entire X-ray spectrum (e.g., 0-20 keV) is collected for a short period of time, and a peak search algorithm operates on the spectrum and stores the statistically determined X-ray peak information (along with the various size parameters) as an object vector for each particle or feature analyzed. Although real-time feature classification is precluded, no previously defined spectral data windows are needed and the opportunity for operator induced error is reduced. Simple computer programs to assist in post-analysis particle classification mechanisms are illustrated in Figures 6a and 6b. X-ray peaks can be sorted in order of decreasing spectral intensity for each object vector. The peak energies of the primary, secondary, or tertiary intensity X-ray lines can be plotted as frequency encoded scatterplots, and using Boolean logic the various combinations of peak energies can be categorized into readily distinguishable particle types.[10,11]

The major advantages of this approach are

(1) no spectral information is lost,
(2) iterative re-estimation of chemical classes/types is not required,
(3) computer graphics can be used to facilitate manipulation of the database, and
(4) small calibration drifts in spectrometer energy scales are quickly recognized as localized clusters.

Digital Imaging Opportunities

Figure 7a is a secondary electron micrograph representative of the particulate loading on one of the surveillance filters. Figure 7b is the backscattered electron image of the same area as Figure 7a. A close comparison of the two images reveals that many of the very small particles (suspected to be auto exhaust) visible in the secondary image are lost in the backscattered image and consequently are missed in the PRC analysis. A digital 256 pixel x 256 pixel secondary electron image of the same region isdisplayed as the pseudocolor image Figure 8a. Note the aspect ratio is square in the digital image rather than rectangular as in the analog representation of Figure 7a. Using the pseudocolor advantage to discriminate between closely related "grey-levels" permits the operator to set a video threshold signal for the PRC control from the digital image (either secondary or backscattered) with greater ease than with the conventional manipulation of the backscattered video line scan signal.

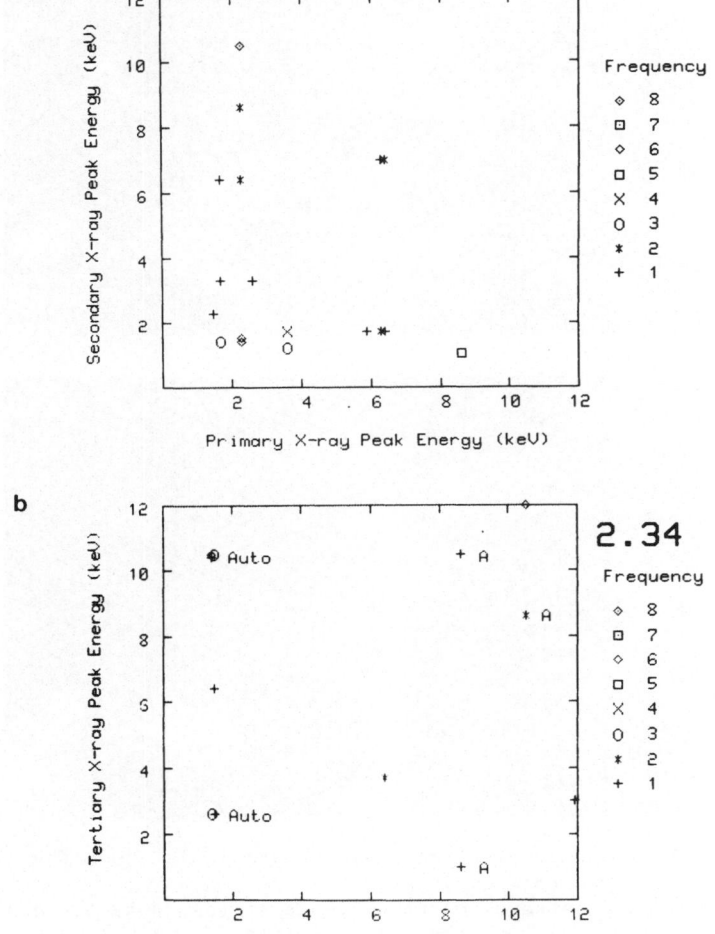

Figure 6. Scatterplots to assist in particle X-ray classifications
(a) the peak energies of the two most intense X-rays emited from the particle are plotted as a function of frequency;
(b) a window has been established at 2.34 keV for the most intense X-ray peak. Viewing within that window the peak energies of the second and third most intense X-ray peaks are plotted as a function of frequency. Various combinations of relative peak intensity are combined using Boolean logic to permit particle classification of "typing".

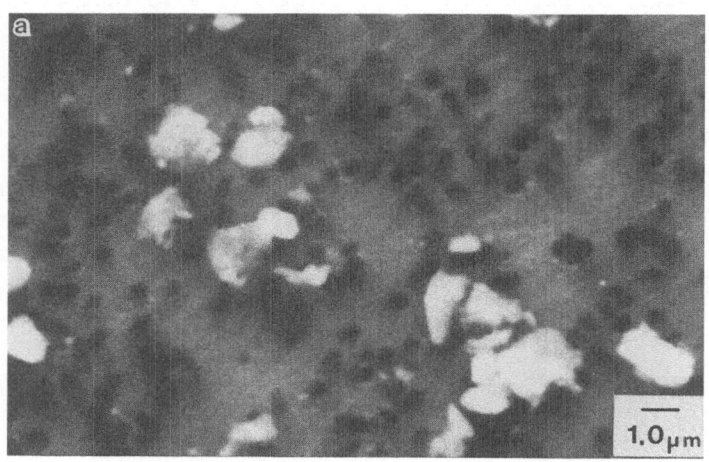

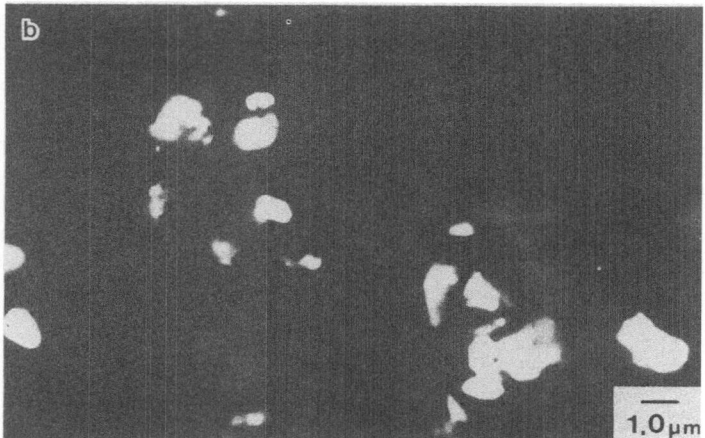

Figure 7. (a) Secondary electron image of particulates on the
            polycarbonate filter.
         (b) Backscattered electron image of (a); note the absence of the
            very small particulates visible in (a).

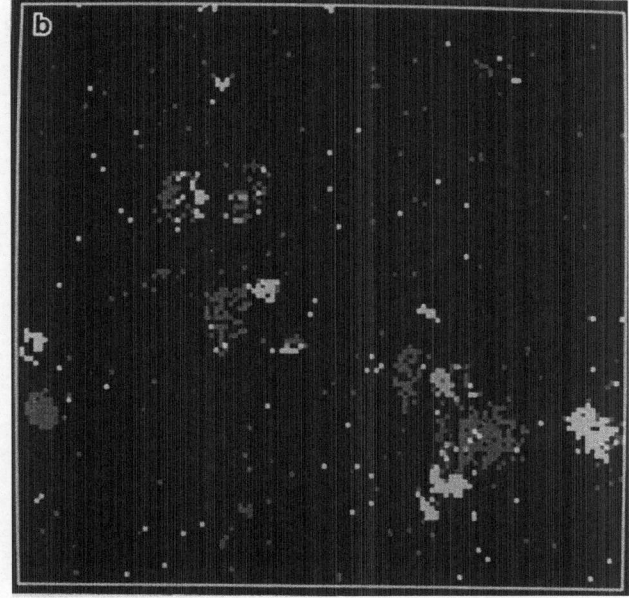

Figure 8. (a) Pseudocolour secondary electron image of the same filter
area as Figure 7 (a); image resolution is 256x256 pixels.
(b) Composite X-ray map of region in 8(a); yellow is indicative
of silicon, green of iron, blue of calcium, red of lead.
The X-ray map resolution is 128x128 pixels.

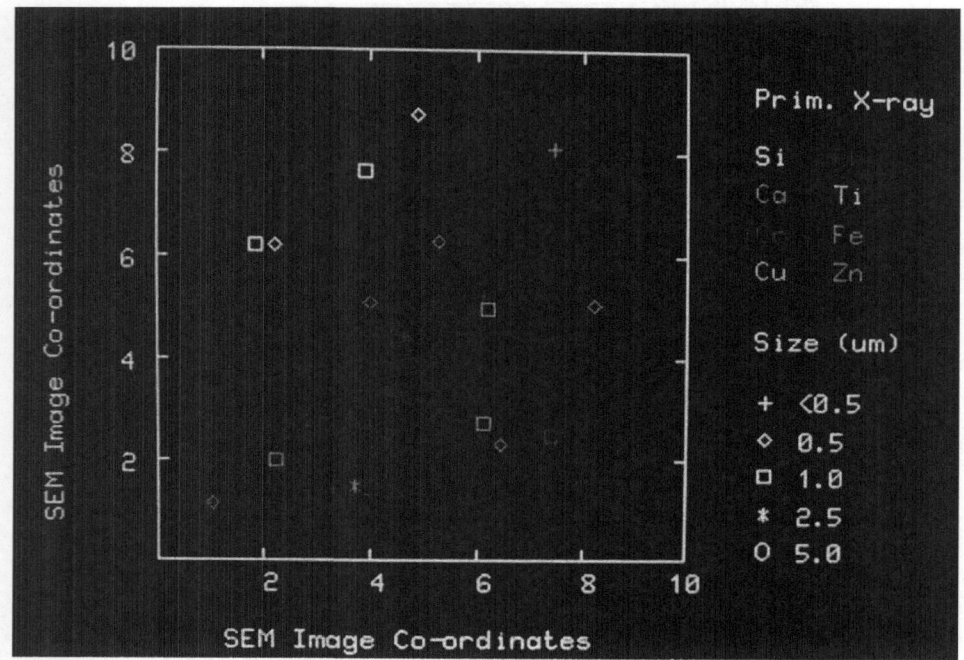

Figure 9. Pseudocolour graphics display of particle classification and size. Colour indicates the particle's chemical class or primary X-ray intensity, the graphic symbol indicates the particle's size category.

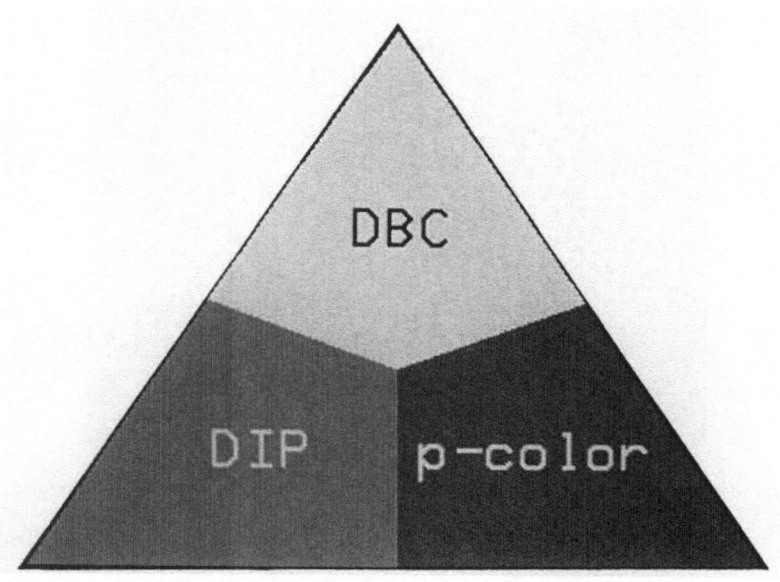

Figure 10. The synergistic relationship between digital image processing, digital electron beam control for feature microanalysis, and pseudocolour computer graphics for data analysis and display.

A composite set of X-ray maps are included as Figure 8b, and show the distribution of some of the major elements in the various features (e.g., particles) of interest. The major impediment to the presentation of data in this format is the substantial investment of instrument time (due to the very low X-ray count rates versus the high secondary or backscattered video signals) required to accumulate multiple X-ray maps of the complete image field containing a sparse distribution of features of interest. Since most of the time is spent analyzing the filter background rather than the particulates, the information return as a function of instrument analysis time is very low. A second limitation of the mapping approach is the limited opportunity to utilize properly the palette of available pseudocolors in displaying composite X-ray maps. Typically only the three primary colors, or perhaps a fourth color are used to avoid confusion in the display of composite maps. X-ray maps are typically accumulated at a lower pixel density (e.g., 128 x 128) to minimize acquisition time, and again the display of small features of interest may be lost in this reduced display.

An alternative to the conventional digital "frame" image analysis or presentation method, is to use a hybrid frame-feature approach. A digital video image of the field is collected in the conventional manner, and a PRC action threshold is determined from information contained within the digital video image; either by levels associated with pseudocolors, or by displaying a video linescan of a raster line in the digital image. The PRC control program then operates on the field of view, and particle characterization data are collected, analyzed, and displayed in the format of Figure 9. The use of pseudocolor computer graphics provides a large palette for display of the various particle chemical (X-ray) classifications, and a choice of graphic symbols can offer a range of particle size parameters not visible in the 128 x 128 X-ray maps. Since PRC typically analyzes a small field of particles in a much shorter time than it takes to acquire the X-ray maps of particular elements, the information return per unit of analysis time is substantially increased by using the digital video signal to capture the "frame" or field of analysis, and presenting the "feature" analysis characteristics of PRC with complementary pseudocolor computer graphics to yield the visual imagery associated with microscopy. This synergistic concept is graphically summarized in Figure 10. The concept incorporates conventional digital imaging schemes with the digital electron beam control capabilities essential to rapid feature analysis, and integrates pseudocolor computer graphics to present the analysis data in a visual format.

Future Developments

Essential hardware for automated particle characterization has been available for a number of years. Current developments in the hardware are related to larger, faster computers and data storage options, fine tuning of digital beam control, and providing peripherals for easier or enhanced computer interaction such as digitization tablets, a computer "mouse", or digital video cameras for light optical microscopy.

The largest area for development opportunities is in software enhancements.[12] Needed are better software "hooks" or integration points to permit use of commercial database or graphics programs, and an extensive library of chemometric or statistical analysis packages.[13,14] The ability to collect and store massive amounts of analysis data is currently available; the abilitiy to condense and efficiently analyze the volume of data is limited and rather lacking in sophistication.

ACKNOWLEDGEMENTS

The authours gratefully acknowledge the assistance of W. A. Reid of
Canadian Bronze Company Ltd. for the financial contribution to the lead
microanalysis program, W. T. Grieve and members of the W. M. Ward Technical Services Laboratory for the sod and soil chemical analyses,
D. C. Jones of Environmental Management Division for the development and
supervision of the sampling programs, S. F. Phillips of Environmental
Management Division and W. H. Hocking of Whiteshell Nuclear Research
Establishment for their evaluation and review of analytical results and
technical manuscripts.

REFERENCES

1. J. A. Cooper and J. G. Watson, Jr., Receptor Oriented Methods of Air
    Particulate Source Apportionment, J. Air Pollution Control Assoc.
    30:1116 (1980).
2. J. E. Core, J. A. Cooper, P. L. Hanrahan, and W. M. Cox, Particulate
    Dispersion Model Evaluation: A New Approach Using Receptor Models J.
    Air Pollution Control Assoc. 32:1142 (1982).
3. F. E. Doern and D. L. Wotton, Microanalysis of Airborne Lead Particu-
    lates in an Urban Industrial Environment, in "Proceed. EMSA 1985",
    G.W. Bailey ed., San Francisco Press, San Francisco, 1985.
4. D. C. Jones and D. L. Wotton, Manitoba Environmental Management Divi-
    sion, Terrestrial Standards and Studies, Report 82-3 (1982).
5. G. S. Casuccio, P. B. Janocko, R. J. Lee, J. F. Kelly, S. L. Dattner,
    and J. S. Mgebroff, The Use of Computer Controlled Scanning Electron
    Microscopy in Environmental Studies, J. Air Pollution Control Assoc.
    33:937 (1983).
6. D. L. Johnson, Automated Scanning Electron Microscopic Characteriza-
    tion of Particulate Inclusions in Biological Tissues, Scanning
    Electron Microscopy/1983/III 1211 (1983).
7. G. Fritz, "PRC-Particle Recognition and Characterization", Tracor
    Northern, Madison, 1982.
8. D. L. Johnson, B. McIntyre, R. Fortmann, R. K. Stevens, and
    R. B. Hanna, Particle Analysis - Bulk Analysis, Chemtech 14:678
    (1984).
9. R. J. Lee, J. S. Walker, and J. J. McCarthy, Micro Imaging: A Link
    Between Microscopy, Image Analysis, and Image Processing, in "Micro-
    beam Analysis-1985", J. T. Armstrong, ed., San Francisco Press, San
    Francisco, 1985.
10. J. Gether and H. M. Seip, Analysis of Air Pollution Data by the Com-
    bined Use of Interactive Graphic Presentation and a Clustering Tech-
    nique, Atmospheric Environment 13:87 (1979).
11. W. S. Cleveland and R. McGill, The Many Faces of a Scatterplot, J.
    Amer. Statistical Assoc. 79:807 (1984).
12. B. Raeymaekers, P. Van Espen, and F. Adams, The Morphological Charac-
    terization of Particles by Automated Scanning Electron Microscopy,
    Mikrochimica Acta - 1984 II 437 (1984).
13. S. Borman, Scientific Software, Anal. Chem. 57:983A (1985).
14. M. F. Delaney, Chemometrics, Anal. Chem. 56:261R (1984).

MEASUREMENT OF SUBCELLULAR IONS

BY X-RAY MICROANALYSIS

FOR EVIDENCE OF HEPATOTOXICITY

J.R. Millette[1,2], A.L. Allenspach[3], P.J. Clark[1], J.A. Stober[1], T. Mills[4], C. Weiler[4], and D. Black[4]

[1] Toxicology and Microbiology Division, Health Effects Laboratory USEPA, Cincinnati, OH 45268
[2] Present address: McCrone Environmental Services, Inc. 200 Oakbrook Business Center, 5500 Oakbrook Parkway, Norcross, GA 30093
[3] Department of Zoology, Miami University, Oxford, OH 45056
[4] Computer Sciences Corporation, Falls Church, VA 22046

PREFACE

This chapter is largely an elaboration of a paper which was presented at the Forensic, Occupational and Environmental Health Symposium at the 1985 joint national meetings of the Electron Microscopy Society of America and the Microbeam Analysis Society in Louisville, Kentucky. An extended abstract of that presentation has been published by San Francisco Press.

INTRODUCTION

The hypothesis that the influx of extracellular Ca^{2+} across a damaged plasma membrane is the final common pathway for chemically mediated cell death[1] has stirred considerable controversy and continued interest in the study of subcellular ion shifts as sensitive indicators of hepatotoxicity caused by environmental pollutants. A number of studies have shown that increases in hepatic mitochondrial calcium result from the ingestion of carbon tetrachloride.[2-4] Quantitative measurement of calcium in subcellular components was accomplished in early studies by using techniques involving whole tissue homogenation and movement of fractionation by centrifugation. In this study x-ray microanalysis of quick-frozen liver tissue, cryo-sectioned for the electron microscope, provided an alternative approach to studying subcellular ion distribution.

METHODS AND MATERIALS

Male CD-1 ICR mice obtained from Charles River Breeding Laboratories, Wilmington, MA, were given 80 mg/kg of carbon tetrachloride (0.5% of the reported LD_{50}) in paraffin oil by oral gavage and sacrificed by decapitation 20 hours later. Immediately after sacrifice four 1 mm cubes shaped like pyramids were taken from the interior of the central liver lobe of each animal. Each cube was mounted on a silver pin with a small drop of glycerol and plunged into a slush of 10% isopentane in liquid propane. The temperature of the mixture was -182 °C. Pieces of whole liver lobe were also successfully prepared by the freezing plier technique.[5] However, although cells on the periphery of the liver lobe showed good cryo-fixation, cells in the interior of the lobe showed evidence of freeze-damage and could not be used in our study comparing interior liver cells.

The cryo-preserved liver samples were sectioned in a Reichert FC4 ultramicrotome with cryo-attachment. Temperature settings were knife = -121°C and specimen = -122°C. The measured temperature at the knife/specimen interface was -133°C. Thin sections (3000 A) were placed on carbon-stabilized formvar-coated grids, pressed and transferred to the JEOL 100CX electron microscope in the frozen state. They were then freeze-dried in the microscope vacuum. The sections were viewed in the JEOL 100CX at an accelerating voltage of 100 KV. Examples of freeze-dried unstained sections of liver cells are shown in Figures 1 and 2.

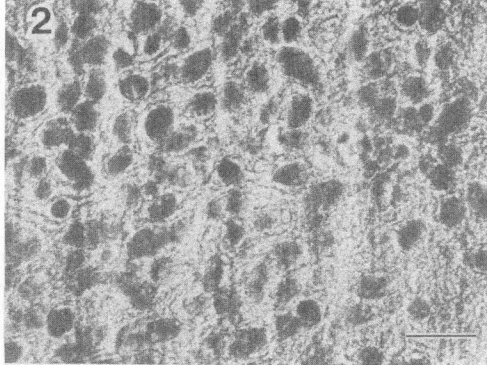

FIG 1.--Low magnification photomicrograph of an unstained, freeze-dried liver section. Nuclei, sinusoids and mitochondria are visible. Bar equals 5 um.
FIG 2.--Higher magnification showing mitochondria and rough endoplasmic reticulum. Bar equals 1 um.

X-ray spectra (200 seconds) were collected from 0.1 um^2 areas within mitochondria, nuclei, or other subcellular components using an ORTEC Si-Li detector and EEDS II multichannel analyzer. Spectra were transferred via floppy disk to a PDP 11/70 computer. Peak-to-continuum values for each element were determined using a version of the Sequential Simplex program developed at the National Institutes of Health.[6,7] The Fortran IV Plus program was run in the IAS Operating System and required 56K bytes of memory. A plot of the best-fit gaussian curve overlayed on the original data was produced for each spectra using Calcomp-Tektronix plotting routines. Normalized Chi-Squared values for each elemental peak were calculated. A standard curve constructed with binary crystalline salts was used to convert peak-to-continuum
ratios to millimole per kilogram dry weight values.[8] The standard curve was calibrated with data from pure organic materials containing chemically

bound ions of interest: glutathion (sulfur) and cyclohelammonium phosphate (phosphorus) (Sigma Chemical, St. Louis, MO). Corrections were made for mass loss with sulfur and chlorine.

Some tissue cubes were prepared by freeze-substitution techniques in order to check the quality of cryofixation. For freeze-substitution, frozen samples were transferred to pre-cooled (-80°C) tetrahydrofuran, (THF, Lot 744613, Fisher Scientific) containing 1% OsO_4 and 0.025% uranyl acetate for a minimum of 70 hours. Following osmication the issues were brought to room temperature by 1 hour steps in -20°C and +4°C freezers, whereupon they were rinsed with acetone and embedded in Epon. Thick (2um) and thin sections were cut on LKB or Reichert ultramicrotomes. Phase contrast microscopy shown in Figures 3 and 4 was performed on a Wild research photoscope. The photomicrographs show both morphological differences for treated and control tissue differences in the freezing preservation of the tissues.

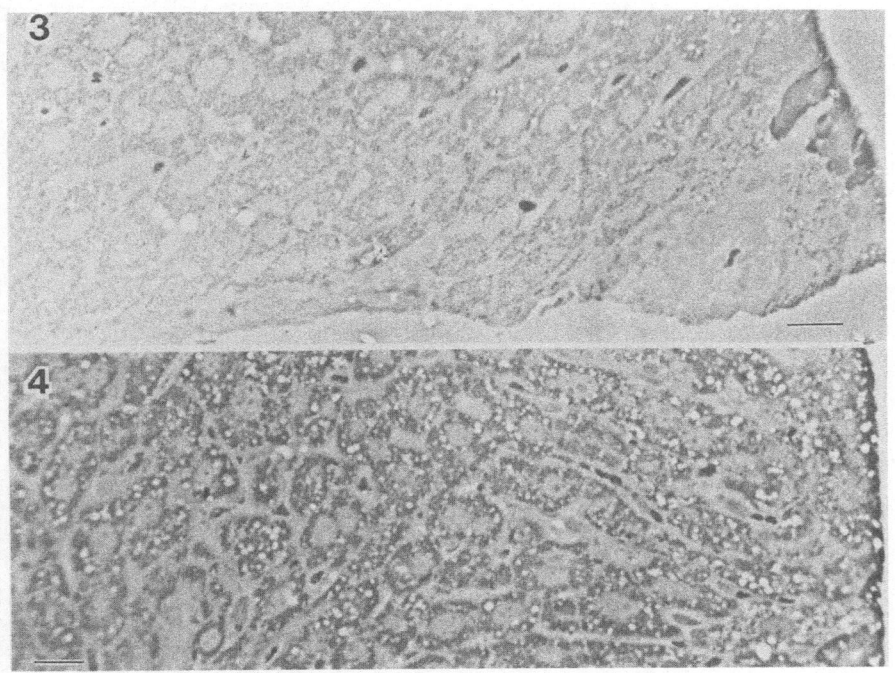

FIG 3.--Phase contrast photomicrograph of control mouse liver tissue which was frozen by plunging into isopentane/propane slush (-182°C). The surface cells exhibit atypical vacuolation and lack of good morphological preservation while deeper cells display good mitochondrial and cytoplasmic preservation. Bar equals 20 um.
FIG 4.--Phase contrast photomicrograph of CCl_4-treated mouse liver tissue, frozen and prepared by methods identical to control tissue (Figure 3). Vacuolation typical of that associated with the effects of CCl_4 is present in deeper cells but the morphological preservation is the same as that shown in the control tissue. Bar equals 20 um.

RESULTS

Based on x-ray microanalysis of mitochondria and nuclei, three populations of parenchymal cells were identified in the livers of mice treated with carbon tetrachloride. One group could not be readily distinguished from cells of control animals, another group of cells had mitochondria with high density calcium granules (as has been reported by others), and a third group had mitochondria which were similar to control in their elemental compositions but had substantially higher (4x) concentrations of chlorine in the nuclei. In some cases all three types of cells were present in the same thin section. Neither high calcium mitochondria nor high chlorine nuclei could be found in any of the cells of control animal liver tissue including cells in sections showing obvious freeze-damage.

Before liver cells can be differentiated on the basis of more subtle subcellular ion concentration differences information must be available about the normal concentration variability present within an individual liver cell. Figure 5 is a photomicrograph of one liver cell and its surroundings. In addition to the nucleus and mitochondria, some rough endoplasmic reticulum and a bile canaliculus are evident. Figure 6 shows a computer assisted drawing of the section and the results of the analyses for one element at 56 sites. This figure gives an indication of the variability in subcellular elemental concentrations within one cell. Data on the other elements detected at these sites and other similar cell analyses were used to develop parameters for statistically representative sampling.

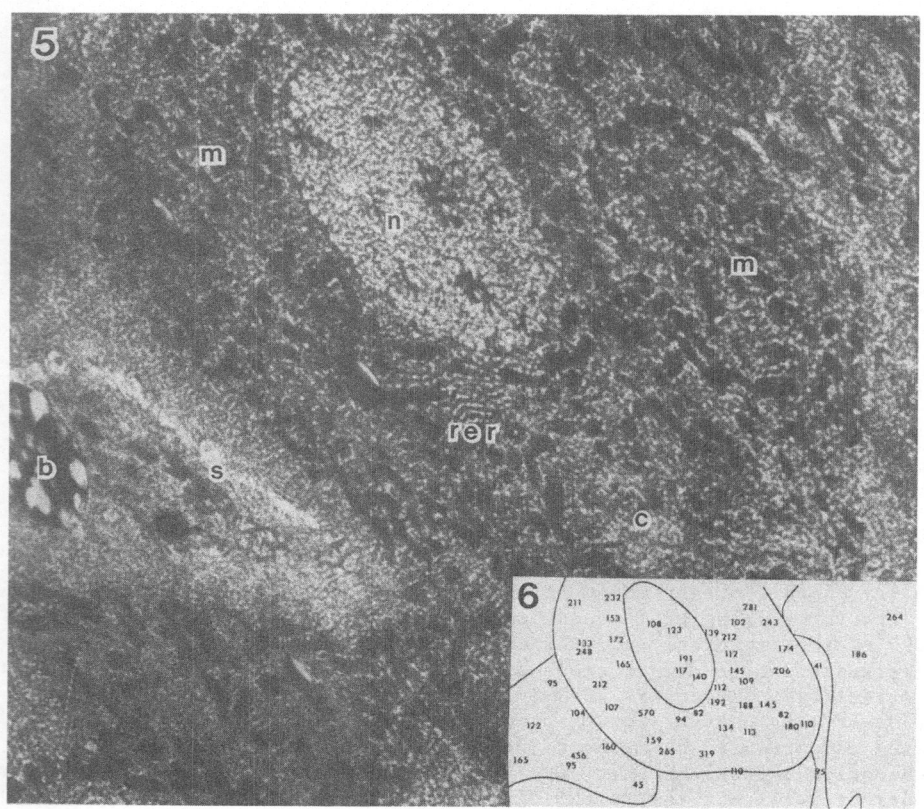

FIG 5.--Photomicrograph of a single parenchymal cell and surrounding area in unstained, freeze-dried liver tissue. The nucleus (N), mitochondria (M), rough endoplasmic reticulum (rer), bile canaliculus (C), sinusoid (S) and blood cell (B) are readily identifiable. Bar equals 1 um.
FIG 6.--Computer-assisted drawing of the cell in Figure 5 showing the location and magnitude of the sulfur concentration as determined by x-ray microanalysis at that spot. Values in millimole per kg/dry weight.

DISCUSSION

The most widely applied technique for quantitative x-ray microanalysis of biological thin sections is the continuum-normalization method proposed by Hall[10]. The basic assumption of the method is that the concentration of an element in the region of the sample can be calculated by comparing the ratio of the peak-to-continuum x-ray intensity of the unknown sample to the same ratio produced by analyzing a standard sample containing a known amount of the element in a matrix similar to that of the unknown.

After an x-ray spectrum has been collected under microscopic conditions which minimize all extraneous signals, the practical problem is to determine the most accurate values for both peak and continuum intensities. The computer program published by Fiori et al.[6] in 1981 is a non-linear technique which fits gaussian profiles to spectral peaks using a sequential simplex matrix optimization procedure for selection of the best fit. The x-ray spectral peaks are represented in a mathematical expression by a set of independent variables for each peak: (1) peak energy (center of the gaussian curve on the x-axis), (2) amplitude (height of the gaussian curve on the y-axis) and (3) peak width (sigma of the gaussian curve). The program compares numerous calculated estimates of the variables representing the peaks with the actual spectrum using the sequential simplex matrix manipulation algorithm until a composite best-fit curve is produced. The area under the fitted gaussian peak of the K-alpha peak for an element is used as the measure of peak intensity. Specifically, the area is calculated using the following equation:

$$Area = 2 \times sigma \times amplitude \qquad (1)$$

Subtracting the best-fit gaussian peaks from the raw spectrum leaves residual counts representing the continuum. A polynomial is fit to the residual data by regression analysis and the value of the polynomial at the central peak energy channel on the x-axis is used as the estimate of the continuum.

In addition to separating the overlapping potassium K-beta and calcium K-alpha peaks (a difference of 0.102 keV), the non-linear fitting program determines the best peak energy and width of the largest peak in the spectrum and subsequently adjusts all other peak energy values. This adjustment corrects for small electronic shifts in energy and resolution of the detector system. This potentially minimizes errors introduced by peak shifts and peak broadening that can create major difficulties in an analysis.[11] Also included in the program is a correction for spectral escape peaks and for a second order effect, incomplete charge collection, when fitting the theoretical gaussian curves to the raw data. This latter effect which is evidenced by a slight distortion on the low energy side of large peaks is important when analyzing small peaks on the low energy side of large peaks. This did not appear to be a problem with normal biological tissue analyses. None of the peaks are what would be considered large.

The Sequential Simplex program published by NIH was considered a research instrument and as such was provided to us with very little documentation. The version we received was not identical to the one published in 1981 and took considerable effort to get up and running. The program is not "user-friendly" and requires specific formats for inputting the parameters used to initialize the simplex matrix. Documentation for the EPA version of the program is available from the first author at his present address. It describes how to run the programs and gives specific examples. As indicated in the methods section, the program takes up considerable computer memory space on the PDP 11/70. A version of the program has been converted to run on an IBM-PC but it is of little use at this time because the ORTEC disks with x-ray data cannot be used directly in the PC. The spectra are converted to fortran data files using disk drives connected to the PDP 11/70. Of greater importance is the lack of sophisticated plotting routines available for our PC. The results of simplex runs could not be plotted with currently available PC software.

The program can take a long time to run - over half an hour for some spectra. The use of so much computer time forced us to develop a batch program where groups of spectra could be run through the program at night when the computer was less in demand. The results of each run was plotted on the next day to visually check the fit of calculated spectrum to the raw data. Running in batch mode requires that certain input parameters be standardized. This did not cause any problems because the spectra were all from similar areas of the liver.

Fiori, Mykelbust and Gorlen (1981)[6] did considerable work testing the effect of varying the initial input parameters on the results of the Simplex runs. Their studies used single pure gaussian peaks to test the program. A spectrum containing peaks corresponding to silicon, phosphorus, sulfur and chlorine as found in our samples was used by our laboratory to test the program for our specific application. A set of 35 different initial estimates of the largest peak amplitude were tested to see the effect on the final product of the program-peak-to-continuum ratio. As with findings of Fiori, Mykelbust and Gorlen (1981) done with a single pure gaussian peak, the initial estimate of amplitude did not have much effect on the final peak-to-continuum ratios. See Table 1.

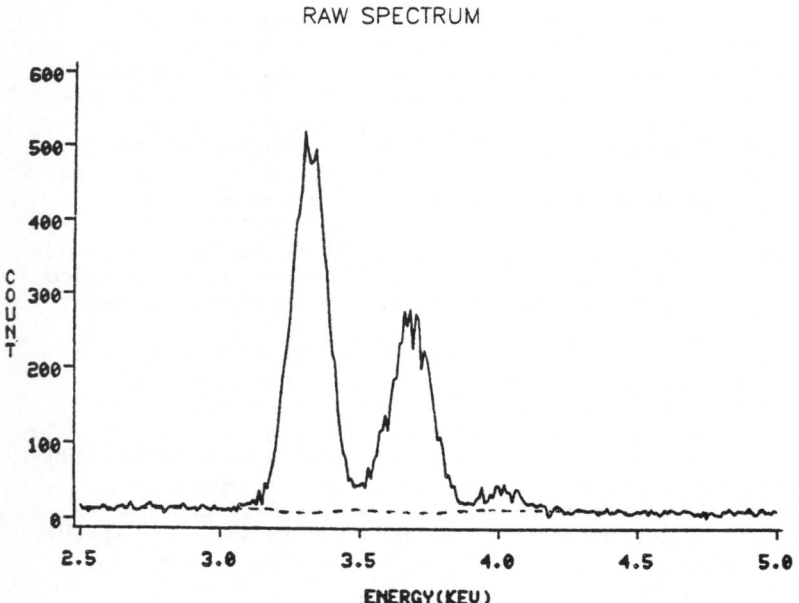

FIG 7.--X-ray Spectrum from standard sample showing curved continuum estimate under principal peak.

The estimation of the continuum was found to be a larger source of variability. The NIH program uses a fourth degree polynomial to fit to the residual values after the best fit gaussian peaks are subtracted from the raw spectrum. A non-standard form of weighting both sets of variables (x and y) is used during the regression. While running standard spectra it was noticed that occasionally the fourth degree regression would produce a continuum which curved downward under the principle peak. See figure 7.

A project to determine which polynomial equation would best model the particular continuum of our spectra was undertaken. All possible regressions from a constant through a fourth degree polynomial, weighted and unweighted, were fit to the residuals of a number of spectra. Based on Chi-square goodness-of-fit tests, consistency of peak-to-background ratios for standards and visual inspection of the fit, four polynomials were chosen as best modeling the continuum. Further work is required to determine which of the 4 polynomial regressions; weighted second degree, unweighted second degree and weighted and unweighted fourth degree, will provide the best continuum estimator for all spectra produced by our analytical system.

Converting from peak-to-continuum ratio values to millimole per kilogram dry weight values requires comparison the unknown spectrum with spectra from standard materials. Pure element materials have been added to gelatin, hardened and sectioned for standards by Hagler et al[5]. Analysis by atomic absorption spectroscopy of available gelatin stocks and agar stocks showed levels of calcium ranging from 14 to 50 mmol/kg dry weight. One of the gelatin materials also contained traces of tin and iodine which have x-ray peaks which overlap with calcium and potassium peak determination. Because a pure gelatin matrix material could not be found, it was decided to construct a standard curve with binary compounds as was done by Shuman et al.[8]. Instead of using the analysis of sulfur in albumin as the standard for calibrating the binary curve as done by Shuman, a pure sample of glutathione (No. G-4251, Sigma Chemical) was ground between glass slides and sprinkled on a carbon-coated grid. The glutathione molecule has one chemically bound sulfur atom in a matrix of light elements. The thin plates of glutathione are therefore chemically well-defined, homogeneous at the molecular level and similar in hydrogen, oxygen and carbon content to biological tissue. For glutathione the factor, Z^2/A, which is used to compare standard and sample compositions is directly calculable and equal to 3.29. The value of Z^2/A for freeze-dried liver tissue has been estimated to be 3.19 by Roomans, 1980.[12] The standard material and the sample material therefore have similar composition.

The work presented in this paper is clearly an initial effort to use the sophisticated x-ray microanalysis technique for studying hepatotoxicity on a subcellular level. A number of areas need further study. The Sequential Simplex program, although in working order, needs to be compared with linear methods of peak fitting. The long execution time required for the program may preclude its use as a routine analysis tool for large numbers of spectra.

Sections of frozen liver tissue can be prepared on nearly a routine basis. Comparisons between the frozen slush technique and freeze plier are needed to determine the advantages of each. Deciding which part of the liver lobe the sections in the microscope represent has not been worked out. A technique for collecting a thick section after thin sections have been cut is being considered as a way to determine whether the cells analyzed are central lobular or periportal.

Despite the initial problems and complexity of the technique it has great potential for comparing subcellular ion concentrations in liver tissue.

TABLE 1

Effect of Varying Initial Amplitude on
Peak-to-Continuum Ratios
(Spectrum BLAV04)

ELEMENT	N	MEAN	STD	95% C.I.	INCR. CHANGE
Silicon	35	484	29	(475, 494)	4%
Phosphorus	35	1058	24	(1050, 1066)	2%
Sulfur	35	414	18	(408, 420)	3%
Chlorine	35	249	8	(246, 251)	2%

Acknowledgement

Figures 1 to 6 have been reproduced from Proceedings 43rd Annual Meeting, Electron Microscopy Society of America, G. Bailey (ed), Lousiville, KY. Aug. 5-9, 1985, San Francisco Press, San Francisco, California, p. 108.

References

1. F. A. X. Schanne, A.B. Kane, E.E. Young and J.L. Farber. Calcium dependence of toxic cell death: A final common pathway. Science 206:700-703 (1979)

2. E. S. Reynolds. Liver parenchymal cell injury. J. Cell Biol. 19:139-57 (1963)

3. E. A. Smuckler. Studies on carbon tetrachloride intoxication. Lab. Invest. 15:57-166 (1966)

4. L. Moore, G. R. Davenport and E. J. Landan. Calcium uptake of a rat liver microsomal subcellular fraction in response to in vivo administration of carbon tetrachloride. J. Biol. Chem. 251:1197-1210 (1976)

5. D. Parsons, D. J. Bellotto, W. W. Schultz, M. Buja and H. K. Hagler. Toward routine cryoultramicrotomy. EMSA Bulletin 14:49-60 (1985)

6. C. E. Fiori, R. L. Mykelbust and K. Gorlen. Sequential Simplex: a procedure for resolving spectral interference in energy dispersive x-ray spectroscopy. NBS Spec. Pub. 604. Proc. Workshop on Energy Dispersive X-ray Spectroscopy, Gaithersburg, MD. Apr. 23-25, 1979. 233-72 (1981)

7. C. E. Fiori, C. R. Swyt and K. Gorlen. Application of the top-hat digital filter to a non-linear spectral unraveling procedure in energy dispersive x-ray microanalysis. In: Microbeam Analysis (R. Geiss, ed.) San Francisco Press, 320- 324 (1981)

8. H. Shuman, A. V. Somlyo and A. P. Somlyo. Quantitative electron probe microanalysis of biological thin sections: methods and validity. Ultramicroscopy 1:317-39 (1976)

9. T. E. Phillips and A. F. Boyne. Liquid nitrogen-based quick freezing: experiences with bounce-free delivery of cholinergic nerve terminals to a metal surface. J. Electron Microsc. Tech. 1:9-15 (1984)

10. T. A. Hall. Biological x-ray microanalysis. J. Microsc. 117:145 (1979)

11. J. M. Tormey. Improved methods for x-ray microanalysis of cardiac muscle. In: Microbeam Analysis (R. Gooley, ed.) San Francisco Press 221-228 (1983)

12. G. M. Roomans. Quantitative x-ray microanalysis of thin sections in M.A. Hayat (ed). X-ray Microanalysis in Biology. Univ. Park Press, Baltimore 401-453 (1980)

USING MICROPARTICLE CHARACTERIZATION TO IDENTIFY SOURCES OF

ACID-RAIN PRECURSORS REACHING WHITEFACE MOUNTAIN, NEW YORK

James S. Webber

Wadsworth Center for Laboratories and Research
New York State Department of Health
Albany, NY 12201

PREFACE

This chapter is largely an elaboration of a paper which was presented at the Forensic, Occupational and Environmental Health Symposium at the 1985 joint national meetings of the Electron Microscopy Society of America and the Microbeam Analysis Society in Louisville, Kentucky. An extended abstract of that presentation has been published by San Francisco Press.[1]

INTRODUCTION

Atmospheric scientists and limnologists no longer monopolize the terms ''acid rain'' or ''acid deposition.'' This phenomenon has become a topic of discussion among the public in general and a source of international debate among scientists, politicians and decision makers. Most concern centers around the severe ecological damage which has occurred in eastern regions of North America where the Canadian Shield is exposed. Airborne acids deposited on this bedrock, which has little buffering capacity, can quickly decrease pH in streams, lakes and groundwater. Adverse effects on aquatic ecosystems have been most widely documented, e.g., eradication or depletion of fish populations in hundreds of lakes in the Adirondack Mountains of New York.[2] Forests may also suffer as lowered soil pH alters metal-ion balance.[3] There is concern that human health may be directly affected by respirable sulfates and associated fly ash[4],[5] or by the increased solubility of toxic trace metals, especially lead, in acidified waters.[6] Even man's monuments are deteriorating as airborne acids etch their metal and stone features. As much as 5 billion dollars' worth of damage may be caused by acid deposition in 17 states from Illinois to Maine.[7]

Sulfates (primarily sulfuric acid and ammonium bisulfate) account for about two thirds of airborne acids in eastern North America with wet and dry deposition contributing equally to acids reaching surface waters or the ground.[8] A recent study of Adirondack lake sediments has revealed that the large increase in atmospheric sulfate is a recent phenomenon.[9] Sulfate loadings in the lake began an 8- to 10-fold increase around 1935. The industrial Midwest is widely assumed to be the source of most sulfates reaching the acid-sensitive Northeast since the amount of sulfur dioxide emitted as a byproduct of midwestern coal combustion dwarfs that emitted

261

from all sources in the Northeast. For example, the seven midwestern states (Illinois, Indiana, Kentucky, Michigan, Ohio, Pennsylvania and West Virginia) located upwind of New York State emit 13.4 million metric tons of sulfur dioxide annually compared to less than 1 million metric tons from New York and the six New England states combined.[10]

Recent studies have revealed with increasing clarity that the Midwest is indeed the primary source of aerosol sulfate. A 1978-9 study based on 18 months' continuous daily data and backward air trajectories concluded that 68% of the high-sulfate air masses reaching Whiteface Mountain in the Adirondacks arrived from the direction of the industrial Midwest.[11],[12] A similar geographic correlation of sulfate in precipitation was reported for Whiteface Mountain.[13] Husain et al.[14] showed that midwestern air masses could be differentiated from East-Coast air masses by their trace-element signatures, specifically manganese/vanadium ratios. Manganese/vanadium ratios are low in air masses from the Northeast because these are highly enriched in vanadium, a byproduct of residual-oil combustion. Ratios in the Midwest are much larger since there is relatively little residual-oil combustion while large amounts of manganese are emitted during industrial operations. Almost all analyzed high-sulfate air masses reaching Whiteface Mountain since 1979 have had typically large midwestern manganese/vanadium ratios. The most direct evidence yet of a coal-burning sulfate source was revealed in an electron microscopical (EM) investigation of a high-sulfate air mass reaching Whiteface Mountain.[15] This paper will summarize that study and continue with results from a second high-sulfate period.

EXPERIMENTAL

Particulate air samples were collected in consecutive 6-hour samples on Whatman 41 filters in high-volume samplers located at Mayville in southwestern New York State and at the summit of Whiteface Mountain during summer 1983. Consecutive 12-hour samples were collected in the same manner at Alexandria Bay in northern New York State and at West Haverstraw in southeastern New York State. Sulfate concentrations were determined by ion chromatography while trace-metal concentrations were determined by atomic absorption spectrophotometry or neutron activation analysis.[14],[16],[17] Particles for EM analysis were collected at Whiteface Mountain on 0.4-μm Nuclepore filters which had been carbon coated. These samples were collected for 3-hour periods in the middle of the 6-hour high-volume sample periods.

The Nuclepore filters were coated again with vacuum-evaporated carbon and prepared for transmission EM (TEM) analysis by a modified Jaffe-wick procedure.[18] Sections were excised, placed on copper finder or nylon TEM grids and placed in a warm chloroform atmosphere for 1 to 3 days. This dissolved the polycarbonate filter, leaving on the grid a carbon film with embedded particles.

Most analyses were done with the Wadsworth Center's AEI EM7 1240-kV high-voltage EM (HVEM) which was equipped with a Kevex Si-Li detector coupled to a Tracor-Northern 2000 multichannel analyzer. Analyses were done at 1000 kV because the high-energy beam yielded higher-order diffraction points and because the forward-peaked bremsstrahlung increased the signal-to-noise ratio in the x-ray spectrum.[19] Discernible x-ray peaks were usually yielded by particles larger than 0.1 μm during the 40-60-second collection period. Collection dead times were generally 35-50%. Specimen morphology and selected-area electron diffraction patterns were collected on Cronex Lo-dose mammography film. Additional filter sections were Au/Pd shadowed at a 27° angle[20] for high-contrast relief characterization with a Siemens 102 TEM or were mounted for analysis by an ETEC Autoscan scanning EM (SEM) which was equipped with a PGT Si-Li detector and Tracor Northern NS880 multichannel analyzer.

The directions from which air masses had traveled to Whiteface Mountain were calculated with the Air Resources Laboratory Atmospheric Transport and Dispersion model.[21] These backward air trajectories were determined for the middle of each 6-hour sampling period, with geographic coordinates calculated for consecutive 6-hour intervals backward in time. Only the first two or three days backward in time were considered significant given the uncertainties of extended trajectory calculations.

ELECTRON-MICROSCOPICAL EVIDENCE

A spike of high sulfate concentration (24 µg/m³) was measured in the last 6-hour sample collected at Whiteface Mountain on 23 June 1983. The high-sulfate period was brief, however, as sulfate concentrations 12 hours earlier and 12 hours later were only 2.7 and 2.2 µg/m³, respectively. Microparticles collected during these three periods were characterized by EM.

Electron-translucent spots ranging from <0.1 µm to 0.8 µm were the most abundant particles in the high-sulfate sample. These yielded only sulfur x rays and were assumed to be sulfates. Two distinct morphological categories are depicted in Figure 1. The first category is typified by the residue H. This was left from a sulfate spot which melted and boiled as soon as it was touched by the electron beam. Sulfate spots in the other category, denoted by N, were more numerous. These spots did not deteriorate as rapidly in the electron beam. It is likely that the more volatile particles were desiccated sulfuric acid and that the stable spots were ammonium sulfate. Because filters from this first episode did not receive a second carbon coat until sections were put through the Jaffe-wick procedure, changes in sulfate morphology were noticed. Within six weeks, most sulfate particles had developed one to several dark spots of condensed sulfate at the margins of the original impaction (Figure 2). The exact nature of the original impaction spot compared to the earlier high-contrast border (as in N, Figure 1) does indicate relaxation of its

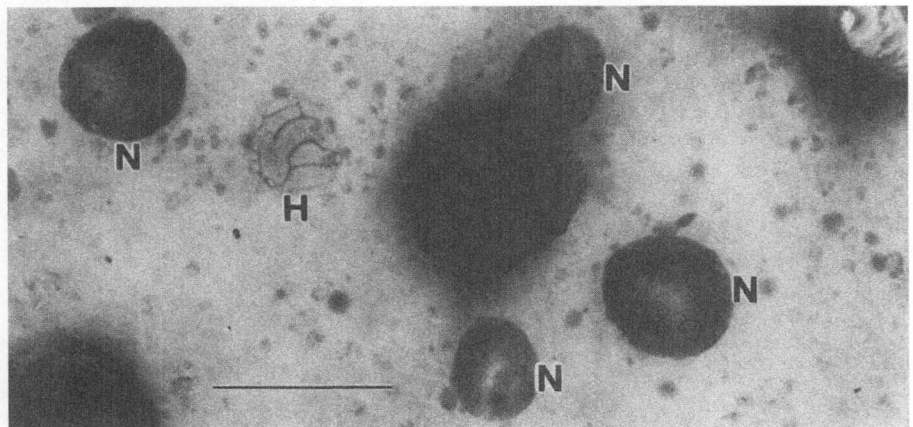

Figure 1. Transmission electron micrograph of sulfate particles collected from 1930–2230 EST on 23 June 1983 atop Whiteface Mountain when sulfate=24 µg/m³. Scale bar=0.5 µm.

three-dimensional morphology (thought to be wart-shaped as described later). Note that both the original spot and the condensed outgrowths had reacted to the 125-kV electron beam after 60 seconds. Because of these alterations, subsequent samples were coated with carbon soon after their return to the laboratory.

Electron-translucent and opaque submicrometer spheres were abundant in the high-sulfate sample but were absent in the pre- and post-episodic samples. Iron-rich spheres or sphere clusters were identified by electron diffraction as magnetite, hematite, maghemite or combinations of these iron oxides. Silicon-rich spheres were also commonly seen in the high-sulfate sample and usually yielded no diffraction patterns. These were consequently classified as glass. In addition, there were spheres with mixtures of iron, silicon, aluminum, manganese, zinc, nickel, chromium, titanium, potassium and calcium.[15] The significance of these spheres will be discussed below.

A more intense high-sulfate episode reached Whiteface Mountain on 28 July. Six samples from this period were examined by EM. As in the earlier episode, submicrometer sulfates were the most abundant. Sulfur x rays from these particles were occasionally accompanied by x rays from smaller amounts of calcium, iron or potassium. These translucent particles were difficult to resolve at high voltage but Au/Pd shadowing revealed interesting morphological details (Figure 3). Common morphologies were domes, flattened circles and clusters. These shapes probably reflected differences in sulfate species. For example, the domes (denoted as H) in Figure 3a melted and boiled readily in the beam and were probably sulfuric acid. The wart-shaped clusters (denoted as N) in Figures 3b and 3c did not boil as readily and were probably ammonium sulfate. The particles with the most nodular aggregations (denoted as S in Figures 3b and 3c) might be carbonaceous soot which was coated with sulfate as discussed shortly. Note that in Figure 3c small domes are concentrated around the rims of the pores. These ~50-nm particles were small enough to be entrained in the accelerating air streams as they approached the pores, but the particles' inertia was too great to allow the particles to make the sharp bend over the pores' rims. Hence the impaction pattern around the pores. The contents of these domes boiled in the beam, indicating a sulfuric acid content.

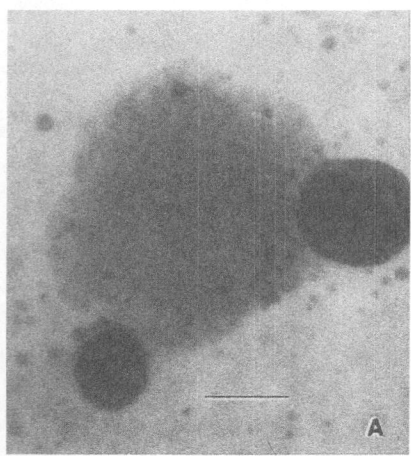 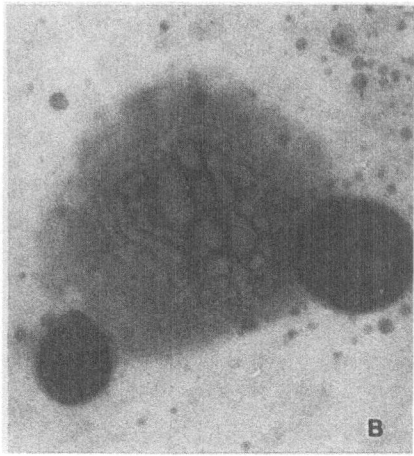

Figure 2. Transmission electron micrographs of a sulfate particle taken 6 weeks after its 23 June 1983 collection atop Whiteface Mountain. Scale bar=0.1 μm. A) During initial beam contact. B) After 60 seconds in beam.

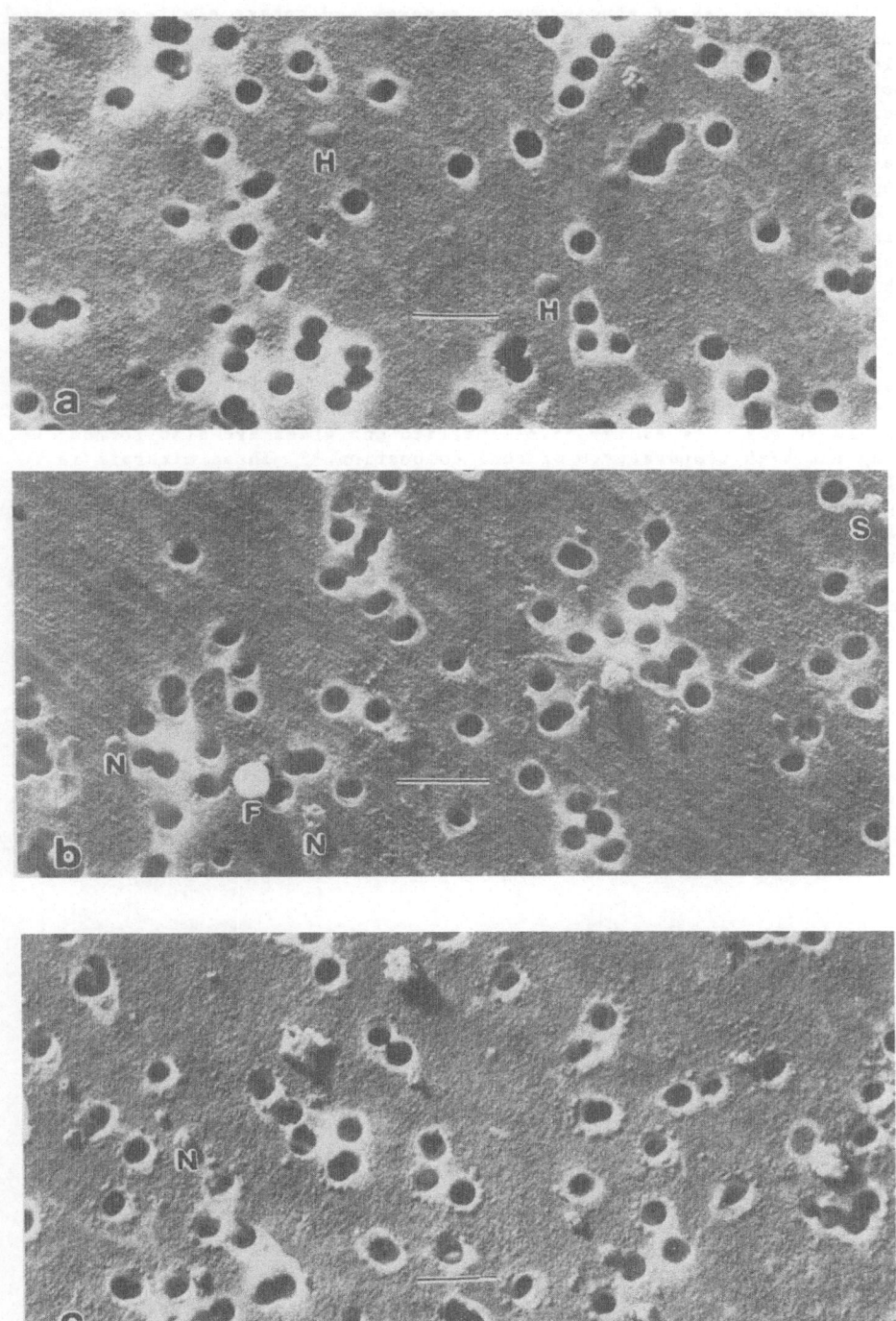

Figure 3. Transmission electron micrograph of Au/Pd-shadowed filters with particles collected atop Whiteface Mountain on 28 July 1983. Scale bars=1 μm.   a) 0130-0430 EST, sulfate=4.9 μg/m³. b) 1330-1630 EST, sulfate=14 μg/m³. c) 1930-2230 EST, sulfate=49 μg/m³.

265

Concentrations of submicrometer spheres and sphere clusters composed primarily of silicon and iron varied almost three orders of magnitude during this second period: abundant when [sulfate] exceeded 10 μg/m³ but almost absent in the low-sulfate samples.[22] The sphere denoted by F in Figure 3b is one of these. Its electron opacity is indicated by the large amounts of white in this negative image. Elemental compositions were the same as in the earlier episode and mineral identities were also similar except that mullite was identified and maghemite rather than magnetite was the dominant mineral in the iron spheres. The iron spheres frequently contained other metals as well (manganese, zinc, nickel, chromium, titanium or potassium). Silicon spheres also contained other elements such as aluminum, calcium, titanium, manganese or iron. Typical spheres and x-ray spectra are shown in Figure 4.

The correlation of these spheres with both high-sulfate air masses is significant in light of previous EM and x-ray diffraction studies of fossil-fuel fly ash. Magnetite, hematite and maghemite are common minerals in coal fly ash.[23,24,25] Mullite and glass are also formed during the high temperatures of coal combustion.[26] These minerals in coal fly ash are usually spherical as a result of the freezing of molten droplets as they cool after combustion. A typical cluster of silicon-rich coal fly ash spheres collected from a midwestern power plant is shown in Figure 5a. Oil fly ash, on the other hand, is enriched in vanadium[27] and morphology is often honeycombed as in Figure 5b. In the nine samples from Whiteface Mountain examined by EM, only one particle with a small

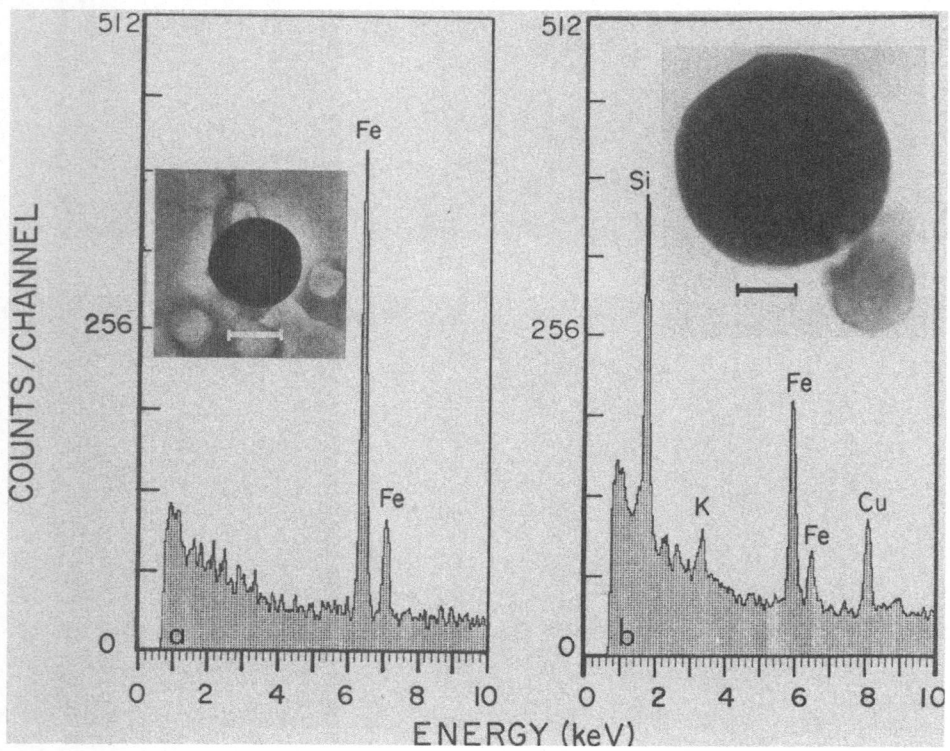

Figure 4. Spheres collected atop Whiteface Mountain from 1930-2230 EST on 23 June 1983 when sulfate=24 μg/m³. Prominent x-ray peaks are labeled in accompanying energy-dispersive spectra. Scale bars=0.25 μm. Cu x-ray peak is from EM grid bar.

vanadium x-ray peak was seen and this was in a post-episodic sample (24 June) when sulfate was only 2.2 μg/m³.

Silicon-rich fragments, ranging from submicrometer to 3 μm, were seen in all samples (Figure 6). Some were identified as quartz, potassium feldspars or plagioclase, common minerals in the Adirondacks.[28] Abundance of these silicon-rich fragments did follow sulfate concentrations, perhaps because some of the fragments were from industrial emissions and were carried along with coal fly ash in the atmosphere to Whiteface Mountain. Concentrations of trace metals were high, commensurate with previous sulfate episodes,[14] and this indicated that favorable conditions existed for atmospheric transport of anthropogenic emissions. Nonetheless, the magnitude of fragment variation was much less than for coal fly-ash spheres.

The types and sizes of the fly-ash particles detected provide interesting insight into the source of the high-sulfate air masses. Although a large (850 MW) oil-fired power plant was operating just 200 km upwind of Whiteface Mountain, there was no microparticle evidence of its emissions during either high-sulfate period. Only spheres characterized as coal fly ash were markedly more abundant during the high-sulfate periods. The lack of spheres larger than one micrometer might point to

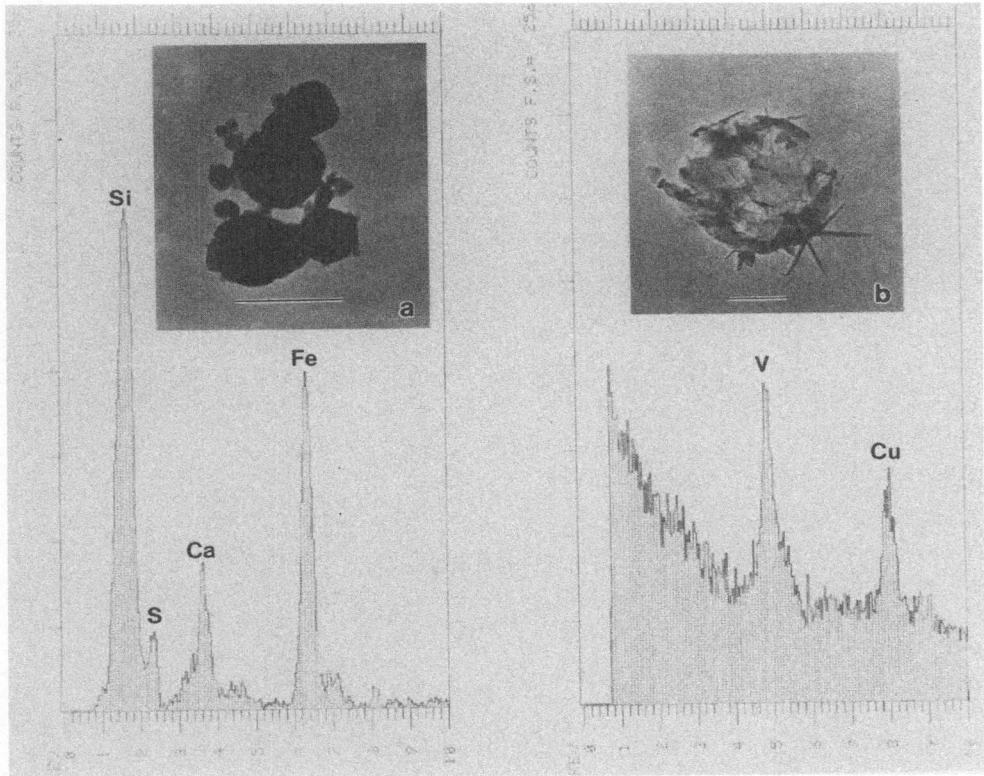

Figure 5. Fossil-fuel fly ash collected in stacks. Prominent x-ray peaks are labeled in accompanying energy-dispersive spectra. Scale bars=1 μm. a) Coal fly ash from a midwestern power plant. b) Fly ash collected in an eastern oil-fired power plant. Cu x-ray peak is from EM grid bar.

distant emission sources since larger spheres would have fallen out during the >500-km transport from the high concentration of coal-fired power plants in the Midwest. Some of the iron spheres might have been fallout from steel-mill emissions, but this would not substantially alter our conclusions. Steel mills are generally concentrated in the same geographic region as the coal-fired power plants.

Detailed features on particle surfaces became evident after angle shadowing by Au/Pd. The accretion of sulfate on the surfaces of both particles in Figure 7 are similar to findings elsewhere. The small (<50nm) sulfate particles on the silicon sphere in Figure 7a were common to many of the fly ash spheres. Less common were the encapsulating sulfate ''pools'' (Figure 7b) which melted and boiled in the electron beam. Cheng[29] reported from laboratory experiments that sulfates were formed on coal fly ash by heterogeneous oxidation of sulfur dioxide in a humid atmosphere. The pool (probably sulfuric acid as mentioned earlier) surrounding the silicon sphere in Figure 7b is similar to observations by Ferek et al.[20] of coal fly ash collected four hours downwind of emissions. They interpreted this as homogeneous oxidation of sulfur dioxide followed by heterogeneous condensation on the fly ash. Although the oxidation mechanism of sulfates on the few particles scrutinized here is uncertain, continued examination of shadowed particles in future episodes may shed light on oxidation pathways and ages of aerosols.

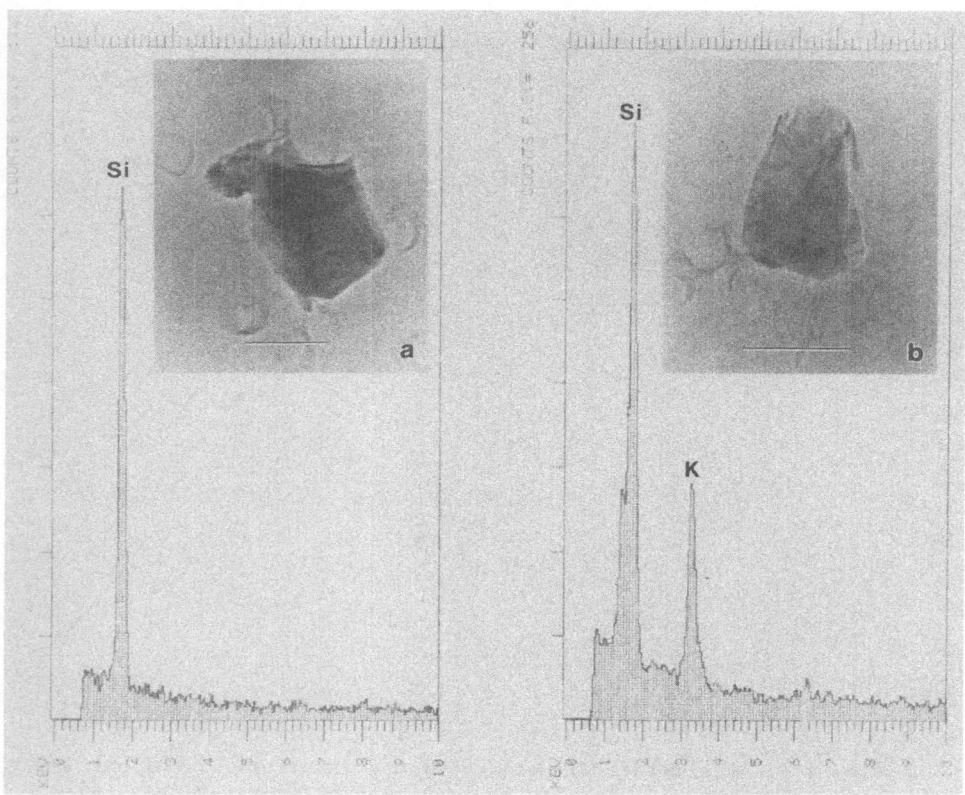

Figure 6. Crustal particles collected atop Whiteface Mountain, 1930-2230 EST, 28 July 1983. Prominent x-ray peaks are labeled in accompanying energy-dispersive spectra. Scale bars=1 μm. a) Quartz. b) Microcline.

SUPPORTING EVIDENCE

Meteorologic conditions and backward air trajectories revealed that both high-sulfate air masses had spent considerable time in the industrial Midwest before reaching Whiteface Mountain. In the June episode, midwestern air rotating around a high-pressure cell in Pennsylvania was funneled through upstate New York by a cold front situated along the lower Great Lakes. When the cold front passed through the state and brought northwesterly winds, sulfate concentrations immediately decreased. The second episode was produced by a more typical pattern where a high-pressure cell passed through the Midwest and stalled off the Atlantic coast, pumping midwestern air into the Northeast. Figure 8 reveals the results quite clearly. On 27 July when air masses were arriving from the northwest, sulfate concentrations were generally low. Only as air trajectories passed through the industrial Midwest did sulfate concentrations increase dramatically. During the last part of 28 July, air masses were coming from the vicinity of the Ohio River Valley. This region has the largest concentration of coal-fired power plants in the United States and produces the largest regional sulfur dioxide emissions in the United States. The close spacing of points on these trajectories indicates that air masses spent a long time in the Midwest before traveling to Whiteface Mountain, a condition which favors sulfate accumulation.[30],[31] The sulfate concentration in the last 6-hour sample of 28 July was 49 $\mu g/m^3$, the highest concentration measured by our group during 10 years of monitoring at Whiteface Mountain. The midwestern origin of these two episodes of high-sulfate air masses repeats the most common pattern seen in previous years.

Chemical signatures at the three upstate sites (Mayville, Alexandria Bay and Whiteface Mountain) were distinctly midwestern during both high-sulfate periods. Manganese/vanadium ratios were consistently >2 when sulfate concentrations were high, indicating that there was little if any aerosol contribution from northeastern sources. Vanadium concentrations at the upstate sites were lower than the double-digit concentrations previously associated with northeastern emissions.[32] In fact, the

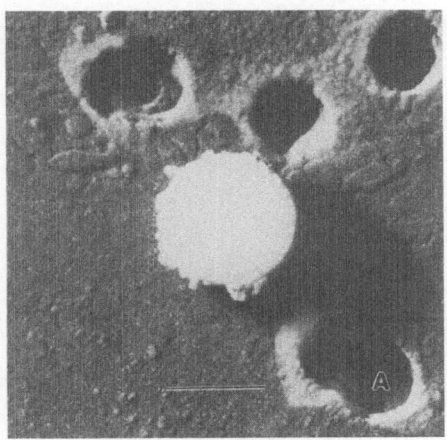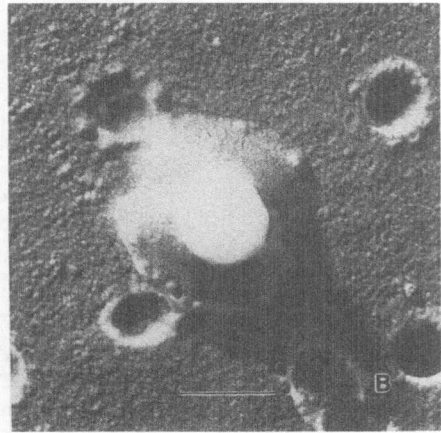

Figure 7. Silicon spheres collected atop Whiteface Mountain on 28 July 1983. Scale bars=0.5 μm. A) 1930-2230 EST. B) 0730-1030 EST. (Figure 5A reprinted with permission from San Francisco Press[1]).

highest vanadium concentration (9.5 ng/m³) at Whiteface Mountain
during the time periods under discussion was measured when air masses
had circled through eastern and central upstate New York. The sulfate
concentration at that time was only 2.7 μg/m³, consistent with the
relatively small sulfur dioxide emissions in upstate New York.   In
general, downstate West Haverstraw was less affected by both midwestern
episodes because it received air masses from different regions.
Manganese/vanadium ratios only exceeded 0.5 when sulfate concentrations
rose from their typical levels of ~5 μg/m³ to 10-15 μg/m³.

Figure 9 shows that selenium, a volatile trace metal which is highly
enriched in coal-combustion emissions, closely followed sulfate
concentrations during both episodes at Mayville and Whiteface Mountain.
Note that because selenium concentrations were measured in
4-sample(24-hour) composites, sulfate concentrations are also plotted as
24-hour averages and hence the variability of individual 6-hour samples is
obscured. The excellent correlation of selenium and sulfate
concentrations has been observed in past years in New York State[17] and
more recently in the Shenandoah Valley[33] and indicated that the major
source of sulfate was indeed coal combustion.

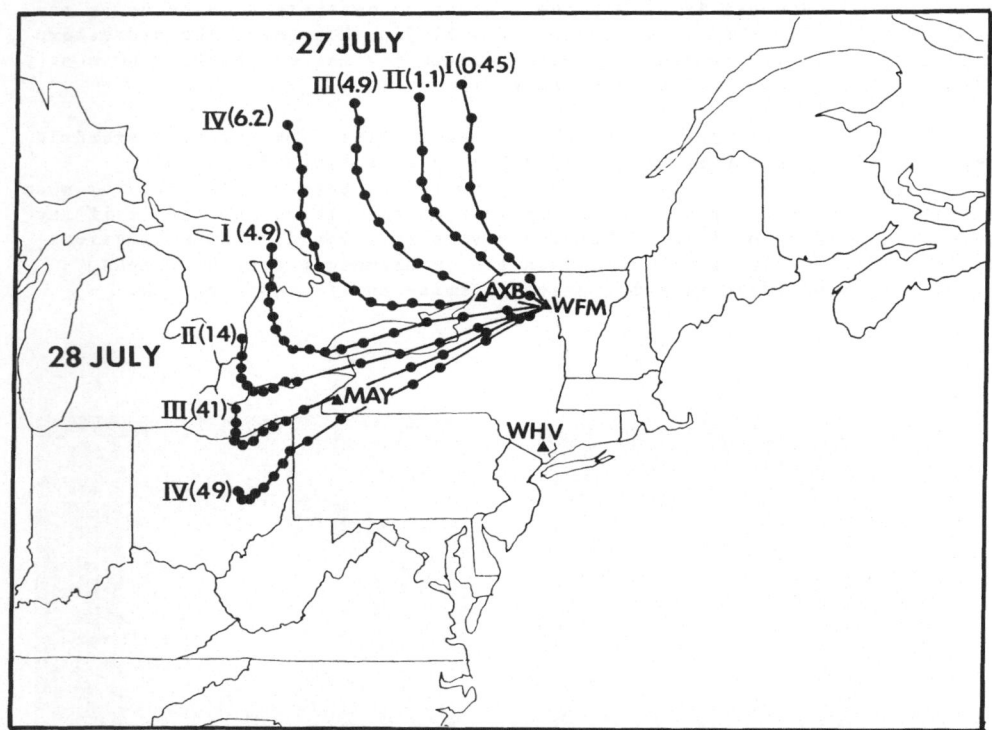

Figure 8. Backward air trajectories reaching Whiteface Mountain at
times where I=0200 EST, II=0800 EST, III=1400 EST, IV=2000
EST.   Numbers in parentheses are sulfate concentrations
(μg/m³). Each dot corresponds to a 6-hour interval
backward in time from the air mass' arrival at Whiteface
Mountain.   Sampling sites were AXB=Alexandria Bay,
MAY=Mayville, WFM=Whiteface Mountain and WHV=West
Haverstraw.

Sulfate concentrations in both episodes were consistent with levels seen in previous midwestern air masses and were greater than sulfate concentrations measured in northeastern air masses reaching Whiteface Mountain. Sources southwest of New York State were indicated during both episodes by initial increases in sulfate at Mayville, the southwestern-most site. A distant source was indicated by the widespread coverage of the entire upstate area during both episodes; a local in-state source would have created high sulfate levels only in the localized area and immediately downwind.

CONCLUSIONS

Electron microscopy combined with trace-element analysis has provided the most direct evidence yet of the source of high-sulfate air masses reaching the acid-stressed Northeast. All lines of evidence point to coal combustion in the Midwest. A distant source was indicated by lack of coal fly ash larger than one micrometer and widespread coverage of the upstate area by the sulfate air masses. A midwestern source was indicated by backward air trajectories, manganese/vanadium ratios greater than 2 and initial increases in sulfate at the southwestern-most site. A coal-combustion source was indicated by markedly increased coal fly ash abundance, increased selenium concentrations and lack of oil fly ash.

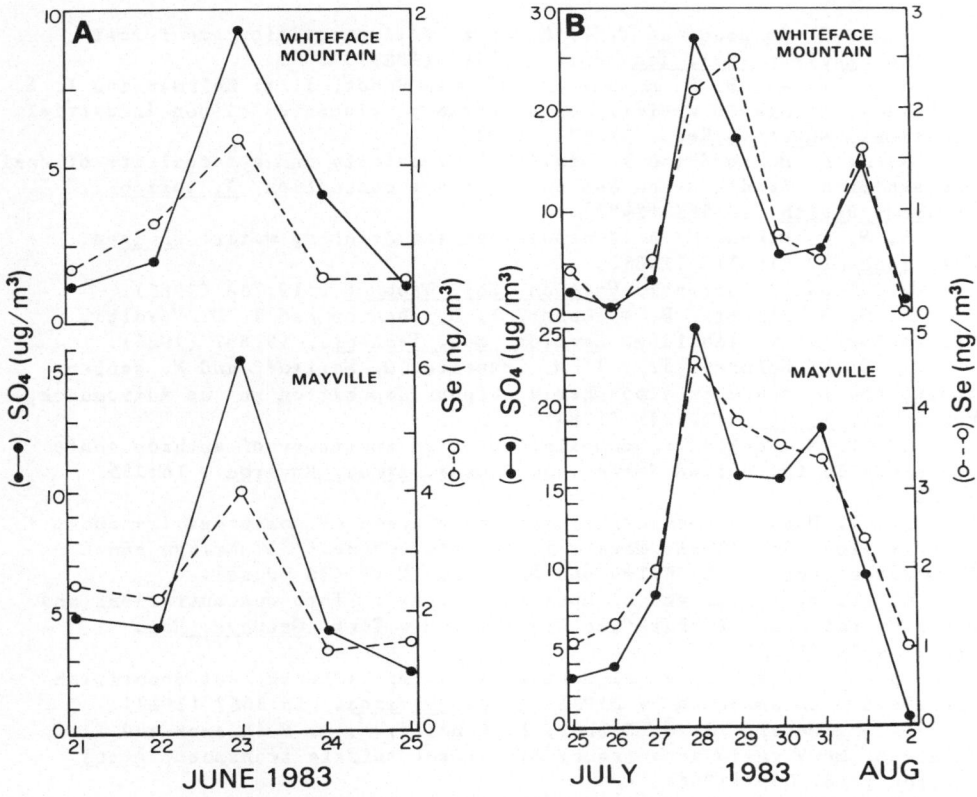

Figure 9. Daily mean sulfate and selenium concentrations at Whiteface Mountain and Mayville, New York, during two 1983 high-sulfate episodes in A) June. B) July/August.

ACKNOWLEDGMENTS

This work would not have been possible without the substantial contributions of Vincent Dutkiewicz and Liaquat Husain. The assistance of Edmondo Canelli, Ed LeGere, Jill Spierre, Pravin Parekh, Adil Khan, Joan Fleser, Jeff Haas, Dan Dickinson, Dan Sullivan and Bob Lincoln in collecting, preparing and analyzing samples is gratefully acknowledged. Analysis was assisted by PHS grant number RR01219 supporting the New York State High-Voltage Electron Microscope as a National Biotechnology Resource, awarded by the Division of Research Resources, DHHS. The Atmospheric Sciences Research Center of the State University of New York at Albany generously provided the use of their field station and personnel atop Whiteface Mountain. Chatauqua County is gratefully acknowledged for the use of facilities and personnel for sample collection in Mayville. Special thanks to Rusty Gates and Roger Cheng for providing source samples of fossil-fuel fly ash. Perry Samson kindly provided the air trajectories.

REFERENCES

1. J. S. Webber, Identifying sources of sulfate aerosols reaching Whiteface Mountain, New York, in: ''Proceedings-Electron Microscopy Society of America, Forty-Third Annual Meeting,'' G. W. Bailey, ed., San Francisco Press, Inc., San Francisco (1985).

2. C. L. Schofield, Acid precipitation: effects on fish, Ambio 5:228 (1976).

3. A. H. Johnson and T. G. Siccama, Acid deposition and forest decline, Environ. Sci. Technol., 17:294A (1983).

4. G. M. Alink, H. A. Smit, J. J. van Houdt, J. R. Kolkman and J. S. M. Boleij, Mutagenic activity of airborne particulates at non-industrial locations, Mutation Res., 116:21 (1983).

5. J. L. Mumford and J. Lewtas, Mutagenicity and cytotoxicity of coal fly ash from fluidized-bed and conventional combustion, J. Toxicol. Environ. Health, 10:565 (1982).

6. M. E. McDonald, Acid deposition and drinking water, Environ. Sci. Technol., 19:772 (1985).

7. ES and T Currents, Environ. Sci. Technol., 19:758 (1985).

8. D. G. Streets, B. M. Lesht, J. D. Shannon and T. D. Veselka, Climatological variability, Environ. Sci. Technol., 19:887 (1985).

9. G. R. Holdren, Jr., T. M. Brunelle, G. Matisoff and M. Wahlen, Timing the increase in atmospheric sulphur deposition in the Adirondack Mountains, Nature, 311:245 (1984).

10. C. M. Benkovitz, Compilation of an inventory of anthropogenic emissions in the United States and Canada, Atmos. Environ., 16:1551 (1982).

11. L. Husain, Chemical element as tracers of pollutant transport to a rural area, in:''Toxic Metals in the Atmosphere,'' J. Nriagu and C. I. Davison, eds., John Wiley and Sons, New York (in press).

12. P. P. Parekh and L. Husain, Ambient sulfate concentrations and windflow patterns at Whiteface Mountain, New York, Geophys. Res. Lett., 9:79 (1982).

13. J. W. Wilson, V. A. Mohnen and J. A. Kadlecek, Wet deposition variability as observed by MAP3S, Atmos. Environ., 16:1667 (1982).

14. L. Husain, J. S. Webber, E. Canelli, V. A. Dutkiewcz and J. A. Halstead, Mn/V ratio as a tracer of aerosol sulfate transport, Atmos. Environ., 18:1059 (1984).

15. J. S. Webber, V. A. Dutkiewicz and L. Husain, Identification of submicrometer coal fly ash in a high-sulfate episode at Whiteface Mountain, New York, Atmos. Environ., 19:285 (1985).

16. E. Canelli and L. Husain, Determination of total particulate sulfur at Whiteface Mountain, New York, by pyrolysis microcoulometry, Atmos. Environ., 16:945 (1982).

17. P. Parekh and L. Husain, Trace element concentrations in summer aerosols at rural sites in New York State and their possible sources, Atmos. Environ., 15:1717 (1981).

18. A. V. Samudra, F. C. Bock, C. F. Harwood and J. D. Stockham, Evaluating and optimizing electron microscope methods for characterizing airborne asbestos, EPA 600/2-78-038 (1978).

19. W. F. Tivol, D. Barnard and T. Guha, Progress in element analysis on a high-voltage electron microscope, Scanning Electron Microscopy 1985:455 (1985).

20. R. J. Ferek, A. L. Lazrus and J. W. Winchester, Electron microscopy of acidic aerosols collected over the northeastern United States, Atmos. Environ., 17:1545 (1983).

21. J. L. Heffter, Air resources laboratories atmospheric transport and dispersion (ARL-ATAD), NOAA Tech. Memo. ERL-ARL-81, Air Resources Laboratory (1980).

22. L. Husain, J. S. Webber, V. A. Dutkiewicz, E. Canelli and P. P. Parekh, Electron-microscopical identification of coal fly ash at a remote site in the northeastern United States, in: ''Environmental Effects of Fossil Fuel,'' R. Markuszewski and B. Blaustein, eds., ACS Symposium Series, (in press).

23. L. D. Hansen, D. Silberman and G. L. Fisher, Crystalline components of stack-collected, size-fractionated coal fly ash, Environ. Sci. Technol., 15:1057 (1981).

24. G. E. Cabaniss and R. W. Linton, Characterization of surface species on coal combustion particles by x-ray photoelectron spectroscopy in concert with ion sputtering and thermal desorption, Environ. Sci. Technol., 18:271 (1984).

25. L. D. Hulett, Jr., A. J. Weinberger, K. J. Northcutt and M. Ferguson, Chemical species in fly ash from coal-burning power plants, Science, 210:1356 (1980).

26. R. J. Lauf, Application of Materials Characterization Techniques to Coal and Coal Wastes, ORNL/TM-7663, Oak Ridge National Laboratory, TN (1981).

27. R. J. Cheng, V. A Mohnen, T. T. Shen, M. Current and J. B. Hudson, Characterization of particulates from power plants, J. Air Poll. Control Assoc., 26:787 (1976).

28. P. Whitney, N.Y. State Geol. Survey, personal communication (1983).

29. R. J. Cheng, Heterogeneous SO_2 oxidation: a laboratory observation, presented at the 1985 Pittsburgh Conference and Exhibition, New Orleans (1985).

30. P. J. Samson, Ensemble trajectory analysis of summertime sulfate concentrations in New York State, Atmos. Environ., 12:1889 (1978).

31. V. A. Dutkiewicz, J. A. Halstead, P. P. Parekh, A. Khan and L. Husain, Anatomy of an episode of high sulfate concentration at Whiteface Mountain, New York, Atmos. Environ., 17:1475 (1983).

32. L. Husain, J. S. Webber and E. Canelli, Erasure of midwestern Mn/V signature in an area of high vanadium concentration, J. Air Poll. Control Assoc., 33:1185 (1983).

33. S. G. Tuncel, I. Olmez, J. R. Parrington and G. E. Gordon, Composition of fine particle regional sulfate component in Shenandoah Valley, Environ. Sci. Technol., 19:529 (1985).

CONTRIBUTORS

Allan L. Allenspach

Department of Zoology
Miami University
Oxford, OH 45056

Istvan Balogh

Department of Forensic Medicine
Semmelweis Medical University, Budapest

Samarendra Basu Ph.D

New York State Police Crime Laboratory
Albany, New York 12226

D. Black

Computer Sciences Corporation
Falls Church, VA 22046

Eric J. Chatfield

Ontario Research Foundation
Sheridan Park Research Community
Mississauga, Ontario, Canada, L5K 1B3

Present Address: Chatfield Technical Consulting Limited
2071 Dickson Road, Mississauga, Ontario, Canada L5B 1Y8

Patrick J. Clark

Toxicology and Microbiology Division
Health Effects Research Laboratory
U.S. Environmental Protection Agency
Cincinnati, OH 45268

Thomas J. David

Georgia Division of Forensic Sciences
2613 Bolton Road
Atlanta, GA 30318

F. E. Doern

Whiteshell Nuclear Research Establishment
Pinawa, Manitoba, Canada

F. H. Y. Green

Division of Respiratory Disease Studies
National Institute for Occupational Safety and Health
Morgantown, Virginia   26505

D. H. Groth

Division of Biomedical and Behavioral Science
National Institute for Occupational Safety and Health
Cincinnati, Ohio   45226

Charles W. Huggins

U.S. BuMines
4900 LaSalle Rd.
Avondale, Md.   20782-3393

Suzanne, J. Johnson

U.S. BuMines
4900 LaSalle Rd.
Avondale, Md.   20782-3393

Arisztid G. B. Kovach

Department of Physiology
Semmelweis Medical University, Budapest

Paul T. McCauley

Toxicology and Microbiology Division
Health Effects Research Laboratory
U.S. Environmental Protection Agency
Cincinnati, OH   45268

James R. Millette

Toxicology and Microbiology Division
Health Effects Research Laboratory
U.S. Environmental Protection Agency
Cincinnati, OH   45268

Present Address: McCrone Environmental Services, Inc.
200 Oakbrook Business Center, 5500 Oakbrook Parkway
Norcross, GA   30093

T. Mills

Computer Sciences Corporation
Falls Church, VA 22046

S. F. Platek

Division of Biomedical and Behavioral Science
National Institute for Occupational Safety and Health
Cincinnati, Ohio 45226

Gabor M. Rubanyi

Department of Physiology
Mayo Clinic, Rochester, MN 55905

Joe M. Segreti

U.S. BuMines
4900 LaSalle Rd.
Avondale, Md. 20782-3393

Sueshige Seta

National Research Institute of Police Science
Tokyo, Japan

R. P. Singh

Department of Science & Technology
Government of India, Technology Bhawan
New Delhi 110 016, India

Janet G. Snyder

U.S. BuMines
4900 LaSalle Rd.
Avondale, Md. 20782-3393

L. E. Stettler

Division of Biomedical and Behavioral Science
National Institute for Occupational Safety and Health
Cincinnati, Ohio 45226

J. A. Stober

Toxicology and Microbiology Division
Health Effects Laboratory
USEPA
Cincinnati, OH 45268

Knud Aage Thorsen

Department of Metallurgy
The Technical University of Denmark
Lyngby, Denmark

V. Vallyathan

Division of Respiratory Disease Studies
National Institute for Occupational Safety and Health
Morgantown, Virginia   26505

Isaac S. Washington

Toxicology and Microbiology Division
Health Effects Research Laboratory
U.S. Environmental Protection Agency
Cincinnati, OH   45268

James S. Webber

Wadsworth Center for Laboratories and Research
New York State Department of Health
Albany, NY   12201

C. Weiler

Computer Sciences Corporation
Falls Church, VA   22046

D. L. Wotton

Province of Manitoba
Environmental Management Division
Winnipeg, Manitoba, Canada